Spatial Complexity

Fivos Papadimitriou

Spatial Complexity

Theory, Mathematical Methods and Applications

Springer

Fivos Papadimitriou
Forschungsbereich Geographie
Mathematisch-Naturwissenschaftliche
Fakultät
Universität Tübingen
Tübingen, Germany

ISBN 978-3-030-59670-5 ISBN 978-3-030-59671-2 (eBook)
https://doi.org/10.1007/978-3-030-59671-2

This Springer imprint is published by the registered company Springer Nature Switzerland AG
The registered company address is: Gewerbestrasse 11, 6330 Cham, Switzerland

To the memory of my parents

Preface

By "spatial complexity" is meant the complexity of surfaces and spatial objects of dimension 2 or higher. Its manifestations in nature are manifold and so are its implications for science and technology. Although it is primarily important for mathematics, geography, ecology, physics, psychology, aesthetics, medicine, engineering, it relates to everyone and everything. This is because it is behind all spatial processes and spatial distributions that give rise to the diversity and heterogeneity observed in spatial forms. It is therefore unsurprising that its assessment is so tricky. Yet, we need indices and methods to assess it for practical purposes.

This book goes beyond the well-known and well-established field of "complex systems" which aims to examine the various behaviours of complex systems. Instead of focusing on the intensely and widely researched "complex dynamics" (involving nonlinear behaviours, chaos, etc.), its focus is on explaining *why* some spatial form, region, figure, object or surface is complex, why is it more complex than another and how much so. Hence, a "grassroots" approach to complexity is presented here, tackling the problem of assessing the complexity of an object or a surface in two (and higher)-dimensional spaces, *as is* (without any change or movement).

Considering these, five key questions about spatial complexity are addressed here:

i. What defines it?
ii. Why and for whom is it important?
iii. How might it be assessed quantitatively?
iv. How can it be created by adopting simple methods?
v. How is it perceived and how is it associated to qualities and meanings?

As concerns the first question, we should not loose sight from the fact that expectedly, there are several alternative approaches to complexity and spatial complexity should not expected to be an exemption. I tried to shed light on as many as possible, in order to eventually derive a general classification of all the basic determinants of spatial complexity.

As for the second question, I aim at offering the possibility to scientists from across diverse scientific domains, if not to carry out further research in spatial complexity, at least to take a moment and think if and how in their own domains they may already be addressing questions of spatial complexity. And if the answer were positive, then they might as well consider whether the research framework proposed here might be useful to them.

With respect to the third question, some simple and easily applicable methods are suggested in this book, enabling oneself to explore spatial complexity quantitatively.

The fourth question is addressed here by adopting a mixed approach, ranging in between mathematics, geography, philosophy and experimentation: I propose a theory for the creation of spatial complexity from simple square maps, by following number-theoretic, algebraic and combinatorial methods, which I call "Spatium Numerorum".

Evidently, tackling the fifth question involves research results and concepts from psychology, aesthetics, epistemology, philosophy. If not anything else, these explorations show how important spatial complexity is, not only to science and technology, but to everyday life also.

The book is articulated in five parts as follows:

Part I *Introducing Spatial Complexity* (Chaps. 1–2) is the reader's first borders crossing into the realm of spatial complexity (Chap. 1), followed by an overview of the main fields in which spatial complexity emerges in scientific research and technological applications (Chap. 2). The most often encountered computational complexity classes are also presented in Chap. 1, as they pop up in various problems related to spatial complexity. The same chapter (1) contains necessary disambiguations from misleadingly similar-looking terms, such as "space complexity", "shape complexity" and "topological complexity". The style of introduction to spatial complexity is intentionally kept informal, so readers from various disciplines understand what spatial complexity is and appreciate its significance. The realm of spatial complexity is immense, and therefore no "complete" presentation of its significance and applications could possibly be made in Chap. 2. Some key areas are outlined there, for which spatial complexity is particularly important, given that it is not feasible to cover within one chapter all the scientific disciplines and domains for which spatial complexity plays a key role. But, hopefully, with this chapter the reader will appreciate the central position of spatial complexity, at least for disciplines for which it is particularly significant.

Part II *The Mathematical Basis of Spatial Complexity* (Chaps. 3–8) presents my basic mathematical theory of spatial complexity, based on three main sets of determinants: geometry (Chap. 3), randomness and entropy (Chap. 4), topology (Chap. 5). Although spatial complexity has several determinants, its assessment will eventually be algorithmic (Chap. 6). The term "algorithmic" here is used sensu lato, meaning the complexity as measured by number of symbols, or number of arithmetic operations (addition, exponentiation), or number of operations on the nature of the object or the surface examined (substitutions, deletions, etc). In this part also (Chap. 6), two metrics of spatial complexity (C_{P1} and C_{P2}) are defined that are extensively used in the subsequent chapters for various calculations. Some ideas for measuring the spatial complexity of 3d objects are also presented in this part (Chap. 7) while briefly examining the main problems encountered in the assessment of spatial complexity in 4d and higher-dimensional domains (Chap. 8).

Part III *Numbers Behind Spatial Complexity* (Chaps. 9–12) constitutes a theoretical and practical framework for assessing spatial complexity of small maps, for creating spatial complexity from numbers and spatial combinatorics, as well as some surprising and amazing aspects of spatial complexity, as they emerge by looking for it in mazes, labyrinths and games (Chap. 9). All these conduce to a "Spatium Numerorum" ("space of numbers") which seems to lie behind spatial forms, distributions, surfaces, objects, defining their shape and mathematical properties. It may sound like a bold assumption, but it nevertheless is possible to test by examining how spatially complex forms can be created from simple square maps. I suggest a method for this, by unveiling the number theory behind geometry and combinatorics, which, in turn, lie behind the spatial complexity of square maps (Chap. 10). The application of the metrics C_{P1} and C_{P2} on 3×3 binary maps (Chap. 11) reveals how the fascinatingly complex interplay between entropy, patchiness, clumpiness and complexity can bring forth both expected and counter-intuitive features of spatial complexity. The complete characterization of spatial complexity of 3×3 binary maps turns out to be a surprise in many ways. It proves, if not anything else, that assessing spatial complexity is, by all means, a difficult undertaking, but it also gives clues for quantitative assessments of spatial complexity. And in this context, mapping out primes and transcendentals as binary square maps (Chap. 12) fosters the need to further explore the "Spatium Numerorum".

Part IV *Understanding Spatial Complexity* (Chaps. 13–14) exposes some issues that can either hinder or facilitate our understanding of spatial complexity. Inevitably, spatial complexity is replete with enigmas that test our mathematical and computational capabilities to their limits (Chap. 13). Yet, there is ground to hope that we can "understand" spatial complexity by giving a fresh look upon some already known results from probability theory and algebra (Chap. 14). Some of the questions addressed in this part are: What is the role of symmetries in understanding spatial complexity? How do singularities affect its assessment? Can spatial complexity be infinite? How might spatial stochasticity be "tamed" by laws of probability and stochastic geometry? Might simple methods (such as the "sudoku method" proposed here) be used to evaluate the complexity of large maps?

Part V *Epistemological, Psychological, Geophilosophical and Aesthetic Perspectives on Spatial Complexity* (Chaps. 15–18) examines spatial complexity from within three domains of the humanities: Psychology (Chap. 15), Aesthetics/Art Theory (Chap. 16), Philosophy and Epistemology (Chap. 17). The central questions addressed to in these chapters are to understand how we perceive spatial complexity (Chap. 15), how we evaluate it aesthetically (Chap. 16), which philosophical approaches are more fitting to examine it and what are the epistemological repercussions of all these (Chap. 17). For instance, to what extent can we hope to understand spatial complexity at large scales with our current mathematical knowledge and computational capabilities? (Chaps. 17 and 18) Finally, an agenda is laid out for future research in spatial complexity (Chap. 18), with the suggestion that a future “Observatory of Spatial Complexity” might be helpful to monitor new developments in spatial complexity, cutting across sciences and disciplines. Such an observatory, which I named “Mens Spatii” (the mind of space), should be strongly interdisciplinary in its methods and scope.

As stated earlier, all throughout the book, and in order to conform with the common perception of the words “space” and “spatial”, these words are taken to refer to dimensions 2 and higher. Indeed, a mathematician may object that lines and points are “spaces” too (of one and zero dimensions, respectively); yet, the analysis of spatial complexity is interdisciplinary and should remain so.

It might be any researcher’s aspiration to possess the scientific means to calculate the spatial complexity of a surface or object in 4, 5, 6, …, n dimensions. However, as the reader will repeatedly discover throughout this book, the assessment (not to mention the measurement) of spatial complexity is hard, even in very simple cases of smooth 2d surfaces such as 3×3 binary square maps. Furthermore, as discussed in the book, the extent to which our 2d methods also apply to 3d surfaces is still poorly known.

Although the aim of this book is not meant to present extensive advanced topics of mathematics or computer science, effort was made to introduce non-mathematicians to some key notions of mathematics that are indispensable for studying spatial complexity. Thus, some key issues are presented step by step, so that, even readers without a particularly strong background in mathematics would be able to follow. Besides, overly detailed analyses of particular technical issues might dissuade some readers with less advanced mathematical skills. Thus, seasoned mathematicians may come across concepts perhaps with only limited explanations, which nevertheless may turn out to be quite promising for future research. But these cases are rare and clearly meant for experts in those fields, so non-mathematicians can safely bypass them and come back to them at some later stage, once they will have acquired further knowledge about those domains. Reversely, when readers feel already familiar with some topics (which nevertheless need to be presented for non-mathematicians), they may as well just skip them.

The world around us is rarely ever “simple” (=“not complex”) as much as isn’t anyone of the objects partaking it. Science has proved this repeatedly: the more we look closer at finer spatial scales, the more complex the world seems to be. But choosing a convenient spatial scale of observation, a suitable spatial representation to code the observations we make and a set of appropriate methods to assess the complexity of this spatial code is necessary and desirable for theoretical and practical needs. Hence, the study of spatial complexity should combine theory and practice, mathematical reasoning and, quite often, experimentation. The quantitative analysis of spatial complexity is a complex problem of its own, of immense breadth and depth; but, as we are biological entities endowed with advanced information processing capabilities, one of our ultimate goals and challenges is to discover how complex is the space we live in—be it the tangible space of our immediate surrounding, the geographical space, the “virtual” space, or even the cosmic space. It would be far too ambitious to assume that one single book would offer a definite "answer" as to how spatial complexity might be measured and perceived. Instead, it is hoped that it might serve as a handbook of research methods in spatial complexity, a compass orienting experts and non-experts alike into the vast realm of spatial complexity, an agenda for exploration and experimentation.

Rio de Janeiro: Part of the Copacabana beach and pavement, against the lights of the Avenida Atlantica at the backdrop, with a view towards Leme. It was there when I decided to write this book on spatial complexity. Now, after having finished writing and recalling Archimedes' "Sand Reckoner", I reckon that, even after years of research effort, those grains of ultra fine sand that were lying in front of me that night, by far outnumbered the sum total of thoughts I made about spatial complexity ever since and, almost certainly, all the thoughts I will be able to make about spatial complexity in the rest of my life.

Tübingen, Germany

Fivos Papadimitriou
geotopia@yahoo.fr

Acknowledgements

I am deeply grateful to the members of my family whose affection and guidance influenced my thinking during all my decades-long trekking in the realm of spatial complexity: my grandfather Christos who first taught me arithmetic and how to love numbers long before I went to school, my grandmother Filanthi (Φιλάνθη, the name means “friend of flowers”) who instructed me how to look for the hidden “flower” (that is the hidden good) in all “difficulties” (I now consider in retrospect she meant “complexities”), my father Theodossios who first initiated me in the challenges of geometric (spatial) thinking, and my mother Athina, for her endless love and support at every step of my life, and for orienting me, time and again, towards the exploration of the mysteries of geographical space. Without her wisdom, sensitivity and rare sense of humour, I would not have dedicated a part of my life to attempt to understand what it means for a spatial entity to be “complex” and why. Although she was a classicist and historian, she firmly believed that space is the ultimate ruler of our lives instead of time (that many people reify nowadays). Being able to see straight through to the meaningful essentials and discard what is unnecessary has always been her key to unlock the meaning of life. This is what one might call “complexity reduction”. Yet, while she always prompted me to simplify the unneccessarily complex, she also taught me the aesthetic pleasures of complexifying the dullness of extreme simplicity. Her guiding principle in life was Pythagoras’ dictum: “What I did wrong? And what did I make well? What I did not do which I should have done? (“Πῆ παρέβην; τί δ’ ἔρεξα; τί μοι δέον οὐκ ἐτελέσθη;”). Metaphorically, this is essentially an application of the Levenshtein algorithm (or “edit distance”, which is adopted here) to everyday life: find a minimum for deletions (of one’s wrongdoings) + additions (of right deeds) + substitutions (of deeds that should replace other ones) so as to have the desired life mode. Without adopting her dual approach and sense of measure, and without her life-long inspiration, this book would have never been written.

Contents

Part I Introducing Spatial Complexity

1 What is Spatial Complexity? 3
1.1 Definition and Disambiguation 3
1.2 Disorder, Asymmetry, Inequality 7
1.3 Spatial Complexity in Three Dimensions 11
1.4 Computational Complexity Classes 13
1.5 Perceiving and Creating Spatial Complexity 14
References 17

2 Spatial Complexity in Nature, Science and Technology 19
2.1 Spatial Complexity in Cosmology 19
2.2 Spatial Complexity in Geography and Ecology 21
2.3 Spatial Complexity in Physics and Electronics 23
2.4 Spatial Complexity in the Life Sciences 28
References 32

Part II The Mathematical Basis of Spatial Complexity

3 The Geometric Basis of Spatial Complexity 39
3.1 Orthogonality 39
3.2 Intersections 41
3.3 Curvature and Non-Euclidean Geometries 44
3.4 Spatial Combinatorics and Polyominoes 46
References 50

4 The Probabilistic Basis of Spatial Complexity 51
4.1 Spatial Entropy Versus Complexity 51
4.2 Spatial Randomness and Algorithmic Complexity 56
References 60

5 The Topological Basis of Spatial Complexity ... 63
5.1 Boundaries ... 63
5.2 Knotting ... 66
5.3 Braiding ... 71
5.4 Writhing and Linking ... 72
5.5 Genus ... 74
5.6 Dimensions ... 74
5.7 Orientability and Immersions ... 76
References ... 77

6 The Algorithmic Basis of Spatial Complexity ... 81
6.1 The Language of Space ... 81
6.2 Metrics of Spatial Complexity ... 86
6.3 Extrema of Spatial Complexities C_{P1} and C_{P2} ... 93
References ... 99

7 Exploring Spatial Complexity in 3d ... 101
7.1 Simplicial Complexes, Betti Groups and Matveev Complexity ... 101
7.2 Evaluation on Möbius Bands, Tori and Multiply-Connected Surfaces ... 103
7.3 Spatial Complexity of Simple 3d Solids and Voxels ... 107
7.4 Spatial Complexity of Reeb Graphs ... 110
References ... 112

8 Spatial Complexity in 4-and-Higher-Dimensional Spaces ... 115
8.1 Hypercubes, Tesseracts, Doxels, Clifford Tori ... 115
8.2 Manifolds, Gropes and Exotic Spheres ... 118
References ... 121

Part III Numbers Behind Spatial Complexity

9 Squares, Cats and Mazes: The Art and Magic of Spatial Complexity ... 127
9.1 Square Partitions ... 127
9.2 Squares, Minimalism and Art ... 131
9.3 Mazes, Labyrinths and Spatial Games ... 133
9.4 The Arnold Cat Map ... 136
References ... 140

10 Entering the "Spatium Numerorum": Creating Spatial Complexity from Numbers ... 143
10.1 Calculating Spatial Partitions ... 143
10.2 Entropy Class ... 153
10.3 Generic Maps and Symmetry ... 154

10.4 Calculating Binary Map Configurations 158
References 160

11 The Spatial Complexity of 3 × 3 Binary Maps 163
11.1 Parameters of Spatial Complexity of Binary Maps 163
11.2 The Spatial Complexity of 3 × 3 Binary Maps 165
11.3 Patchiness, Adjacency and Clumpiness. 169
11.4 Generic 3 × 3 Binary Maps and Their Multiplicities 171
11.5 Spatial Analysis at "King's Neighborhood" 175
References 178

12 Complexity of Binary Maps of Primes and Transcendentals 179
12.1 Numbers Defining Spaces 180
12.2 Complexity of Binary Square Maps of Primes and Composites 183
12.3 Square Maps from Transcendental Numbers 187
References 191

Part IV Understanding Spatial Complexity

13 Enigmas of Spatial Complexity 195
13.1 Geometrization, Singularities and Immersions 195
13.2 Spatial Complexity and Infinity 201
13.3 Same Numbers—Different Spatial Complexities 205
References 206

14 Taming Spatial Complexity 209
14.1 Taming Spatial Randomness? 209
14.2 Taming Spatial Complexity with Symmetries? 214
14.3 Taming Spatial Complexity with Grothendieck's Inequality? 218
14.4 Taming Spatial Complexity with a "Sudoku Method" 221
References 224

Part V Epistemological, Psychological, Geophilosophical and Aesthetic Perspectives on Spatial Complexity

15 Spatial Complexity, Psychology and Qualitative Complexity 229
15.1 Gestalt Psychology and Spatial Complexity 229
15.2 Scale-Dependence of Spatial Complexity 233
15.3 Perception of Spatial Randomness and Spatial Complexity 236
15.4 Qualitative Complexity Versus Spatial Complexity 239
References 241

16 Spatial Complexity, Visual Complexity and Aesthetics 243
16.1 Visual Complexity 243
16.2 Visual Complexity Versus Spatial Complexity of 3d Objects 246
16.3 Landscape Aesthetics and Spatial Complexity 248
16.4 Spatial Complexity and Aesthetic Evaluations 251
References 258

17 Geophilosophy and Epistemology of Spatial Complexity 263
17.1 From Difference to Complexity 264
17.2 Curvature and Subjectivity 266
17.3 A Philosophical Ladder to Spatial Complexity 268
17.4 Experimenting in the Spatium Numerorum 271
17.5 Large-Scale Spatial Complexity: Limits to Spatial Analysis? 273
References 276

18 Spatial Complexity and the Future 279
18.1 Key Determinants of Spatial Complexity 279
18.2 Spatial Complexity, "Com-Possible" Worlds and Quantum Computation 283
18.3 Large-Scale Spatial Complexity, Spatial Computing and Planetary Futures 286
18.4 *Mens Spatii*: Towards an Observatory of Spatial Complexity 289
References 291

Index 293

Symbols

A	Area
a	Parameter (real number)
ai	Parameter (real number)
$A_k(n)$	Kloosterman sum
A_d	Adjacency of square cells
B	Black square cell
B	Clumpiness
b	Parameter (real number)
b_i	Betti number
B_p	Boundary points
c	Parameter (real number)
C_{LN}	Spatial complexity metric defined by the author for number of crossings
C_{P1}	First Fivos Papadimitriou metric of spatial complexity
C_{P2}	Second Fivos Papadimitriou metric of spatial complexity
C_R	Spatial complexity metric defined by the author for Reeb graphs
$C(s)$	Spatial complexity metric defined by the author for 3d solids
D	Distance
D_i	Diffusion coefficient along the coordinate i
$E_2(q)$	Eisenstein series
${}_2F_1$	Hypergeometric function
G	Permutations group
g	Generic maps
H	Shannon entropy
H^2	The hyperbolic plane
h	Parameter (real number)
i	The imaginary number
Im( )	The imaginary part of a complex number
I_p	The number of internal points
$I_{3/2}$	Modified Bessel function of the first kind
k	Positive integer

k_G	Grothendieck's constant
L	Number (quantity) of lines
$L(\gamma_1, \gamma_2)$	Linking number for curves γ_1, γ_2
m	Positive integer
mod	Modulo
n	Positive integer
N_g	Number of generic maps
$N(n)$	Number of possible square map configurations of size n
$O(\ )$	Order of magnitude
P	Patchiness
p	Prime number
$P(n)$	Partition function of the number n
P_r	Probability
q	Nome of a Dedekindt eta or Jacobi theta function
$\wp(z)$	Weierstrass elliptic function
Q_i	Percentage of a map covered by cover i
R	The set of real numbers
r	Entropy class of a map
$\mathrm{R_d}$	The ring of algebraic integers
Re( )	Real part of a complex number
S^2	The 2-sphere
T_i	Quadrant i
t	Variable (real number)
U	The number of boundaries in a square map
u	Variable (real number)
V	The number of colours or thematic categories on a map
v	Variable (real number)
W	White square cell
$W(\ )$	Lambert series
w	Variable (real number)
x	Variable (real number)
y	Variable (real number)
Z	The set of integers
z	Variable (complex number)
mod	Modulo
$\Gamma(\ )$	Gamma function
γ	The Euler-Mascheroni constant
γ_i	Curve
δ_{mn}	Kronecker delta
$\varepsilon(\kappa)$	Crossings of two curves
$\eta(q)$	The Dedekind eta function
θ	Angle
∂_i	Boundary between i square cells
$\vartheta_n(z, q)$	Jacobi theta function with nome q
$\zeta(\ )$	Riemann function

k_e	The elliptic modulus
λ	Block of consecutive symbols
ξ	Multiplicity class of a generic map
ξ_r	Multiplicity class of a generic map for entropy class r
$\sigma_k(n)$	The divisor function
τ	The half-period ratio of the Dedekind eta function
φ	Angle
ϕ	The golden section
χ	The Euler characteristic of surface s
ψ	Angle
$\|\psi>$	The quantum state vector in the Hilbert space
$\Psi(g)$	Number of colourings
$\lceil\,\rceil$	The ceiling function
$\lfloor\,\rfloor$	The floor function

Part I
Introducing Spatial Complexity

Chapter 1
What is Spatial Complexity?

Existence, Space/Land and Becoming is a triad
of discrete elements, which preexisted the origin of the skies.
"ὄν τε καὶ χώραν καὶ γένεσιν εἶναι,
τρία τριχῇ καὶ πρὶν οὐρανὸν γενὲσθαι"
(Plato, 428–348 b.C., "Timaeus", 52d)

Abstract Spatial complexity is defined here as the difficulty to simplify the structure or form of a 2-and-higher-dimensional surface or object. The study of spatial complexity refers to the geographical space, to mathematically abstract spaces, to physical objects, or to any surface or object, in a n-dimensional space with n equal to two or higher. Spatial complexity should not be confused with "space complexity", "topological complexity", "shape complexity" or "complex stystems". Spatial complexity is scale-dependent (it changes according to the level of generalization at which it is examined and is, under certain conditions, perception-dependent also.

Keywords Spatial complexity · Psychology and complexity · Topology and complexity · Map complexity · Computational complexity · Simplicity · Geography and Complexity

1.1 Definition and Disambiguation

The end is in the beginning and yet you go on
(Samuel Beckett, 1906-1989, "Endgame", 1957)

Plainly put, "spatial complexity" is exactly what its two constituent words suggest: the complexity of a spatial object.

Any spatial object, be it two-dimensional (i.e. a surface) or three-dimensional, or even n-dimensional, large or small in size, on a plane or on a curved surface, compact or with holes, rugged or smooth, can be more or less complex in comparison to another.

In terms of "spatial science", the most characteristic cases of spatially complex objects or settings can be identified from the perception of geographical spaces or outdoor environments (i.e. landscapes), or even from representations of spaces by

F. Papadimitriou, *Spatial Complexity*, https://doi.org/10.1007/978-3-030-59671-2_1

maps, photographs etc. To other sscientists, a machine constituting in many parts, the shapes and colors of a garment, a painting, and innumerable other objects and surfaces may be spatially complex.

Let us consider what comes to mind when we think that some object of the real world is "complex", in contrast to another one that is "simple". If something is more "complex", it means that it is more difficult to understand, to learn how it works, to break down in pieces, to re-create from its basic constituents, to explain to ourselves and to others. All these difficulties are summarized by the word "complexity" and if the "complex" object is spatial, then we talk about "spatial complexity", that is the complexity of a spatial object.

Let us consider these in a practical way: A room is certainly more complex than another, if there are more objects in it (might as well be so for several other reasons, as will be examined later). Contrast, for instance, the simplicity of a zen monk's room with one of the royal halls of the palace of Versailles. Certainly, the fewer items or any other categories of *distinct* objects are confined within a spatial extent, the simpler the space is. For instance, the floor of a 30 m^2 room with three chairs on it is more complex than if only one chair was there, and even more complex than if the room had been completely empty. Carrying on along the same line of thought, the fewer items or any other categories of *different* classes of objects are found within a strictly confined spatial extent, the simpler this spatial extent will be: if, in addition to the three chairs, the floor of the same room had a table, a sofa and nineteen books on it (that is 24 objects belonging to 4 different categories), then it would be more complex than if it had all its objects belonging to the same category, i.e. 24 books or 24 chairs.

Intuitively also, we understand that some space is more complex, if the objects in it are in a state of disorder. As we all know, disorder increases the difficulty of understanding something. If there were only 24 books in the room and they were all packed together as a single stack, then the room would simply consist in two halves: the area of the room covered by the books (the stack) and the uncovered area. If the 24 books were scattered all over, then the rooms' tenant would need to brace herself to "arrange" the place, or "put things in order". But what if these 24 books were not placed disorderly on the floor? What if they were all stacked together on a table instead? Then the room would have an area of low complexity (where there are no tables, books, chairs, or sofas) and an area where there is a table with a heap of books thrown randomly on it. What would the complexity of that room be then? While complexity may be localized and concentrated within a restricted area of space, it may as well be absent in other areas.

Hence, spatial complexity is the degree of difficulty to describe (computationally, linguistically) or to code (i.e. algorithmically) a spatial object, surface, arrangement, assortment, or piece of space containing objects or surfaces. As such, its study is part of the study of "complexity" as it has been developed in mathematics and computer science over the last decades, with the peculiarity that it is focused on the complexity of spatial entities only, without any restriction as to the nature, the physical or chemical constitution or functions of these spatial entities. Whatever their nature may be, we need tools to evaluate the spatial complexity of ordinary objects: the spatial complexity of a balloon, of a landscape, of one square centimetre of one's skin etc.

Some notices of disambiguation are due here.

(i) Objects, sets of objects or systems that *behave* in a complex manner are beyond the scope of the present book. Complex behaviors, processes and dynamics constitute the research subject of the already fairly advanced discipline of "*complex systems*", which studies phenomena such as chaos, bifurcations, unpredictability etc.

(ii) "*Spatial* complexity" should be clearly distinguished from the term "*space* complexity" that is used in informatics to denote the amount of space resources required for the computation of a problem's solution. Consequently, the complexity of processes, behaviors, situations, relationships, temporal changes, of any process that presents changes in time or changes in contexts other than spatial, or can not be brought into spatial form *only*, lies beyond the scope of "spatial complexity".

(iii) The notion of "*topological complexity*" only partly relates to spatial complexity, because it means quite different things in different scientific contexts, and, as such it can not replace the significance of the term "spatial complexity", nor should it be confused with it. It was defined by Farber (2003, 2004) as the complexity of the problem of constructing a motion planning algorithm in the 3d space. This term has been adopted mainly in robotics, and particularly in the study of the complexity of trajectories and motion planning (Grant 2007), but it has also appeared in various contexts in other domains also: by Finkel et al. (2006), in the context of diffeomorphisms of 2-dimensional manifolds, by Martensen (2003) and Godefroy et al. (2001) in the context of Banach spaces, by Grigoriev (2000) and by Souvaine and Yap (1995) in range searching. Besides these, it is also encountered in biology and DNA analysis (Hertling 1996; Martin-Parras et al. 1998), and in neural networks (Chapline 1997). It seems that in all these cases the second component of the term ("complexity") was apparently unrelated to any other domain of complexity analysis (i.e. algorithmic complexity).

(iv) Spatial complexity is a much wider concept than "*shape complexity*" for which the reader may consult the relevant literature (e.g. Chazelle and Incerpi 1984; Catrakis and Dimotakis 1998; Rossignac 2005; Joshi and Ravi 2010; Chambers et al. 2016). The term "shape complexity" has so far been poorly related to algorithmic complexity and there are no methods from algorithmic complexity theory to calculate it, nor is there a universally accepted measure of "shape complexity". Furthermore, spatial complexity refers not only to complexity of shapes (outlines/forms), but also to the spatial allocation and/or distribution of colors/covers/classes/types/categories over surfaces or spatial objects; it therefore also heavily relies on entropy and probabilities of distributions (while "shape complexity" does not).

(v) Conforming with the common understanding of the word "space", a two-dimensional space is the least-dimensional space that can, without any doubt, be considered as a "space"; as points and lines alone can not be considered as "spaces" in the broad *non-mathematical* sense of the word "space". This convention is followed here, and therefore the examination of strictly-less-than-two-dimensional cases is beyond the scope of "spatial complexity", unless they form indispensable constituents of spatial complexity. Spatial complexity is encountered in innumerable forms, in two, three and higher spatial dimensions, although it *can* be generated by objects with dimensions lower than two (i.e. lines). While mathematicians normally deal with "spaces" of dimension strictly less than two also, the 0-and-1-dimensional spaces do not conform with what most people call "space" (lines and points are not "spaces" in the non-mathematical sense) and, as spatial complexity is a fundamentally inter-disciplinary subject, the common meaning attributed to the word "space" is respected, so here we mean the complexity of objects and surfaces of *(at least) two dimensions.*

Spatial complexity may be charmingly beautiful, puzzling, or even disdainful (often infuriatingly so). It can be a source of frustration (i.e. to scientists and engineers who seek simplicity and efficiency) and a source of inspiration to philosophers and artists. Whatever the attitude towards it, all living beings have to cope with it in different forms, time and again, from their birth until their passing away. Bees seek flowers among the plants' leaves, predatory animals assess every tiny change in a small or large spatial area in order to spot and stalk their prey. Humans seek to conquer and understand the entirety of space that is available to them on the surface of this planet and to expand their quest for complex forms of existence in the outer space.

Distinguishing the interesting from the uninteresting, the useful from the useless, the certain from the uncertain, eventually involves taking snap decisions that are based on one's spatial perception. Taking decisions on whether to stay or leave a particular location in space, evaluating the aesthetic appeal of an image, deciding whether a shape in space is purposeful or not, all these and countless more decisions inevitably involve some kind of assessment of spatial complexity. Quite often, processing such assessments quickly can be a matter of life and death. Generally, the more complex the spatial area or spatial object surveyed, the more difficult it is to reach a decision about it. Some successful professionals however, are often able, with admirable effectiveness, to assess the complexity of spatial arrangements and take correct snap decisions accordingly (i.e. some military officers). Others still (i.e. visual artists), are able to create pleasant forms, by masterly exploiting the aesthetic characteristics of spatial forms, by either enhancing or downsizing spatial complexity in their artworks. And then, there are scientists (mathematicians, geologists, ecologists, engineers among many others), who seek to understand how spatially complex a two-dimensional surface (such as a map or an image) or even three-dimensional object may be and why. As a general rule, the more advanced a life form is, the more complex its internal structure.

Aside of the complexity of bodily functions and operations, the human body organs, tissues and cells display increasingly complex forms if examined at finer spatial scales. Furthermore, the rise of increasingly more spatially complex forms throughout the earth's history is manifested not only biologically, but also geologically (in the separation of landmass from the oceans and then the breaking up of Pangea into several pieces and continents). The incessant processes of spatial complexification are difficult to grasp in their immensity and multiplicity, as we possess only limited tools to investigate them to their full breadth and depth. But striking differences of spatial simplicity vs. complexity are all around us. Given this, it should be rather easy to start grasping the basics of spatial complexity by using common knowledge and everyday experience, as will be seen in the following section.

1.2 Disorder, Asymmetry, Inequality

> He had brought a large map representing the sea,
> without the least vestige of land: and the crew were much pleased
> when they found it to be. A map they could all understand...
> Other maps are such shapes, with their islands and capes!
> but we've got our brave captain to thank that he's brought us the best:
> a perfect and absolute blank!
> (Lewis Carroll, "The Hunting of the Snark", 1876)

Let us see some examples of spatial complexity, as contrasted to spatial simplicity. A barren landscape bearing no plants, no water and virtually no life is spatially less complex than a landscape in which many different plant species cover its surface (Fig. 1.1).

A two-dimensional object, such as a photograph of the sky may have almost zero variability among its cells (and thus low spatial complexity), while another may display a great diversity in its color palette (Fig. 1.2). In the animal kingdom, genetic rules prescribe remarkable differences in the spatial complexities of the appearances

Fig. 1.1 Part of a dry and desert-like landscape, which can not bear vegetation (left) contrasted to a complex landscape (right) that is endowed with many kinds of different plants within a small area

Fig. 1.2 A sky with only one color contrasted to a sky with many colors: more colors, higher spatial complexity

of living beings of the same species (Fig. 1.3). Expectedly, a piece of space is more difficult to understand, if the objects in it are in a state of *disorder* (Fig. 1.4).

Besides disorder, also characteristic is the absence of symmetries that would qualify the space as ordered. Spaces with *asymmetries* are more difficult to decode and require more spatial information than spaces endowed with symmetric or repetitive patterns (Fig. 1.5). Yet, as often happens, order may occur side by side with asymmetries within the same space (Fig. 1.6).

Obviously, the whole problematic of spatial complexity can easily spiral out of any possible computational control and this compels us to seek as simple approaches to it as possible. That is why, in so many cases, we need to reduce spatial complexity. *Simplifying* details in either *geometric* features or *categories* of spatial elements (i.e. colors) in a spatial object results in reducing its complexity. For instance, a regular

Fig. 1.3 Two stray cats (left and middle) that the author used to take care of. On the right, one of the author's cats, Flashy, enjoying getting herself entangled in complex spatial settings. Notice the varying levels of spatial complexity on the cats' fur: an entirely black one (less complex), a black-and-white (more complex), and a multicolored one (even more complex). Having closely observed these three cats' characters, the author has concluded that the more spatially complex their appearance, the more complex their behaviors also (an observation that applied to these particular ones only!)

Fig. 1.4 An ordered space (left) requires less effort to describe it and therefore has lower spatial complexity in comparison to a space that hosts disorder (right)

Fig. 1.5 Symmetries in space (left) are indicators of lower spatial complexity, in contrast to asymmetries (right) that (most probably) imply higher spatial complexity

hexagon is a more complex form than a square. Similarly, an irregular hexagon is even more complex than a regular hexagon (Fig. 1.7). This is because the hexagon has more sides and angles than the square and the irregular hexagon has different angles and side lengths than the regular hexagon.

Simplicity does not only depend on order, symmetry and shape. It has to do with *numbers* of objects and shapes also: quantity matters for complexity. It is easier to draw a square (either by hand or with the aid of a computer), than to draw 439

Fig. 1.6 The symmetric design of a building, contrasted with the spontaneously spreading branches of a tree

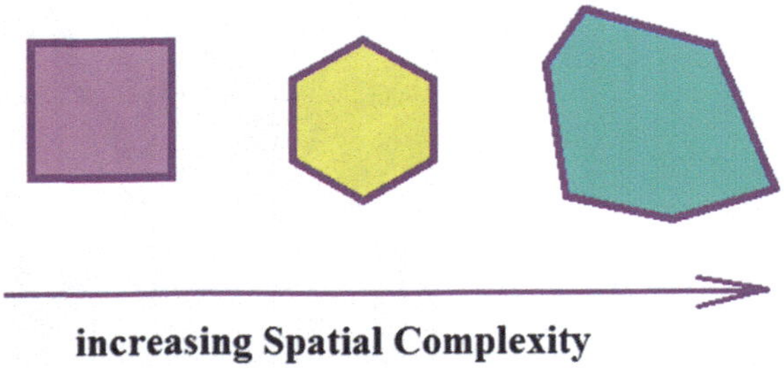

Fig. 1.7 Three shapes compared with respect to their spatial complexity: A regular hexagon is more complex than a square and an irregular hexagon is more complex than a regular hexagon

squares of the same side length. One step further: it is easier to perceive 439 equal-sized squares than 9 such squares. And, eventually, it is easier to perceive 9 squares of the same size than 9 squares of unequal sizes (spatial inequalities increase spatial complexity). To make things even more "complex", it is easier to perceive 39 equal-sized squares all red than 39 squares of unequal size *and* of different color each (can anyone easily figure out 39 different colors in 39 unequal squares?). So same class (represented i.e. by the same color), same size, and same geometry mean lower spatial complexity. Plainly put, spatial dissimilarities increase spatial complexity. Besides geometric simplification however, thematic simplification is commonly applied in order to reduce the spatial complexity of an image. In Fig. 1.8 for instance, while the original picture requires 545 kbytes memory, the simplified one needs only 4 kbytes.

Fig. 1.8 A photograph of the skyline of Rio de Janeiro (above) and a simplification of this picture (below). The geometric features (points, lines, areas) of the original picture have been regrouped, so as to become as simple as possible and the same was done with the colors. Simplification in either geometric features or spatial elements reduces spatial complexity: the original photograph (above) requires 545 kb memory while the simplified (below) only 4 kb

1.3 Spatial Complexity in Three Dimensions

> "Complex" is a transition that comes
> with a reversal or an adventure, or both
> "Πεπλεγμένην δὲ ἐξ ἧς μετὰ ἀναγνωρισμοῦ
> ἢ περιπετείας ἢ ἀμφοῖν ἡ μετάβασίς ἐστιν"
> (Aristotle, 384-322b.C.,"Poetics", 1452a)

While 2d square cells constitute the basic spatial element for the analysis of 2d surfaces in Z^2, *voxels* (a composite word from volume and pixel) are the 3d equivalent of pixels in the digital topology Z^3. A 3d surface (even curved surface) can be voxelized in the same way that a 2d surface is pixelized (Fig. 1.9) and thus, voxelized landscapes can be created to model the surface of a 3d object or its internal structure. Although several algorithms have been devised for voxelization, the calculation of the complexity of voxelized spatial forms remains rather poorly studied to date.

And yet, surfaces and objects may not appear straightly stretched in space: they can be knotted, linked, braided, writhed (Fig. 1.10) and *topological differences* among

Fig. 1.9 Voxeliation of a cube (left) and of an irregular surface (right)

Fig. 1.10 The "*unknot*" (that is a trivial knot, homeomorphic to the circle, on the left) and two knots made from rubber that can easily be transformed to the unknot on the left

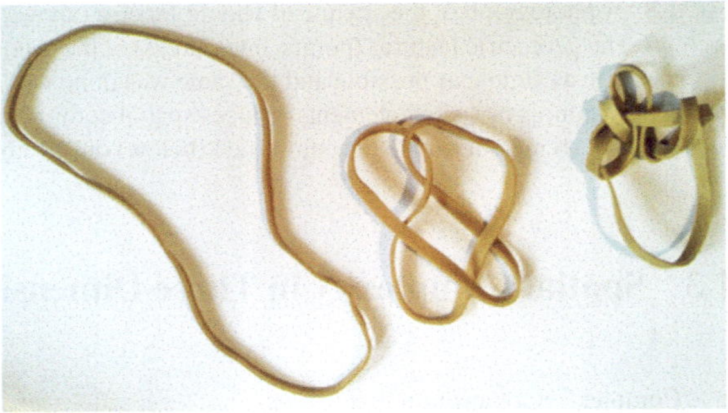

knots can help us understand whether an object in 3d space is more complex than another (Fig. 1.11). A central question in knot theory is whether and how a knot can be untied (would Alexander the Great be able to solve the Gordian knot if he lived in the twenty-first century without eventually cutting it off?).

Fig. 1.11 Increasing surface complexity in the 3d space is reflected by knotting and linking: it is the same rubber strand that is simple (left) or increasingly complex (middle and right)

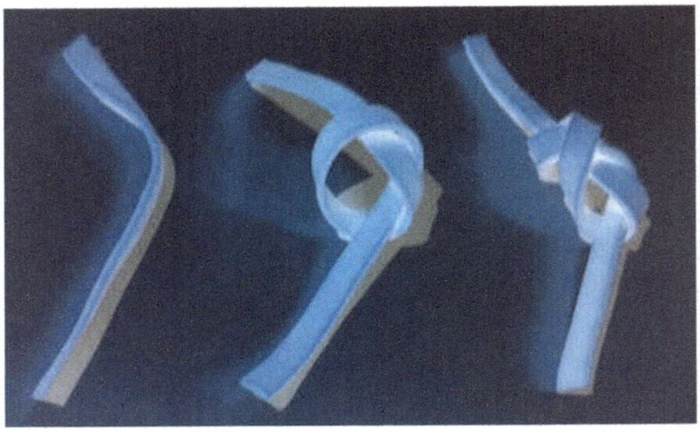

Beyond the third dimension, the equivalent of the voxel is the "*doxel*" in the 4d space (dynamic voxel), for which we are still short of satisfactorily efficient algorithms for complexity estimates. Unsurprisingly, as regards spaces of even higher dimensions (fifth and higher), although our knowledge from topology is fairly advanced, our methods for estimating spatial complexity are still very poor. Yet, and as will be explained in the next chapters, severe difficulties are encountered in encoding, decoding, measuring and perceiving spatial complexity even in two dimensions only and even in cases of simple small binary maps.

1.4 Computational Complexity Classes

"A short piece of work means as much to me as a long piece of work"

(Harold Pinter, 1930–2008)

No doubt, many problems of spatial complexity involve some kind of computation. Eventually, the complexity of a spatial object reflects the effort or resources (measured in terms of time, energy, computational power, either consumed by a human or a machine) required to fully decipher an object by using a sequence of symbols or operations. Whether these computations are easy, difficult or even impossible to carry out is a question that falls in the field of "*computational complexity*". The "computational complexity" of a problem concerns the computational difficulty of solving a certain problem or a class of problems. For further information, the reader may consult anyone of the classic texts on this subject (i.e. Garey and Johnson 1979; Lawler et al. 1985; Rayward-Smith 1986; Papadimitriou 1994; Van Leeuwen 1998). Although several "computational complexity classes" have been identified, some categories will be briefly presented next, as they sporadically appear in assessments of spatial complexity. The complexity class "*P*" contains all problems that can be solved by a polynomial algorithm (one such is to determine whether a given number is a prime or not). The class *PSPACE* is the set of decision problems that can be solved in polynomial space and a polynomial number of bits of space or memory (to by used/occupied) are used in any number of time steps. The nondeterministic variant of *PSPACE* is *NPSPACE* and the equivalence between the two is guarranteed by Savitch's theorem (Savitch 1970). For other interesting equivalences of *PSPACE*, the reader is referred to Immerman (1988) and Szelepcsényi (1988). The class *EXPTIME* is the set of problems that are solvable in exponential time, that is by an order of magnitude $O(2^{p(n)})$ time, where $p(n)$ is a polynomial function of n, and much alike them, *EXPSPACE* is the set of problems that are solvable by an order of magnitude $O(2^{p(n)})$ space, where $p(n)$ is a polynomial function of n.The often-encountered class "*NP*" contains all problems of which the solution can be verified by a polynomial algorithm (i.e. the problem of deciding whether two graphs are the same). The class "*NP*-complete" contains all those "difficult" problems, which nevertheless have the characteristic that if one of them could be solved, then all the other ones of the same class might as well. A known example of this class from the spatial sciences is the problem of deciding whether it is possible to color any map of different regions (i.e. countries) with three colors only, in a way that that no two adjacent countries are assigned the same color. Finally, the problems of the "*NP*-hard" class are even more difficult to solve and most often involve some optimisation. Also from the spatial sciences, a typical such is problem is the "travelling salesman", that consists in finding an optimal route between points without repeating any part of the itinerary. Despite the fact that many problems of 2d spatial analysis such as those involving raster image analysis (analysis of imagery on the basis of orthogonal grids) are expected to be *NP*-hard, our knowledge of their computational complexity remains restricted (Coeurjolly et al. 2008; Sivignon and Coeurjolly 2009).

1.5 Perceiving and Creating Spatial Complexity

"The transcendental topography of the mind"

(Georg Lukacs 1994, p. 29)

Understanding the way we perceive spatial complexity is an issue of its own. As repeatedly proven experimentally with the use of various strings of symbols, the perception of *randomness* by humans is skewed and hardly (seldom) accurate (e.g. Brugger 1997; Falk and Konold 1997; Kahneman and Tversky 1972; Kareev 1992; Lopes and Oden 1987; Nickerson 2002). This is partly due to the fact that the theoretical concepts of probability may not coincide with subjective views of what is random and what is regular (see Beltrami 1999), to the extent that the term "*subjective* complexity" (Falk and Konold 1997) has been proposed, while the "*qualitative* complexity" (Papadimitriou 2010) refers to the meanings conveyed and the semantics associated with the spatial complexity of a spatial object (Papadimitriou 2012).

Plausibly, a deceivingly innocent question emerges (which, as will be seen later, presents enormous difficulties to answer): what kind of properties a two-dimensional object has that make it to be perceived as more complex than another? Take, for instance, two images (Fig. 1.12), both of the same size (367×375 pixels). On the left side, the photograph of a floor with its orderly arrangement of square tiles contrasts the dense branches of a natural Mediterranean bush (*Spartium junceum*) displayed at the photograph on the right. The latter picture displays a very high spatial complexity (at least as perceived visualy) that is created by the curved, interwoven thin branches of the bush. It would be hard to believe that, as a matter of fact, the image on the right rquires only twice (394 kb) the memory required for the storage of the picture on the left (193 kb). So kilobytes of computer memory are *not* always suggestive of spatial complexity of the object examined, particularly the *perceived* spatial complexity. To make this point more explicit, consider yet another pair of imagery (Fig. 1.13), in which a tree is contrasted to a shop selling pottery. In this case, both images are of the same size (367×375) and require the same storage memory also (334 kb). The point here is to observe the context associated to this

a *b*

Fig. 1.12 Two images of the same size (367 × 375). A floor (**a**) and dense branches of *Spartium junceum* (**b**). Image **b** requires only the double (393 kb) of the storage memory required by image **a** (194 kb), although it is *visually* a lot more complex

a *b*

Fig. 1.13 Two snapshots of the same size (367 × 375). A tree (**a**) and the façade of a shop (**b**). Although both images require the same storage memory (334 kb), image **a** gives a sense of "natural order" as all branches lean towards the same direction) while image **b** gives a sense of disorder

difference in computer memory requirements for each image: the fact alone that the shop's picture is more complex (and hence, it requires more memory) coincides with the *meaning* it conveys to the viewer. Evidently, there is a huge difference associated to meanings conveyed by these two images: the natural order of the orientation of the tree's branches may be a false assurance of low complexity, while the shop's facade is an example of diversity, asymmetry and disorder. And yet, both images require exactly the same memory storage capacity. Hence, bytes of information are not always suggestive of the meanings associated to what the images display. In fact, memory sizes derived from the application of the lossless compression method png on each image do not constitute a spatial complexity *measure* on its own: a compression process or method is not necessarily a measure of complexity and, as a matter of fact, several compression methods exist. The difference between any two such images is in memory bytes. But memory bytes measure information; not complexity (although information may serve as an estimator of complexity of an image lacking other appropriate measures). And here enter psychology and art theory to explain what is the difference between *visual complexity* and the spatial determinants that define the spatial complexity of an image, object or setting.

Yet, nothing precludes the possibility that spatial complexity be *concealed* right before our eyes. This is because the mathematical proof of existence of spatial complexity and its visual perception can be poorly related with these being two almost completely disjoint processes. Consider, for instance, the image of Fig. 1.14: all odd-numbered rows have 6 colored cells each, all even-numbered 8 and all columns have

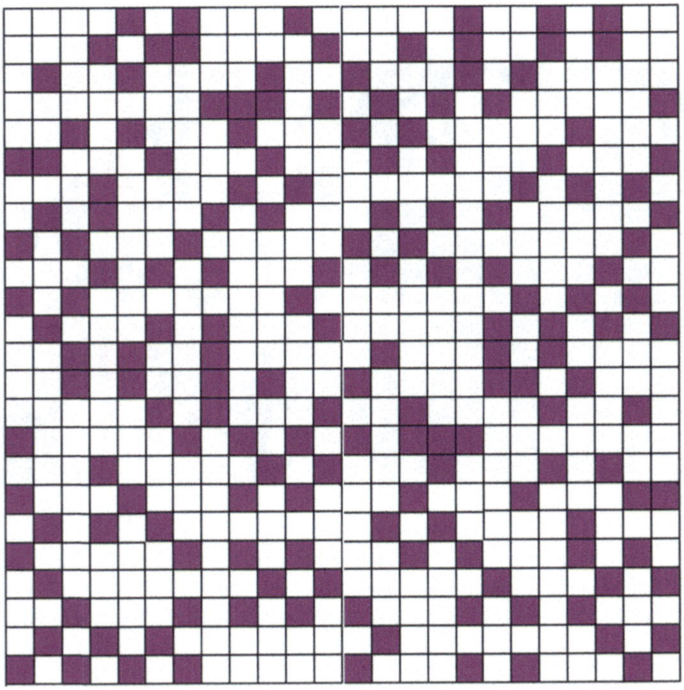

Fig. 1.14 Spatial complexity may be concealed or misperceived. This map has some remarkably simple regularities, which nevertheless evade the reader's attention at first sight (unless one is told how to unveil them): all odd-numbered rows have 6 dark cells each, all even-numbered ones have 8, while each and all columns have 7 colored cells

7 colored cells. Despite this mathematical regularity however, there is no easily discernible pattern in it.

References

Beltrami, E. (1999). *What is random? Chance and order in mathematics and life.* New York: Springer-Verlag.

Brugger, P. (1997). Variables that influence the generation of random sequences: An update. *Perception and Motor Skills, 84,* 627–661.

Catrakis, H. J., & Dimotakis, P. E. (1998). Shape complexity in turbulence. *Physical Review Letters, 80*(5), 968–971.

Chambers, E., Emerson, T., Grimm, C., & Leonard, K. (2016). Exploring 2D Shape Complexity. In A. Genctav, K. Leonard, S. Tari, E. Hubert, G. Morin, N. El-Zehiry & E. Chambers (Eds.) Research in Shape Analysis (pp. 61–83). Springer.

Chapline, G. (1997). Spontaneous origin of topological complexity in self-organizing neural networks, *Network: Computation in Neural Systems, 8*(2), 185–194.

Chazelle, B., & Incerpi, J. (1984). Triangulation and shape-complexity. *ACM Transactions of Graphics, 3,* 135–152.

Coeurjolly, D., Hulin, J., & Sivignon, I. (2008). Finding a minimum medial axis of a discrete shape is NP-hard. *Theoretical Computer Science, 406*(1–2), 72–79.

Farber, M. (2003). Topological complexity of motion planning. *Discrete Computational Geometry, 29,* 211–221.

Farber, M. (2004). Instabilities of robot motion. *Topology Applications, 140,* 245–266.

Falk, R., & Konold, C. (1997). Making sense of randomness: Implicit encoding as a basis for judgment. *Psychological Review, 104*(2), 301–318.

Finkel, O., Ressayre, J.-P., & Simonnet, P. (2006). On infinite real trace rational languages of maximum topological complexity. *Journal of Mathematical Sciences, 134*(5), 2435–2444.

Godefroy, G., Yahdi, M., & Kaufman, R. (2001). The topological complexity of a natural class of norms on Banach spaces. *Annals of Pure and Applied Logic, 111*(1), 3–13.

Grigoriev, D. (2000). Topological complexity of the range Searching. *Journal of Complexity, 16* (1), 50–53(4).

Garey, R. M., & Johnson, D. S. (1979). *Computers and Intractability: A guide to the theory of NP-Completeness.* New York: W.H. Freeman.

Grant, M. (2007). *Topological complexity of motion planning and Massey products.* In Proceedings of the M. M. Postnikov Memorial Conference 2007, Banach Centre Publications [arXiv:0709. 2287].

Hertling, P. (1996). Topological complexity with continuous operations. *Journal of Complexity, 12*(4), 315–338.

Immerman, N. (1988). Nondeterministic space is closed under complementation. *SIAM Journal of Computation, 17*(5), 935–938.

Joshi, I., & Ravi, B. (2010). Quantifying the shape complexity of cast partsDurgesh. *Computer-Aided Design and Applications, 7*(5), 685–700.

Kahneman, D., & Tversky, A. (1972). Subjective probability: A judgment of representativeness. *Cognitive Psychology, 3*(3), 430–454.

Kareev, Y. (1992). Not that bad after all: Generation of random sequences. *Journal of Experimental Psychology: Human Perception and Performance, 18*(4), 1189–1194.

Lawler, E. L., Lenstra, J. K., Rinnooy-Kan, A. H. G., & Shmoys, D. B. (1985). *The Traveling Salesman Problem: A Guided Tour of Combinatorial Optimization.* Chichester: Wiley.

Lopes, L. L., & Oden, G. C. (1987). Distinguishing between random and nonrandom events. *Journal of Experimental Psychology: Learning Memory and Cognition, 13*(3), 392–400.

Lukacs, G. (1994). *The theory of the Novel: A historico-philosophical essay on the forms of great epic literature.* Cambridge MA: MIT press.

Martensen, B. F. (2003). The local topological complexity of *Cr*-diffeomorphisms with homoclinic tangency. *Nonlinearity, 16*(1), 161–186.

Martín-Parras, L., Lucas, I., Martínez-Robles, M. L., Hernández, P., Krimer, D. B., Hyrien, O., & Schvartzman, J. B. (1998). Topological complexity of different populations of pBR322 as visualized by two-dimensional agarose gel electrophoresis. *Nucleic Acids Research, 26*(14), 3424–3432.

Nickerson, R. S. (2002). The production and perception of randomness. *Psychological Review, 109*(2), 330–357.

Papadimitriou, C. H. (1994). *Computational Complexity*. New York: Addison-Wesley.

Papadimitriou, F. (2010). Conceptual modelling of landscape complexity. *Landscape Research, 35*(5), 563–570.

Papadimitriou, F. (2012). Artificial intelligence in modelling the complexity of mediterranean landscape transformations. *Computers and Electronics in Agriculture, 81,* 87–96.

Rayward-Smith, V. J. (1986). *A First Course in Computability*. Oxford: Blackwell.

Rossignac, J. (2005). Shape complexity. *the Visual Computer, 21*(12), 985–996.

Savitch, W. J. (1970). Relationships between nondeterministic and deterministic tape complexities. *Journal of Computation and Systems Science, 4*(2), 177–192.

Sivignon, I., & Coeurjolly, D. (2009). Minimum decomposition of a digital surface into digital plane segments is NP-hard. *Discrete Applied Mathematics, 157*(3), 558–570.

Souvaine, D. L., & Yap, C.-K. (1995). Combinatorial complexity of signed discs. *Computational Geometry, 5*(4), 207–223.

Szelepcsényi, R. (1988). The method of forced enumeration for nondeterministic automata. *Acta Informatica, 26*(3), 279–284.

Van Leeuwen, J. (Ed.) (1998). Handbook of Theoretical Computer Science. Vol. A,*Algorithms and complexity.* Amsterdam: Elsevier.

Chapter 2
Spatial Complexity in Nature, Science and Technology

Science! True daughter of Old Time thou art!
Who alterest all things with thy peering eyes
(Edgar Allan Poe, 1809–1849, "Science")

Abstract Some characteristic domains of science and technology for which spatial complexity is significant are examined here. In cosmology, spatial complexity relates to the large-scale spatial inhomogeneities of the universe. In geography, earth sciences and ecology, it appears as one of the two major components of landscape complexity and thus affects ecosystem managment and landscape protection. In physics, complex systems and fractals are two of the most characteristic fields of science for which spatial complexity is important, while in electronics, spatial complexity is of paramount importance, for the miniaturization of electronic devices, for broadening the capacity of electronic systems, processing times of spatial datasets, QR technologies etc. In the bio-medical sciences, spatial complexity is a key descriptor of melanomas and emerges as a decisive factor in some interpretations of MRI imagery, while the complexity of knotting of proteins is important for DNA analyses.

Keywords Spatial complexity · Geographical complexity · Landscape complexity · complexity and cancer · Fractals and Complexity · Complex systems · Geocomputation

2.1 Spatial Complexity in Cosmology

Nature in the universe has been articulated from infinites and finites.
"Ά φύσις δ' ἐν τῷ κόσμῳ ἁρμόχθη ἐξ ἀπείρων τε καὶ περαινόντων"
(Philolaus, 470-385 b.C.,F1,Diog.Laert. 8.85)

Spatial complexity lies at the heart of the problem of the exploration of the large-scale structure of the universe, and of the "cosmic web" of spider web-like conglomerates of galaxies that are developed in it (Liebeskind et al. 2017). In cosmology,

F. Papadimitriou, *Spatial Complexity*, https://doi.org/10.1007/978-3-030-59671-2_2

the "horizon problem" consists in explaining the absence of a generally convincing explanation of why the universe is homogeneous at its very large scales (that is why its matter is allocated more or less homogenously in space). According to the measurements carried out so far, the large-scale topology of our universe presents a "flatness", in the sense that angles between any two intersecting lines are preserved (which would not be anticipated if it had a hyperbolic or a spherical shape). Yet, following string theory, our universe is a 2d surface (which may as well have infinite extent), that is called a "brane" (abbreviation for "membrane"). In some theories (i.e. the Horava-Witten), there are several branes, with dark matter possibly present in some of them. As one brane moves towards another, it is thought it can modify the other brane's physical laws. Thus, the anticipated set of stable states of a "string landscape" is estimated to range in between 10^{100} and 10^{1000} and each such state is thought to be governed by different physical laws. The theory that different possible worlds might be emanating from simple geometric structures is examined in the context of string theory, particularly in the context of a multiplicity of universes, the "multiverse", which, if exists, should have a truly unfathomable spatial complexity.

At a more detailed spatial scale however, the form, structure and distribution of galaxies in the universe are intricately related to spatial complexity, of which the importance for the development of large-scale structures in the universe is evident, given the presence of areas with large masses (i.e. $\sim 10^{15}$ solar masses of the "Coma" cluster) where "clusters of galaxies" prevail in sharp contrast to other large empty spaces. This *spatial inhomogeneity*, in tandem with the matter-antimatter asymmetry in our universe, has puzzled cosmologists. If a spatial object is homogeneous, then it is not as complex as an inhomogeneous one (of the same size). Perhaps, the large-scale spatial inhomogeneities that have been observed in the universe have never been better observed than from the spatial distribution of the cosmic microwave background radiation (CMB) which was emitted since the "big bang" and was dispersed all over the universe ever since.

Probably, the increasing complexification in the universe occurred in tandem with an increase in its entropy. In turn, the cause of the increase in entropy is believed to be the expansion, which, in turn, is related to the gravitational field. As the universe expands, its entropy increases and so does its complexity, but as it approaches its heat death, its complexity is expected to diminish dramatically. Alternatively the roots of changes in both energy and complexity may be sought in Helmholtz's free energy, in the sense that the entropy difference increases, free energy decreases according to the change in temperature, so free energy grows entropy (Lineweaver et al. 2013). Understanding the gap between actual and maximum entropy in the universe is essential in order to explain the growth of the amazingly complex forms in it and the interplay between entropy and complexity manifests itself in the history of life on earth, the multiplicity of forms of life, and eventually, the rise of civilization itself.

Spatial complexity might also hold the key to unlock further secrets of the cosmic space. Specifically, the possibility of existence of life in extraterrestrial environments depends on particular geoindicators, one of which is the geological differentiation of a planet's surface, which can be detected by means of remote sensing techniques (Jovian satellites for instance, display significant surface fragmentation and hence higher spatial complexity than other satellites in our planetary system). These indicators resemble those used in spatial analysis and landscape ecology for the analysis of complex structures of terrestrial landscapes.

Eventually, understanding how entropy and spatial complexity change due to the expansion of the universe is a big question of its own, although the interplay between entropy and spatial complexity reverberates across spatial scales, from the large-scale structure of the universe down to shapes and patterns, both natural and human-made, on the face of the earth.

2.2 Spatial Complexity in Geography and Ecology

> Vigorous branches falling one on another,
> Complexifications of petals, covers of leaves,
> complexes of fruits: thus was the language of plants
> "Ἔθαλλον οἱ κλάδοι, συνέπιπτον ἀλλήλοις ἄλλος ἐπ᾽ ἄλλον,
> ἐγίνοντο τῶν πετάλων περιπλοκαί, τῶν φύλλων περιβολαί,
> τῶν καρπῶν συμπλοκαί: τοιαύτη τις ἦν ὁμιλία τῶν φυτῶν"
> (Achilles Tatius, 1st-2nd cent. b.C., "Leucippe and Clitophon", 1.15)

Measuring the complexity of a surface constitutes a major challenge for many fields of the earth sciences, such as geospatial technologies and GIS. Geography, landscape analysis and landscape ecology constitute the prominent fields of "visible" applications of spatial complexity analysis (Papadimitriou 2010a, b; Papadimitriou 2012a, b, c), with repercussions for land use planning (Papadimitriou 2013). In geography, spatial landscape complexity can be perceived in various ways, by means of maps, satellite imagery or aerial photographs etc. and specific mathematical methods have been devised for this purpose (Papadimitriou 2002, 2009, 2012a) to measure this type of landscape complexity. In geomorphology, given the puzzling spatial complexity of landforms, it has also been suggested that *landscape complexity* should be one of the highest research priorities in geomorphology (Werner 1999; Fonstad 2006; Murray and Fonstad 2007; Goehring 2013; Tlidi et al. 2018).

Landscape complexity has also been identified as one of the key research priorites in landscape ecology (Wu and Hobbs 2002) and several studies have appeared over the last years in the literature, aiming at quantitative assessments of landscape complexity, while research has produced evidence of non-linear interactions in landscapes (Turchin and Taylor 1992; Pahl-Wostl 1995).

Recognizing the importance of the complexity for landscapes however, a classification of types of landscape complexity was made by Papadimitriou (2002; 2010a), who suggested that landscape complexity is of three basic types: (a) *spatial* (or "structural"), (b) *functional* and (c) *qualitative* (or semantic). The words "structure" and "function" have precise meanings in landscape ecology and the reader is referred to classic texts of landscape ecology (i.e. Forman and Godron 1986) for their explanation.

Landscape ecology often resorts to combinations of already known landscape-ecological indices, such as "landscape patchiness", "landscape diversity" and "landscape fragmentation". Expectedly, such metrics should affect decisions related to land management. The "European Landscape Convention" requires assessments of "landscape fragmentation" (Llausas and Nogue 2012). This can be attributed to the widely held view that landscape fragmentation is held responsible for negative effects on ecosystems' function, as several studies have documented in the literature of landscape ecology, from across different bioclimatic settings (Pütz et al. 2011; Gao and Li 2011; Bassa et al. 2012). Besides, landscape fragmentation often occurs in areas of rapid urbanization (Shrestha et al. 2012). Fragmentation is important in the complexity of predator–prey systems (Morozov et al. 2006), it is usually caused by human activity, and affects complexity not only structurally, but sometimes even functionally so (Briefer et al. 2010). Forest fragmentation poses considerable problems to successful landscape management plans, and constitutes an important factor in land use/land cover change analyses of its own (Andrieu et al. 2011; Mas et al. 2012). Besides, evidence from marine biological observations suggests (Hovel and Lipcius 2002; Hovel 2003) that some marine species (i.e. the juvenile blue) depend on environmental fragmentation.

The complexity of landscapes needs to be assessed for practical applications, such as land management, land use planning and forecasting changes (Papadimitriou 2012a, b, 2013). In geo-ecological spatial analysis, we need to explore spatially structured ecological interactions, but we also need methods to simplify spatial complexity, while taking into account combinations of nonspatial along with spatially explicit approaches. This is particularly interesting, considering the strong evidence that self-organization at the landscape level may emerge even from homogeneous landscapes, therefore forcing us to examine spatial complexity also in the context of its changes with time (Papadimitriou 2002, 2009, 2010a, b). One way to analyze spatial complexity is by means of spatial ecological networks in which the nodes of the network can be species or individuals and links can represent the direction of flows of mass or energy. In these cases, the complexity of a network can be measured i.e. by the number of connections per node or the randomness of connections per node (Fig. 2.1). The extent to which an ecosystem can maintain its functions even if some of its nodes are removed from it, reveals its stability (Papadimitriou 2013).

Spatial complexity also enters as an important parameter in geographical information systems (GIS) and geospatial technologies and in various practical applications (Batty 2005), because it directly impacts time and resources necessary to process geospatial data (although this impact may not always be measurable); relief modelling for instance, depends on terrain complexity. Spatial complexity is also important in qualitative spatial reasoning and in efforts to develop new generation GIS, as it modulates the efficiency of handling large repositories of big spatial data. For these reasons, we sometimes need to reduce the spatial complexity of large spatial databases in order to speed up computations.

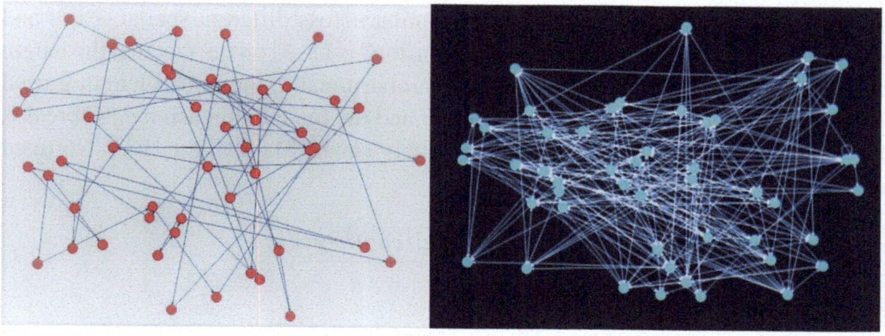

Fig. 2.1 Two networks, with the same number of nodes each (50), but with different number of links per node: there are two connections per node for all nodes in the network on the left. The network on the right is more complex, as it has both a higher number of links per node and a random number of links per node

Thus, comparing landscape ecology and geospatial technologies, it becomes evident that whether high spatial complexity is desirable and beneficial or not varies and depends on particular situations and conditions. In ecology and biogeography, high spatial complexity is desirable for ecosystems, as it is associated with "information-richness" in the geographical space (White and Engelen 1994) and "species richness" in ecology. In contrast to these, when we need to handle data *representing* the environment, ecosystems and geographical spaces, a high complexity of these data may not always be desirable (it can be so only up to an acceptable level that will allow us to process those data at the desired spatial resolution in order to fulfill our practical requirements).

2.3 Spatial Complexity in Physics and Electronics

> You still have Chaos in you
>
> "Ihr habt noch Chaos in euch"
>
> (Friedrich Nietzsche, 1844–1900, "Also sprach Zarathustra" part 5)

It is common place to say that the earth looked from above a spaceship appears like a "blue planet", but zooming in to this all-blue image would reveal the immense variety of landscapes our blue planet is endowed with. Mapping out this variety of natural and artificial settings would certainly not result in "blue" color in them all. In fact, the color of each patch would depend on how far or how close we look at it from. The dependence of spatial analysis (and, inevitably, complexity also) on the spatial scale of observation has given rise to the growth of "fractal geometry", "fractal landforms", "fractal dimensions" etc. that have attracted great interest in physics and other sciences.

Spatial forms are often "*fractal*", self-similar across different spatial scales and non-differentiable. Measuring "fractal dimension" is based on the fact that the length of the contour of a planar fractal object is proportional to the ruler's length *g* used to measure the length of the object's contour and the fractal dimension *d*, where *d* is the limit as the ruler's length *g* grows infinitely small (Mandelbrot 1983; Normant and Tricot 1991):

$$d = 1 - \lim_{g \to 0} \frac{\log[length(g)]}{\log g} \tag{2.1}$$

A common fractal is the "*Koch curve*" which has a fractal dimension of $d = \log4/\log3 = 1{,}262$ (Fig. 2.2).

Fig. 2.2 The "Koch curve" (or "Koch island" or "Koch snowflake") has a fractal dimension of $d = 1{,}262$: a non-integer dimension, classifying this shape as an object with dimension in between 1 (lines) and 2 (planes)

Beyond physics, fractals have been examined in earth sciences, geography and ecology (among many other disciplines). Although they present a useful approach to describe how complex a boundary or a spatial form is with measurements becoming progressively more detailed, they serve as *indicators* of spatial complexity, but not as its *measures*. However, the fact that increasingly more refined spatial scales of analysis reveal higher complexity is important and concerns many domains of scientific inquiry. Assessing the complexity of higher than one-dimensional forms is also a key issue in software measurement, computer graphics, information theory, neural networks, pattern recognition, materials science, physics of fractal objects and many more. For all these domains, cellular automata occupy a prominent position, because discrete models of complex processes have traditionally been based on cellular automata. These are automatic cellular evolutionary processes depending on a set of "states" $S_1, S_2, \ldots, S_n$ and a set of "transition rules" $T_1, T_2, \ldots, T_m$, acting on these states. Each cell is found in one state only and that state is determined by the rules and the states its surrounding cells are in. Consequently, at time $t + 1$, the state of each cell, S_{t+1}, is determined from the transition rule T_i acting on the state of that cell at time t: $S_t \xrightarrow{T_i} S_{t+1}$. The "rules" are simple algorithms acting on an array of cells in two dimensions. Thus, all cells interact with their neighbouring ones, either in the "rook" sense or in the "king" sense (as the rook's or the king's movements in chess). These neighborhood types are defined as neighborhoods around a point (x,y), defined by the sets of cells surrounding the central cell. In the case of a von Neumann's neighbourhood the interactions are described as: $N_4(x; y) = \{(x+1; y); (x\text{-}1; y); (x; y\text{-}1); (x; y+1)\}$ and in the case of Moore's neighbourhood: $N_8(x; y) = \{(x+1; y); (x\text{-}1; y); (x; y\text{-}1); (x; y+1); (x+1; y+1); (x+1; y\text{-}1); (x\text{-}1; y+1); (x\text{-}1; y\text{-}1)\}$. Yet, there are other possibilities (Fig. 2.3) for constructing cellular automata,

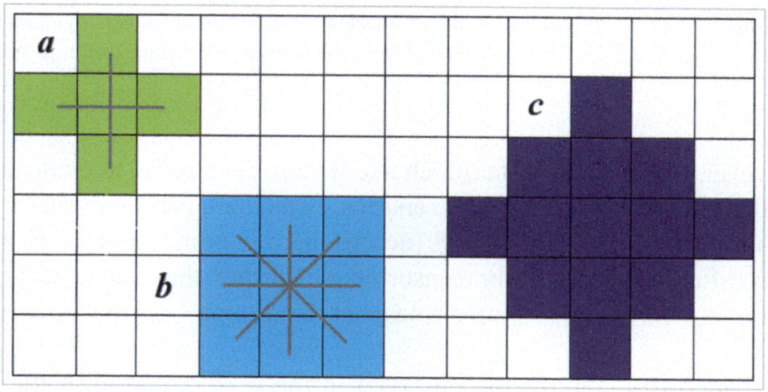

Fig. 2.3 Some common types of spatial arrangements in cellular automata: The central cell interacts with its four surrounding cells in "rook's case" (**a**), its nine surrounding cells in "king's case" (**b**) and with twelve cells around it in the case of a von Neumann neighbourhood (**c**)

with longer interactions than in the 9 cells surrounding the immediate neighbourhood of the central cells (as, i.e. in the case of a "5x5 Von Neumann neighbourhood").

Cellular automata have been intensely explored in order to simulate the emergence of spatially complex forms in geography and ecology. In landscape research for instance, such applications range from the possibility to establish general algorithmic ecological laws, to the exploration of ecological processes such as niches, industrial ecologies, interspecies competition, latitudinal gradients and species diversity (Rohde 2005; Baynes 2009). In geography, they are used to model the progress of urbanization in the course of time, geomorphological processes such as run-off and soil erosion and landscape evolution. On the inteface between ecology and geography, cellular automata can be used to simulate the ways by which landscapes change with the spatial propagation of natural hazards (i.e. forest fires) and to explore the complexity of spatial synchronization and self-organization. With cellular automata, among many applications, it is also possible to simulate the complexity of spatial synchronization processes (Satulovsky 1997) as well as phenomena of self-organization in space (Manrubia and Sole 1996; Malamud and Turcotte 1999).

Changes over time that are simulated by means of cellular automata may unveil pattern formation and self-organization, which are highly significant facets of complex system behaviors. Evidently, these changes emerge if the cellular automata are allowed to run for long times and thus become able to produce aggregates in

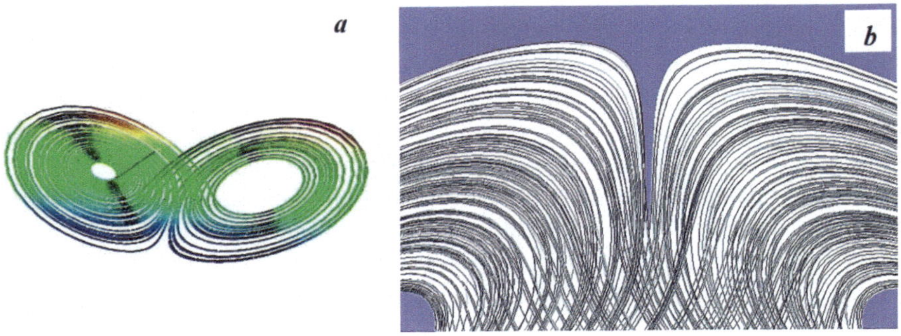

Fig. 2.4 Details (**b**) of the orbits (**a**) of the phase space of a widely known dynamical system, the *Lorenz attractor*. The finer the resolution of the observations, the more different orbits can be discerned

space. However, cellular automata (much like fractals) are useful to model and simulate spatially complex developments, senarios, behaviors, processes and dynamics, but not to measure spatial complexity (describing a complex process as it evolves with time is different than actually measuring it). Further, they can be used to create spatially complex forms from simple rules, but this is a process that only evolves in time.

Spatial complexity can also be observed in the representation of the orbits of dynamical systems in the *phase space*. In this case however, we do not deal with spaces of the physical world but with mathematical constructs that are used to plot and represent the orbits of dynamical systems in time. Phase spaces are not spaces of the real world; they are mathematical constructs used to represent behaviors of systems of the real world. Yet, in terms of spatial complexity, it is interesting to notice that as some orbits are observed at increasingly finer scales, the denser they may appear. As an example, consider the detail in the orbits of the phase space of the system of differential equations that is famously known as the *Lorenz attractor* (Fig. 2.4), which represents (with some detail of abstraction) the atmospheric circulation by means of the "Lorenz system" of nonlinear differential equations (Lorenz 1963):

$$\begin{aligned}\frac{dx_1}{dt} &= a(x_2 - x_1)\\ \frac{dx_2}{dt} &= bx_1 - x_2 - x_1x_3\\ \frac{dx_3}{dt} &= x_1x_2 - cx_3\end{aligned} \tag{2.2}$$

The variables x_1,x_2,x_3 are *not* coordinates of space; they are meteorological variables and therefore the space they define is a mathematical construct devised to help us visualize the system's behavior.

Spatial complexity also manifests itself in phenomena of "nonlocal coherence" (communication of behaviors, properties and processes by distance) that are sometimes observed in natural and social systems. Besides, coherence, coupling and synchronization emerge from models of nonlinear spatial dynamics (Casado 2001; Berglund 2007a, b) but the mechanisms by which these properties emerge in spatial ecosystems are largely unknown. It is probable however, that there exist processes leading prey-predator interactions to the "*edge of chaos*" through oscillators (Rai 2004), by making dynamical systems to appear as "*riddled systems*". In such systems, as their naming suggests, it is difficult to distinguish basins of attraction and orbits in the phase space. If this is the case (as currently appears more likely), then we should not lose sight from the fact that such riddled systems may demand extremely more elaborate models than we already have at our disposition, particularly if they are spatially explicit.

Besides physics, spatial complexity is important in the analysis of "big geospatial data" (in image analysis, most commonly). In fact, the need to reduce image complexity has led to the development of various image compression protocols, such as jpeg. Other fields of technological applications are QR encodings, that have now become ubiquitous in algorithmics (Park et al. 2011), microchip construction, cryptography (Alvarez et al. 2012), materials science (Hyde and Schroder-Turk 2012) etc. The more complex a space is within the confines of a QR square, the more information it contains. Compare, for instance, the QR encoding of the author's name and surname with another QR representing additionally his degrees and mailing address (Fig. 2.5).

Fig. 2.5 QR-representations. Left: the author's name and title. Right: the author's name and title, along with his degrees and mailing address. The finer the (binary) detail in a QR, the more information is stored in it and, the finer the detail, the higher the spatial complexity

Differences in spatial complexity are also important for various functions of opto-electronic systems (radars, scanners etc. technologies), which aim to solve problems of detection, recognition, classification and identification of targets by using opto-electronic tracers (either multispectral or monochrome), radiometry in automatic target recognition, chromatic filters bearing Foreon technology etc., whatever the scanning method is used and embedded in them (i.e.serial "raster" scanning, parallel "linear array" scanning, "rosette scanning").

2.4 Spatial Complexity in the Life Sciences

"All that man is, all mere complexities,
the fury and the mire of human veins"
(W.B. Yeats, 1865-1939, "Byzantium", 1930)

One of the precursors of modern medicine, Paracelsus, in his "Labyrinthus Medicorum Errantium" (1538) used the term "complex" to mean the mixture or coexistence of properties and qualities, while correctly pointing out that the basic elements of nature are not complex, but simple (the "temperamenta"). Perhaps, nowhere in the bio-medical sciences becomes the significance of spatial complexity more explicit than in the identification of skin malignancies; melanomas in particular. The widely used "ABCDE method" is characteristic for spotting a dangerous melanoma (Fig. 2.6): Asymmetric form, rough Borders, many Colors, larger than 6 mm Diameter and Evolution (change in any of the above). Essentially, all these criteria (except for the last one that relates to time) are criteria of spatial complexity. Similar criteria are

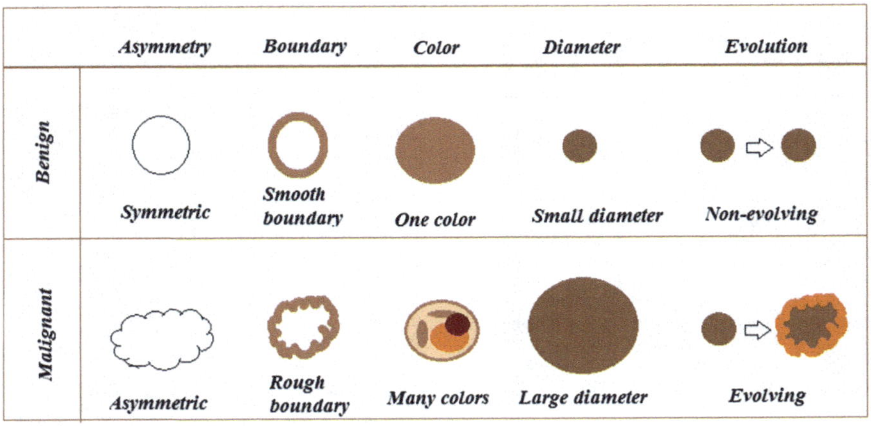

Fig. 2.6 The "ABCDE" criteria for identification of melanomas. A malignancy has a higher spatial complexity than a benign skin spot

used in other cancer malignancy categorization schemes (i.e. the Gleason 5-degrees scale for prostate cancer).

Yet, the significance of spatial complexity for medicine becomes explicit if one considers even more domains of the biomedical sciences, such as MRI imagery, histology, and DNA sequencing. Analyzing ultrasound imagery is an important field of everyday medical practice, more recently also combined with applications in telemedicine. Before proceeding to image analysis, histological observations (images) are converted to binary or multicoloured maps of raster (square) grids to make them suitable for further analysis. Then, successive images of the same tumour revealing its changes in spatial complexity indicate the degree of cellular differentiation and hence malignancy, with a predictability that can mount up to 90% (Tambasco et al. 2009). The potential of fractal geometry has also been explored as an indicator of cancer cell differentiation with high degree of success (Capri et al. 2006; Timbo et al. 2009). Besides, one of the most exciting fields of applications of fractal patterns seems to be the detection of tumour growth (Cross 1994, 1997; Cross et al 1994; Bash and Jain 2000; Esgiar et al. 2002). Estimating the increasing roughness of borders of cells and tissues as a discriminator of pathological situations has found applications in cardiology, osteoporosis detection and diagnosis of pulmonary diseases (Heymans et al. 2000).

But the usefulness of the concept of spatial complexity in biomedical sciences extends well beyond the domain of oncology. For instance, spatial complexity has been identified as a key issue in understanding brain function (Jia et al. 2018; Schulz et al. 2018), so another domain of application is brain MRI analysis. Neuroimaging using MRI (magnetic resonance imaging) is now ubiquitous and aims at detecting changes in texture in the spatial distributions of cell types and biochemical substances in the brain. The applications of such analyses are wide ranging, although the main body of research has thus far focused on neurodegenerative diseases, such as Alzheimer's disease and dementia, particularly with respect to cortical thinning. In such cases, spatial complexity has been correlated with a number of neurodegenerating diseases (Singh et al. 2006; Young and Schuff 2008).

From a topological perspective, knottedness and entanglement serve as measures of complexity of folded proteins (Taylor 2007). Some proteins form links that can be even more complex than knots and, interestingly, the link topology is characteristic of eukaryotic organisms only (Dabrowski-Tumanski and Sulkowska 2017). Furthermore, sequence complexity relates to the topology of proteins (Romero et al. 2001; Park and Levitt 1995; Edgar 2004) while various topologically interesting structures have been identified in them (toroids, solenoids etc.). Interestingly, the higher the knot complexity, the higher the probability of knotting of a physical substance. This has been experimentally been proved in the case of ring polymers (Shimamura and Deguchi 2002). The importance of spatial complexity for DNA topology is exemplified by the role of solenoidal and plectonemic supercoils, involving topoisomerase enzymes, histones etc. (i.e. Vologodskii 1992; Champoux 2001; Bar et al. 2011).

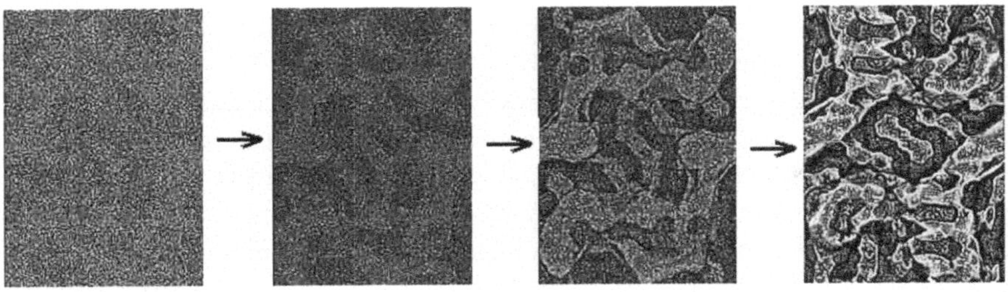

Fig. 2.7 A morphogenetic process resulting from Turing's reaction–diffusion equations. The amorphous landscape at time 1 (top left) soon becomes slightly patterned after 5 time steps (top right), then more so after 10 steps (bottom left) and even more so after 100 time steps (bottom right)

Yet, the extent to which knot invariants can be used in order to derive estimations of complexity remains an open problem. In some cases, it does seem possible (Ricca 2012), as the number of crossings of a knot (a measure of a knot's complexity) remains as the handiest simple measure of knot complexity that is also easily applicable to chemical and biochemical analyses (Vargas-Lara et al. 2017).

But spatial complexity does not leave its imprint on living beings at their DNA level only; patterns that appear on the skin or on the fur of animals are quite often interesting for their spatial complexity also (e.g. zebra strips, puma spots). Trying to explain how such patterns can be modelled and how spatially complex forms can be derived from simple differential equations has led to many researches ever since Turing proposed his set of equations of morphogenesis (Turing 1952):

$$\begin{aligned}\frac{dx_1}{dt} &= (a_1x_1 + a_2x_2^2 + a_3x_3^2) + D_1\nabla^2 x_1\\ \frac{dx_2}{dt} &= (a_4x_2 + a_5x_2x_3 + a_6x_1x_2) + D_2\nabla^2 x_2\\ \frac{dx_3}{dt} &= (a_7x_3 + a_8x_2^2 + a_9x_1x_2) + D_3\nabla^2 x_3 \end{aligned} \tag{2.3}$$

(where a_i are parameters and D_i are the diffusion coefficients) which has been the oldest but also one of the most successful models of self-organization in space (Fig. 2.7).

Other equations model other complex spatial pattern formation processes, i.e. spiral forms that are produced by the Complex Landau-Ginzburg equation (Fig. 2.8) of amplitude $f(u,t)$:

$$\frac{\partial f(u,t)}{\partial t} = (1 + ic_1)\frac{\partial^2 f(u,t)}{\partial x^2} + \varepsilon f(u,t) - (1 - ic_2)|f(u,t)|^2 f(u,t) \tag{2.4}$$

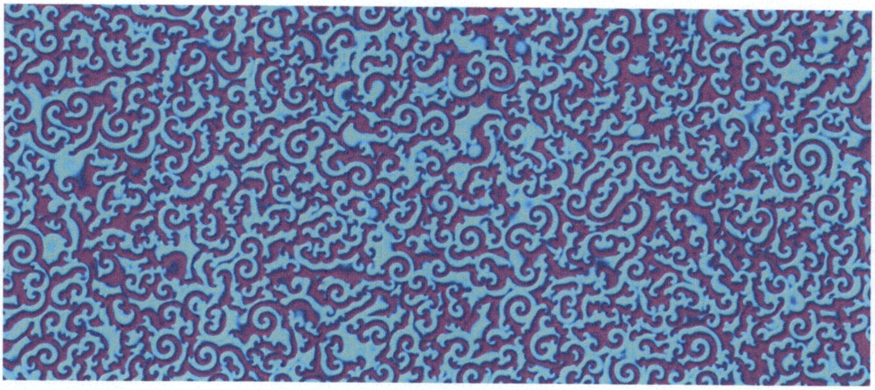

Fig. 2.8 Spirals resulting from modeling spatial diffusion, as calculated from a complex Landau Ginzburg equation

Fig. 2.9 Dendrites are a typical example of natural morphogenesis. Here, a digitally processed picture of a tree in the vicinity of the author's residence in Athens

where ε is the control parameter (instability incurs when $\varepsilon > 0$), c_1 stands for a coefficient measuring linear dispersion (the dependence of the wave frequency on the wave number) and c_2 represents the measure of nonlinear dispersion. This equation has been effectively used to describe ecological invasions affecting the parameters of oscillatory ecological systems (Reichenbach et al. 2008; Sherratt et al. 2009; Smith and Sherratt 2009).

Self-organization produces many spatially complex forms of plants and animals, both regular (geometric) and irregular (dendrites, filaments, aggregates etc.), whereas branching processes can also yield spatially complex dendritic forms of morphogenesis (Fig. 2.9).

References

Álvarez, R., Martínez, F., Vicent, J.-F., & Zamora, A. (2012). Cryptographic applications of 3x3 block upper triangular matrices. *Lecture Notes in Computer Science (including subseries Lecture Notes in Artificial Intelligence and Lecture Notes in Bioinformatics) 7209* LNAI (PART 2), 97–104.

Andrieu, E., Ladet, S., Heintz, W., & Deconchat, M. (2011). History and spatial complexity of deforestation and logging in small private forests. *Landscape and Urban Planning, 103*(2), 109–117.

Bar, A., Mukamel, D., & Kabakçoğlu, A. (2011). Denaturation of circular DNA: Supercoil mechanism. *Physical Review E, 84*(4), 041935.

Bassa, M., Chamorro, L., & Sans, F. X. (2012). Vegetation patchiness of field boundaries in the Mediterranean region: The effect of farming management and the surrounding landscape analysed at multiple spatial scales. *Landscape and Urban Planning, 106*(1), 35–43.

Bash, J. W., & Jain, K. J. (2000). Fractals and cancer. *Cancer Research, 60,* 3683–3688.

Batty, M. (2005). *Cities and Complexity*. Cambridge, MA: MIT Press.

Baynes, T. M. (2009). Complexity in urban development and management: Historical overview and opportunities. *Journal of Industrial Ecology, 13*(2), 214–227.

Berglund, N., Fernandez, B., & Gentz, B. (2007a). Metastability in interacting nonlinear stochastic differential equations: I. From weak coupling to synchronization. *Nonlinearity, 20*(11), 2551–2581.

Berglund, N., Fernandez, B., & Gentz, B. (2007b). Metastability in interacting nonlinear stochastic differential equations: II. large-N behaviour. *Nonlinearity, 20*(11), 2583–2614.

Briefer, E., Osiejuk, T. S., Rybak, F., & Aubin, T. (2010). Are bird song complexity and song sharing shaped by habitat structure? An information theory and statistical approach. *Journal of Theoretical Biology, 262*(1), 151–164.

Capri, A., Vincent, N., Vieyres, P., Poisson, G., & Makris, P. (2006). Interface areas of complexity characterization of echographic images. *Nuclear Instruments and Methods in Physics Research a, 569,* 640–644.

Casado, J.M. (2001). Coherence resonance in a washboard potential. *Physics Letters A, 291*(2,3), 82–86.

Champoux, J. (2001). DNA topoisomerases: Structure, function, and mechanism. *Annual Review of Biochemistry, 70,* 369–413.

Cross, S. S. (1994). The application of fractal geometric analysis to microscopic images. *Micron, 25*(1), 101–113.

Cross, S. S. (1997). Fractals in Pathology. *Journal of Pathology, 182,* 1–8.

Cross, S. S., Bury, J. P., Silcocks, P. B., Stephenson, T. J., & Cotton, D. W. (1994). Fractal geometric analysis of colorectal polyps. *Journal of Pathology, 172*(4), 317–323.

Dabrowski-Tumanski, P., & Sulkowska, J. I. (2017). Topological knots and links in proteins. *PNAS, 114*(13), 3415–3420.

Edgar, R. C. (2004). MUSCLE: A multiple sequence alignment method with reduced time and space complexity. *BMC Bioinformatics, 5,* 113.

Esgiar, A. N., Naguib, R. N. G., Bennett, M. K., & Murray, A. (2002). Fractal analysis in the detection of colonic cancer images. *IEEE Transactions in Information Technology in Biomedicine, 6*(1), 54–58.

Fonstad, M. (2006). Cellular automata as analysis and synthesis engines at the geomorphology-ecology interface. *Geomorphology, 7*(7), 217–234.

Forman, R. T. T., & Godron, M. (1986). *Landscape Ecology*. New York: Wiley.

Gao, J., & Li, S. (2011). Detecting spatially non-stationary and scale-dependent relationships between urban landscape fragmentation and related factors using geographically weighted regression. *Applied Geography, 31*(1), 292–302.

Goehring, L. (2013). Pattern formation in the geosciences. *Philosophical Transactions of the Royal Society a, 371,* 20120352.

Heymans, O., Fissete, J., Vico, P., Blacher, D., Mosset, D., & Browers, F. (2000). Is Fractal geometry useful in medicine and biomedical sciences? *Medical Hypotheses, 54,* 360–366.

Hovel, K. A., & Lipcius, R. N. (2002). Effects of seagrass habitat fragmentation on juvenile blue crab survival and abundance. *Journal of Experimental Marine Biology and Ecology, 271*(1), 75–79.

Hovel, K. A. (2003). Habitat fragmentation in marine landscapes: Relative effects of habitat cover and configuration on juvenile crab survival in California and North Carolina seagrass beds. *Biological Conservation, 110*(3), 401–412.

Hyde, S. T., & Schroder-Turk, G. E. (2012). Geometry of interfaces: Topological complexity in biology and materials. *Interface Focus, 2,* 529–538.

Jia, H., Li, Y., & Yu, D. (2018). Normalized spatial complexity analysis of neural signals. *Scientific Reports, 8,* 7912.

Libeskind, N. I., van de Weygaert, R., et al. (2017). Tracing the cosmic web. *Monthly Notices of the Royal Astronomical Society, 473*(1), 1195–1217.

Lineweaver, C. H., Davies, P. C. W., & Ruse, M. (Eds.). (2013). *Complexity and the Arrow of Time*. Cambridge: Cambridge University Press.

Llausàs, A., & Nogué, J. (2012). Indicators of landscape fragmentation: The case for combining ecological indices and the perceptive approach. *Ecological Indicators, 15*(1), 85–91.

Lorenz, E. N. (1963). Deterministic non-periodc flow. *Journal of Atmospheric Science, 20,* 130–141.

Malamud, B. D., & Turcotte, D. L. (1999). Self-organized criticality applied to natural hazards. *Natural Hazards, 20,* 93–116.

Mandelbrot, B. (1983). *The Fractal Geometry of Nature*. New York: Freeman.

Manrubia, S. C., & Sole, R. V. (1996). Self-organized criticality in rainforest dynamics. *Chaos Solitons and Fractals, 7,* 523–541.

Mas, J. F., Pérez-Vega, A., & Clarke, K. C. (2012). Assessing simulated land use/cover maps using similarity and fragmentation indices. *Ecological Complexity, 11,* 38–45.

Murray, B., & Fonstad, M. (2007). Preface: Complexity (and simplicity) in landscapes. *Geomorphology, 91*(3–4), 173–177.

Morozov, A., Petrovskii, S., & Li, B. L. (2006). Spatiotemporal complexity of patchy invasion in a predator-prey system with the Allee effect. *Journal of Theoretical Biology, 238*(1), 18–35.

Normant, F., & Tricot, C. (1991). Method for evaluating the fractal dimension of curves using convex hulls. *Physical Review A, 43*(12), 6518–6525.

Pahl-Wostl, C. (1995). *The Dynamic Nature of Ecosystems: Chaos and Order Entwined*. New York: Wiley.

Papadimitriou, F. (2002). Modelling indicators and indices of landscape complexity: An approach using GIS. *Ecological Indicators, 2,* 17–25.

Papadimitriou, F. (2009). Modelling spatial landscape complexity using the levenshtein algorithm. *Ecological Informatics, 4*(1), 51–58.

Papadimitriou, F. (2010a). Geo-mathematical modelling of spatial ecological complex Systems: An evaluation. *Geography Environment Sustainability, 1*(3), 67–80.
Papadimitriou, F. (2010b). Conceptual modelling of landscape complexity. *Landscape Research, 35*(5), 563–570.
Papadimitriou, F. (2012a). Artificial intelligence in modelling the complexity of mediterranean landscape transformations. *Computers and Electronics in Agriculture, 81,* 87–96.
Papadimitriou, F. (2012b). Modelling landscape complexity for land use management in Rio de Janeiro Brazil. *Land Use Policy, 29*(4), 855–861.
Papadimitriou, F. (2012c). The algorithmic complexity of landscapes. *Landscape Research, 37*(5), 599–611.
Papadimitriou, F. (2013). Mathematical modelling of land use and landscape complexity with ultrametric topology. *Journal of Land Use Science, 8*(2), 234–254.
Park, B. H., & Levitt, M. (1995). The complexity and accuracy of discrete state models of protein structure. *Journal of Molecular Biology, 249,* 493–507.
Park, K.-S., Park, R.-H., & Kim, Y.-G. (2011). Face detection using the 3x3 block rank patterns of gradient magnitude images and a geometrical face model. *Digest of Technical Papers - IEEE International Conference on Consumer Electronics*, art. no. 5722867, 793–794.
Pütz, S., Groeneveld, J., Alves, L. F., Metzger, J. P., & Huth, A. (2011). Fragmentation drives tropical forest fragments to early successional states: A modelling study for Brazilian Atlantic forests. *Ecological Modelling, 222*(12), 1986–1997.
Rai, V. (2004). Chaos in natural populations: Edge or wedge? *Ecological Complexity, 1*(2), 127–138.
Reichenbach, T., Mobilia, M., & Frey, E. (2008). Self-organization of mobile populations in cyclic competition. *Journal of Theoretical Biology, 254,* 368–383.
Ricca, R. L. (2012). Tackling fluid tangles complexity by knot polynomials. *AIP Conference Proceedings, 1479*(1), 646–649.
Rohde, K. (2005). Cellular automata and ecology. *Oikos, 110*(1), 203–207.
Romero, P., Obradovic, Z., Li, X., Garner, E. C., Brown, C. J., & Dunker, A. K. (2001). Sequence complexity of disordered protein. *Proteins, 42,* 38–48.
Satulovsky, J. E. (1997). On the synchronizing mehcanism of a class of cellular automata. *Physica A: Statistical Mechanics and Its Applications, 237,* 52–58.
Schulz, L., Ischebeck, A., Wriessneggera, S. C., David Steyrla, D., & Gernot, R. (2018). Action affordances and visuo-spatial complexity in motor imagery: An fMRI study. *Brain and Cognition, 124,* 37–46.
Sherratt, J. A., Smith, M. J., & Rademacher, J. D. M. (2009). Locating the transition from periodic oscillations to spatiotemporal chaos in the wake of invasion. *Proceedings of the National Academy of Sciences of the USA, 106,* 10890–10895.
Shrestha, M. K., York, A. M., Boone, C. G., & Zhang, S. (2012). Land fragmentation due to rapid urbanization in the Phoenix metropolitan area: Analyzing the spatiotemporal patterns and drivers. *Applied Geography, 32*(2), 522–531.
Shimamura, M. K., & Deguchi, T. (2002). Knot complexity and the probability of random knotting. *Physical Review E, 66*(4), 4.
Singh, V., Chertkow, H., Lerch, J. P., Evans, A. C., Dorr, A. E., & Kabani, N. J. (2006). Spatial patterns of cortical thinning in cognitive impairment and Alzheimer's disease. *Brain, 129*(Pt11), 2885–2893.
Smith, M. J., & Sherratt, J. A. (2009). Propagating fronts in the complex Ginzburg-Landau equation generate fixed-width bands of plane waves. *Physical Review E, 80,* 046209.
Tambasco, M., Costello, B. M., Kouznetsov, A., Yau, A., & Magliocco, A. M. (2009). Quantifying the architectural complexity of microscopic images of histology. *Micron, 40,* 486–494.
Taylor, W. R. (2007). Protein knots and fold complexity: Some new twists. *Computational Biology and Chemistry, 31*(3), 151–162.
Timbo, C., da Rosa, L. A. R., Goncalves, M., & Duarte, S. B. (2009). Computational cancer cells identification by fractal dimension analysis. *Computer Physics Communications, 180,* 850–853.

Tlidi, M., Clerc, M. G., Escaff, D., Couteron, P., Messaoudi, M., Khaffou, M., & Makhoute, A. (2018). Observation and modelling of vegetation spirals and arcs in isotropic environmental conditions: Dissipative structures in arid landscapes. *Philosophical Transactions of the Royal Society A, 376,* 20180026.

Turchin, P., & Taylor, A. D. (1992). Complex Dynamics in ecological time series. *Ecology, 73,* 289–305.

Turing, A. (1952). The chemical basis of morphogenesis. *Philosophical Transactions of the Royal Society of London. Series B, Biological Sciences, 237*(641), 37–72.

Vargas-Lara, F., Hassan, A. M., Mansfield, M. L., & Douglas, J. F. (2017). Knot energy, complexity and mobility of knotted polymers. *Scientific Reports, 7*(1), 13374.

Vologodskii, A. (1992). *Topology and Physics of Circular DNA*. Boca Raton, FL: CRC Press.

Werner, B. T. (1999). Complexity in natural landform patterns. *Science, 284*(5411), 102–104.

White, R., & Engelen, G. (1994). Cellular dynamics and GIS: Modelling spatial complexity. *Geographical Systems, 1*(3), 237–253.

Wu, J., & Hobbs, R. (2002). Key issues and research priorities in landscape ecology: An idiosyncratic synthesis. *Landscape Ecology, 17,* 355–365.

Young, K., & Schuff, N. (2008). Measuring structural complexity in brain images. *Neuroimage, 39,* 1721–1730.

Part II
The Mathematical Basis of Spatial Complexity

Chapter 3
The Geometric Basis of Spatial Complexity

Classifying geometrical objects by their degrees of symmetry represents a sharp departure from the traditional classification of geometrical figures by their essences
(*Manuel De Landa* 2002, *p. 17*)

Abstract Spatial complexity emerges even from simple geometric objects, once they are arranged at non-trivial geometric positions. The geometric context of spatial complexity depends on the presence (or absence) of symmetries, orthogonality, number of intersections and geometry type (Euclidean or other). A simple spatial relationship (i.e. orthogonality of two sides of the triangle) makes the calculation of various geometric features (i.e. area, volume) less demanding in terms of operations required and hence, computation time and resources. Thus, key spatial details such as the relative position of two or more geometric objects and their intersections (regardless of their sizes) result in substantial differences in spatial complexity. Beyond these, research in polyominoes has furnished various computational complexity results, that are useful for the analysis of spatial complexity on squared surfaces.

Keywords Spatial complexity · Combinatorial complexity · Geocomputation · Coputational Geometry and Complexity · Polyominoes · Map Complexity

3.1 Orthogonality

Inequality is the cause of anomalies in nature
Thus we have gone through the origin of inequality
"Αἰτία δὲ ἀνισότης αὖ τῆς ἀνωμάλου φύσεως
ἀνισότητος δὲ γένεσιν μὲν διεληλύθαμεν"
(Plato, 428–348 b.C., "Timaeus", 58a)

In fact, even the tiniest spatial differences matter a lot for spatial complexity. In landscape analysis this has been examined before (Papadimitriou 2002), but to see why this is so, consider the calculation of the area of the simplest 2d spatial shape (the triangle) from the lengths of its sides: it is much simpler to calculate the area

F. Papadimitriou, *Spatial Complexity*, https://doi.org/10.1007/978-3-030-59671-2_3

of a right triangle with side lengths measured as *a*, *b* and *c*, than that of a scalene triangle. As known from high school, the area of a right triangle is calculated from Pythagoras' rule: *ab*/2, but the area of the scalene triangle is calculated from Heron's rule:

$$\sqrt{\left(\left(\frac{a+b+c}{2}\right)\left(\frac{a+b+c}{2}-a\right)\left(\frac{a+b+c}{2}-b\right)\left(\frac{a+b+c}{2}-c\right)\right)} \quad (3.1)$$

or (most often expressed in an abbreviated form) as:

$$\sqrt{\tau(\tau-a)(\tau-b)(\tau-c)}, \text{ where } \tau = (\alpha + \mathrm{b} + \mathrm{c})/2 \quad (3.2)$$

A simple spatial relationship (orthogonality of two sides of the triangle) therefore makes the calculation significantly less consuming in terms of computation time and resources. Thus key spatial details, such as the *relative position* of two or more geometric objects regardless of their sizes, result in substantial differences in spatial complexity. To verify this, one need only consider the algorithmic side of this calculation. A simple measure of complexity can be the number of arithmetic operations required to measure the area included within each triangle. The algorithm calculating a right triangle's area consists in *two* algebraic operations only (one multiplication followed by one division), but the algorithm for the calculation of the area of a scalene triangle requires as many as *ten* such operations (two additions, one division, three subtractions, three multiplications and one square root). One might be tempted to consider that calculating the area of a scalene triangle is more computationally expansive, because it is more "irregular" than that of a right triangle. This is true, but how much more irregular is it? Slightly so. In fact, only two of the three lines defining each triangle have a special relative position (a right angle) but precisely this special geometric relationship suffices to make a significant difference in the algorithmic process of the calculation of the area defined by each triangle: differences in geometric properties therefore imply differences in algorithmic procedures.

"Small" but key spatial details may be responsible for significant differences in spatial complexity: the calculation of the area of any scalene triangle is always more complex than that of a right triangle. And this is not because of the triangles' particular sizes, locations or orientations in space: it is a general property that applies to all triangles, however large or small they may be and whatever Euclidean space they may be embedded in: calculating the area of the tiniest scalene triangle will always be more computationally expansive than that of the hugest right triangle.

Expectedly, as the geometry of a surface changes, the calculation of spatial complexity may become more demanding. Indeed, the calculation of the area *A* of a spherical triangle with sides *a*, *b*, *c* is possible from a variant of Heron's formula for planar triangles as follows:

$$\tan\left(\frac{A}{4}\right) = \sqrt{\sin p \sin(p-a) \sin(p-b) \sin(p-c)}, \quad (3.3)$$

where $p = (a + b + c)/2$.

Hence, in the case of the spherical triangle, there are more complex operations to be made compared with the planar triangle and the calculation of the area of a planar triangle, however large it is, is always easier compared with that of a spherical triangle, however small that is. Changes in geometry therefore induce changes in spatial complexity.

3.2 Intersections

5. A surface is what has only length and breadth.

6. The extremities of a surface are lines.

"ε΄. Ἐπιφάνεια δέ ἐστιν, ὃ μῆκος καὶ πλάτος μόνον ἔχει.

ς΄. Ἐπιφανείας δὲ πέρατα γραμμαί".

(Euclid, fourth century b.C.,"Elements", Book A)

Let us now consider another example, showing how simple geometric elements can generate spatial complexity. We know that L lines on the plane intersect at $p_{\max}$ points:

$$p_{\max} = \frac{L(L-1)}{2} \tag{3.4}$$

Notice that this is *not* the number of intersection points generated by L lines in all cases. Specifically, the number u of possible intersections can be calculated from the formula

$$u_{L,k} = \binom{L}{k} \tag{3.5}$$

with k lines intersecting the other L-k lines.

For instance, if there are $L = 3$ lines, then the cases that $k = 0,1,2,3$ lines intersect with the remaining L-k lines are given in Fig. 3.1 and if, i.e. there are only one line ($k = 1$) intersecting the remaining L-k $= 2$ lines, then there are three cases of intersection:

$$u_{3,1} = \binom{3}{1} = \frac{3!}{2!} = 3 \tag{3.6}$$

In exactly the same way, if $L = 4$ and $k = 2$, then the number is (Fig. 3.2):

$$u_{4,2} = \binom{4}{2} = \frac{4!}{2!2!} = 6 \tag{3.7}$$

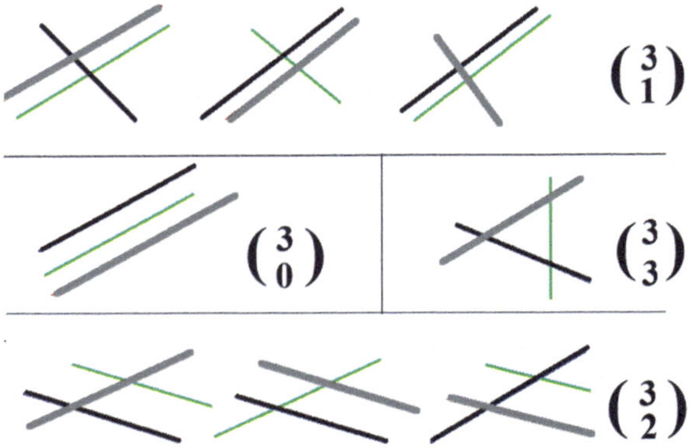

Fig. 3.1 Number of cases of intersections for $L = 3$ lines, when $k = 0$, or $k = 1$, $k = 2$ or $k = 3$ of them intersect the remaining L-k lines

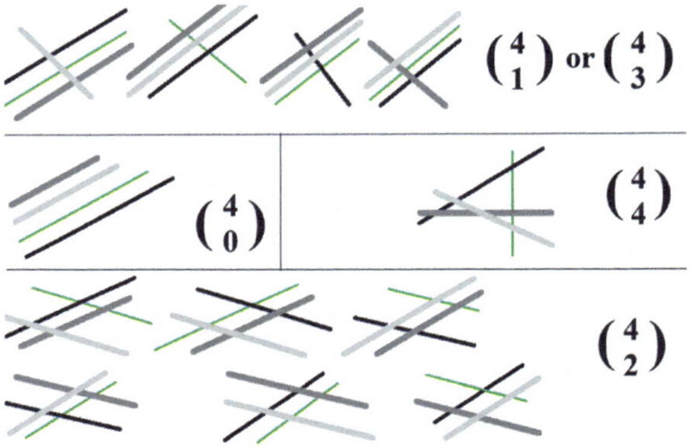

Fig. 3.2 Alternative cases of intersection of $L = 4$ different lines, when $k = 0$, 1, 2, 3, 4

Notice that the number of intersection points $p_{\max}$ result in only one case:

$$u_{L,L} = \binom{L}{L} = 1 \tag{3.8}$$

Proceeding from lines to areas, we may observe that intersecting lines define intersecting areas on the plane. Beginning with $L = 2$ lines, observe that these define one intersection point. Continuing with more lines i.e. up to $L = 7$ (as shown in Fig. 3.3), then progressively more areas are defined in between the intersecting lines and the maximum number of possible such areas is given by the formula:

$$A_{\max} = \frac{(L-1)(L-2)}{2}. \tag{3.9}$$

Now a natural question to ask is how the ratios among lines, intersection points and areas grow with increasing number of lines (L). The answer essentially lies in calculating the following limits:

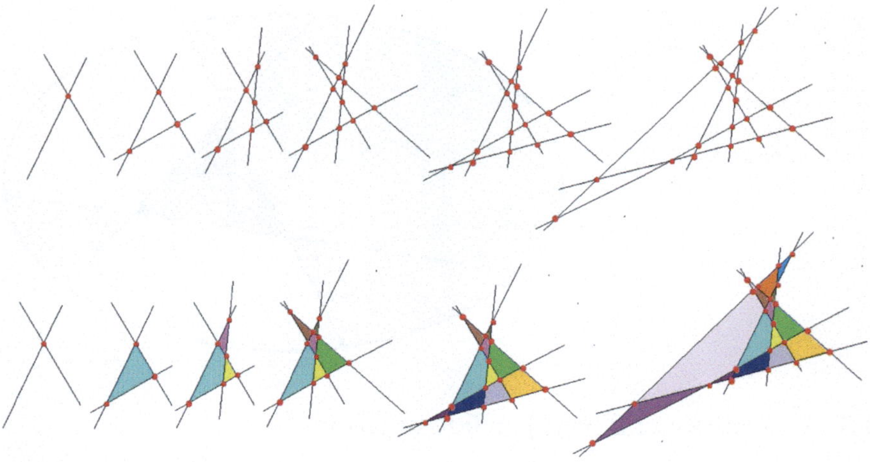

Fig. 3.3 Maximum intersection points (p_{max}) defined by 2 to 7 intersecting straight lines (up, from left to right) and regions (a_{max}) defined by these maximum intersection points (bottom, from left to right)

$$\lim_{L\to\infty} \frac{A_{max}}{p_{max}} = \lim_{L\to\infty} \left(1 - \frac{2}{L}\right) = 1 \quad (3.10)$$

$$\lim_{L\to\infty} \frac{p_{max}}{L} = \lim_{L\to\infty} \left(\frac{L}{2} - \frac{1}{2}\right) = \infty \quad (3.11)$$

$$\lim_{L\to\infty} \frac{A_{max}}{L} = \lim_{L\to\infty} \left(\frac{L}{2} - \frac{3}{2} + \frac{1}{L}\right) = \infty \quad (3.12)$$

Thus, as the number of lines tends to infinity, the number of intersection points approximates the number of areas defined in-between lines. While the complexity of calculating the maximum number of intersection points is polynomial (a function of L^2), the number of areas generated tend to infinity with respect to the number of intersecting lines. This means that as more intersecting lines are drawn on the plane, a lot more areas can be defined by their possible intersections. Otherwise stated, the more 1d objects are used to shape 2d spaces, the more such 2d spaces emerge. The previous formula was for the definition of areas only inside the intersections. If areas outside the intersection regions are also considered, then the formula giving the number of regions A is (Fig. 3.4):

$$A_{max} = \binom{L+1}{2} + 1 \quad (3.13)$$

Thus, spatial complexity can be created by simple geometric objects, once they are arranged at non-trivial geometric positions.

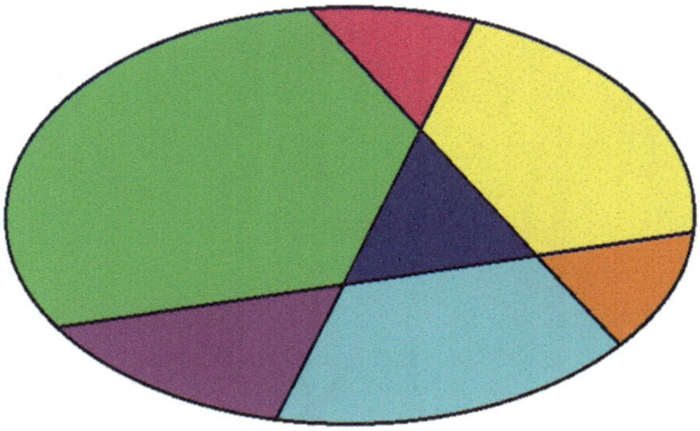

Fig. 3.4 The maximum number of areas *A* defined by *L* lines both inside and outside the lines' intersections. In this case, $L = 3$, so there are $A = 7$ regions

3.3 Curvature and Non-Euclidean Geometries

"On Red Square the earth is roundest,
its slope more firm,
on Red Square the earth is roundest,
and its slope suddenly undolds"
(Osip Mandelstam,1891–1938,"Children's Haircut", 1935)

Curvature has been considered as a determinant of complexity for curves and curved surfaces (Ujiie et al. 2012; Matsumoto et al. 2019). Among all formulas of geometry, the "isoperimetric inequality" is probably the more appropriate one to describe the degree of complexity of a shape *C* that is circumscribed by a curve $\gamma(t) = (x(t),y(t))$:

$$L^2 \geq 4\pi A \tag{3.14}$$

where *L* is the curve length and *A* is the area defined by the curve γ:

$$A = \oint_C F(x, y) \cdot d\gamma = \int_\alpha^\beta x(t)y'(t)dt \tag{3.15}$$

It is easy to verify that when the shape is a circle, then $L^2 = 4\pi A$ and, hence, the ratio:

$$L^2/4\pi \mathrm{A} \tag{3.16}$$

is a geometric assessor of the complexity of a 2d-shape: the higher the ratio, the more complex the shape is (Fig. 3.5).

The mean curvature of a surface is an old problem in geometry, which consists in seeking the surface of the largest area among all compact surfaces in the Euclidean

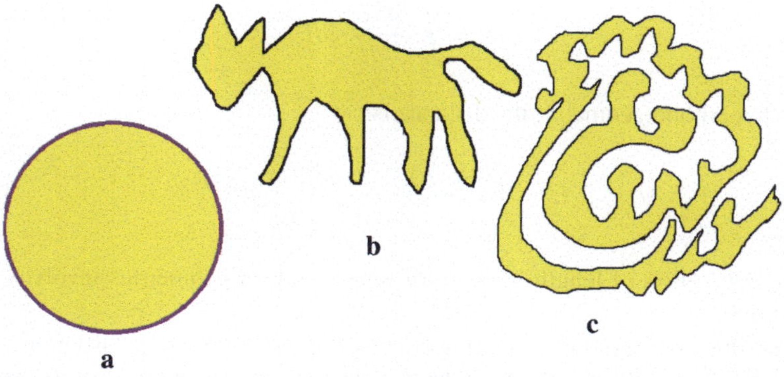

Fig. 3.5 Using the isoperimetric inequality to assess spatial complexity: Among shapes with the same area, but with different perimeter lengths, more spatially complex is the one with the higher perimeter length (**c** is more spatially complex than **b**, and **b** more complex than **a**)

space enclosing a fixed volume. The obvious solution to this problem is the sphere. But keeping the geometric properties as elementarily simple as those of a triangle and changing the geometry type induces considerable changes in spatial complexity.

In assessing the impact of curvature on spatial complexity, one needs to recall that the Gaussian curvature K_G of a surface is calculated on the basis of the three variables (E_C, F_C, G_C) of the "first fundamental form", which, as known from differential geometry, is equal to

$$E_C du^2 + 2F_C du dv + G_C dv^2 \tag{3.17}$$

leading to the well known (and cumbersome) formula:

$$K_G = \frac{\begin{vmatrix} -\frac{1}{2}E_{vv} + F_{uv} - \frac{1}{2}G_{uu} & \frac{1}{2}E_u & F_u - \frac{1}{2}E_v \\ F_v - \frac{1}{2}G_u & E_C & F_C \\ \frac{1}{2}G_v & F_C & G_C \end{vmatrix} - \begin{vmatrix} 0 & \frac{1}{2}E_v & \frac{1}{2}G_u \\ \frac{1}{2}E_v & E_C & F_c \\ \frac{1}{2}G_u & F_C & G_C \end{vmatrix}}{\sqrt{E_C G_C - F_C^2}} \tag{3.18}$$

As a result of Gauss's "Theorema Egregium", there is no planar map representing the curved surface of the earth, without distorting distances. This is because this theorem guarantees that Gaussian curvature K_G must be the same so long as there are local isometries. As the surface of a sphere has a non-zero curvature and the planar map has zero curvature, the respective K_G curvatures are different, and therefore there are no local isometries. Thus, curved surfaces generate objects of higher spatial complexity since they require more calculations to measure lengths and areas. The effect of curvature becomes more decisive for spatial complexity in the case of non-Euclidean geometries. In the case of Euclidean geometry, the length of the curve connecting points x and y is:

$$L = \int |\frac{dx(t)}{dt}| dt \tag{3.19}$$

but in a hyperbolic geometry, the distance is:

$$L = \int \frac{1}{1 - \frac{1}{4}|x(t)|^2} |\frac{dx(t)}{dt}| dt \tag{3.20}$$

so the computation of length in cases of non-euclidean geometries involves more calculations.

Increasing complexity by mounting dimension from two to three further increases the spatial complexity of a shape. Take, for instance, the calculation of the volume of a spherical tetrahedron. This is tantamount to changing the geometry to spherical and adds up more complexity to the calculations. It may come as a surprise that there is no simple formula available for the spherical tetrahedron. In fact, the volume of an arbitrary tetrahedron in a space of nonzero curvature was first calculated only shortly before the beginning of this century (Cho and Kim 1999) for spaces of hyperbolic curvature. This was later also calculated by means of other formulas (Murakami and Ushijima 2005; Murakami and Yano 2005) for the same type of curvature. Murakami (2011) obtained a formula for the volume of the spherical tetrahedron. This is given here without detailed explanation; simply for the sake of illustration of how a lot more complex the calculation of volumes gets when it extends to geometries of non-zero curvature:

$$\begin{aligned} V = {} & \mathrm{Re}\Big(\tilde{L}(b_1, b_{2,}, \ldots, b_6, \tilde{z}_0)\Big) - \pi \arg(-\tilde{q}_2) \\ & - \sum_{j=1}^{6} \frac{\partial \mathrm{Re}\Big(\tilde{L}(b_1, b_{2,}, \ldots, b_6, z)\Big)}{\partial l_j} |_{z=\tilde{z}_0} - \frac{\pi^2 \bmod (2\pi^2)}{2} \end{aligned} \tag{3.21}$$

Notice that, if not anything else, this calculation can not be effectuated without using complex numbers and the dilogarithm function.

3.4 Spatial Combinatorics and Polyominoes

Precise ideas often lead to doing nothing

"Les idées précises conduisent souvent à ne rien faire"

(Paul Valéry, 1871–1945, Mélange, 1934)

Spatial complexity also emerges from geometric combinatorics. Enumerating possible compositions of square cells resulting from geometric arrangements of simple geometric objects is a straightforward method to evaluate the combinatorial complexity of spatial patterns.

The "map coloring" problem consists in the determination of the least number of colors to color any map. In 2d, the "*Four colors problem*" was to prove that any map can be colored with no more than 4 colors, whatever the spatial arrangement of the regions shown on it. After perplexing too many people, it was solved in 1976 by K. Appel and W. Haken. A generalization was achieved with the *Colin de Verdiere number* μ(G), which is an invariant for a graph G (Colin de Verdiere 1990) and any graph with a CdV invariant μ may be colored with at most $\mu + 1$ colors. Planar graphs have $\mu = 3$ and, by means of the "Four Color Theorem" can be colored by at most $\mu + 1 = 4$ colors. Disjoint unions of paths ("linear forests") have $\mu = 1$ and therefore can be colored by at most 2 colors. Thus four colours are enough to color any map, with the exception of some exceptionally complex cases (one example of which is the "*Wada lakes*", that is regions sharing the same boundary). In the case of Wada lakes, the number of colors required is as number as the number of regions defined by the "lakes". In some sense therefore, the minimum number of colors necessary to color just *any* map remains an unsolved problem.

Aside of the colors problem, an example of combinatorial complexity emerging from spatial problems is the problem of covering the plane with polyominoes. A polyomino is a plane geometric shape formed by joining one or more squares of equal area, edge to edge (Fig. 3.6) and is classified according to how many square cells it consists in (with 2-cells it is called "domino", with three cells "triomino", with 4 cells "tetromino" and so on).

Several interesting results relate to tiling squares with polyominoes. For instance, it is known that square maps can (or can not) be tiled by *n*-ominoes depending on whether they are even–numbered or odd-numbered, i.e. a 7 × 7 square map can not be tiled by dominoes (Fig. 3.7).

The complexity of spatial arrangements of polyominoes is interesting because several problems involving them are intractable. It suffices to observe that the number of polyominoes with *n* cells increases fast, giving a glimpse of the high combinatorial complexity involved: 1, 2, 5, 12, 35, 108, 369, 1285, 4655, 17,073, 63,600, 238,591, 901,971, 3,426,576, 13,079,255, 50,107,909... (corresponding to the Sloane sequence A000105). But high combinatorial complexity in spatial

Fig. 3.6 Polyominoes of various sizes

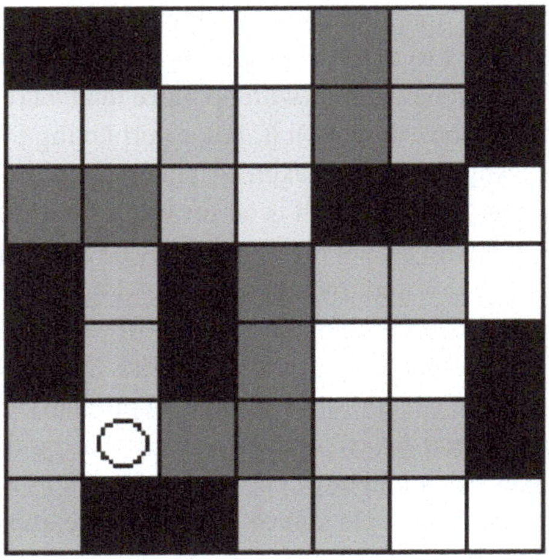

Fig. 3.7 Odd-numbered square maps (7 × 7 here) can *not* be tiled by dominoes

arrangements may emerge even from much simpler cases, i.e. when trying to practically solve simple spatial combinatorial problems.

In this respect, there already are some interesting results about combinations of particular types of 3d polyominoes (Fig. 3.8). In two dimensions, we know that a *m*x*n* rectangle can be tiled with O-tetrominoes, if and only if m and n are even, if a $n \times n$ square can be tiled with T-tetrominoes, then n^2 is divisible by 8, a $n \times n$ square can be tiled with L-trominoes with fourfold rotational symmetry if and only if n is divisible by 6, a $n \times n$ square can be tiled with L-tetrominoes with fourfold

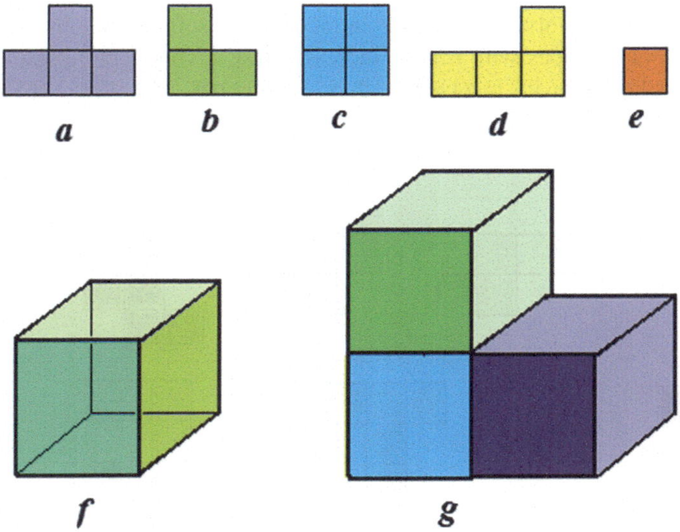

Fig. 3.8 Some types of 2d-and-3d-polyominoes: T-tetromino (**a**), L-tromino (**b**), O-tetromino (**c**), L-tetromino (**d**), Monomino (**e**), 3d monomino (**f**) and 3d L-tromino (**g**)

Fig. 3.9 A solution to the spatial combinatorial problem (see text for explanation): there are 24 alternative land cultivation schemes if one of the four pieces of land is left uncultivated (too many spatial alternative configurations to decide, even for such a simple problem)

rotational symmetry if and only if n is divisible by 4 and any $2^n \times 2^n$ square can be tiled by a monomino and trominoes (Golomb 1954; 1996).

Equivalently, there are similar (unfortunately fewer) results for polyominoes in 3d: any $2^n \times 2^n \times 2^n$ cube with $n = 1(\mathrm{mod}3)$ can be tiled with a 3d monomino and 3d L-trominoes (Starr, 2008) and a *m*x*n*x*k* parallelepiped can be packed with 3d L-trominoes if and only if *mnk* is divisible by 3 (Soifer 2010).

Spatial combinatorial problems are well known for producing big numbers quickly. Consider, for instance, the following spatial planning problem:

A farmer owns a square lot of side length x, which can be divided in four equal squares, of side length $x/2$ each. He has three types of cultivations to allocate on these four squares: arable crops, tree crops and orchards. How many alternative land cultivation schemes are there for the three crop types on the four squares pieces of land, provided that all three cultivations types should be used in that lot? The farmer also needs to know the possible spatial arrangements if one quarter of the land is left uncultivated each time.

There are 24 alternative land cultivation schemes if one of the four squares is left uncultivated (Fig. 3.9), and as many as 36 different land cultivation schemes if all squares are occupied by cultivations. Even in this simple land allocation scheme of only 4 spatial regions and 3 types of spatial entities, it is evident that the number of possible combinations becomes rapidly high; may be unacceptably high for ordinary (i.e. everyday) spatial decision-making.

References

Cho, Y., & Kim, H. (1999). On the volume formula for hyperbolic tetrahedra. *Discrete Computational Geometry, 22*(3), 347–366.

Colin de Verdière, Y. (1990). Sur un nouvel invariant des graphes et un critère de planarité. *Journal of Combinatorial Theory B, 50*(1), 11–21.

De Landa, M. (2002). *Intensive Science and Virtual Philosophy*. London: Continuum.

Golomb, S. W. (1954). Checker boards and polyominoes. *American Mathematical Monthly, 61,* 675–682.

Golomb, S. W. (1996). *Polyominoes: Puzzles, Patterns, Problems, and Packings*. Princeton NJ: Princeton University Press.

Matsumoto, T., Sato, K., Matsuoka, Y., & Kato, T. (2019). Quantification of "complexity" in curved surface shape using total absolute curvature. *Computers and Graphics, 78,* 108–115.

Murakami, J. (2011). Volume formulas for a spherical tetrahedron. arXiv:1011.2584v4 [math.MG] 2 May 2011.

Murakami, J., & Ushijima, A. (2005). A volume formula for hyperbolic tetrahedra in terms of edge lengths. *Journal of Geometry, 83,* 153–163.

Murakami, J., & Yano, M. (2005). On the volume of a hyperbolic and spherical tetrahedron. *Communications in Analytic Geometry, 13,* 379–400.

Papadimitriou, F. (2002). Modelling indicators and indices of landscape complexity: An approach using GIS. *Ecological Indicators, 2,* 17–25.

Soifer, A. (2010). *Geometric Etudes in Combinatorial Mathematics*. New York: Springer.

Starr, N. (2008). Tromino tiling deficient cubes of side length $2n$. *Geombinatorics, XVIII, 2,* 72–87.

Ujiie, Y., Kato, T., Sato, K., & Matsuoka, Y. (2012). Curvature entropy for curved profile generation. *Entropy, 14,* 533–558.

Chapter 4
The Probabilistic Basis of Spatial Complexity

From Chaos became Erebus and the black Night;
From Night were born the Aether and the Day,
"ἐκ Χάεος δ' Ἔρεβός τε μέλαινά τε Νὺξ ἐγένοντο:
Νυκτὸς δ' αὖτ' Αἰθήρ τε καὶ Ἡμέρη ἐξεγένοντο"
(Hesiod, c.750 b.C., "Theogony" 104,123-125)

Abstract This chapter sets the basis for a probabilistic approach to spatial complexity, by focusing on: (a) species richness (the number of different qualitative classes /species/ population types/ categories /covers/colours/types) occupying the spatial extent of a surface, (b) entropy (measured i.e. by Shannon's formula, and reflecting the degree of equidistribution of these classes on the basis of their relative participation percentages) and (c) randomness of allocations of these classes. Evidently, the higher the number of different classes within a surface, space, or spatial object, the higher its spatial complexity. Further, the more disordered their allocation, the higher the spatial complexity also. These considerations however, ramify to a series of questions, i.e.: when is a string of symbols a random one? what alternative definitions of random strings are there and how do they impact spatial randomness?

Keywords Spatial complexity · Complexity and randomness · Entropy and Complexity · Landscape Diversity · Algorithmic complexity · Map Complexity · Landscape complexity

4.1 Spatial Entropy Versus Complexity

"A bit of entropy is a bit of ignorance"
(Seth Lloyd 2007, p.80)

While analyzing spatial complexity, one should always look for the simplest methods possible, despite the fact that spatial settings are "interesting" exactly when they are *not* simple (although they may be pleasant nevertheless). Understanding the relationship between entropy and complexity is a multi-faceted issue cropping up in many domains of scientific enquiry. This relationship has attracted the attention of

F. Papadimitriou, *Spatial Complexity*, https://doi.org/10.1007/978-3-030-59671-2_4

physicists and computer scientists time and again. Generally, the higher the number of different classes (species/populations/categories/covers/colours/types...) within a surface or spatial object, the higher the spatial complexity. But this also depends on the scale under which the object is examined, as different classes or objects can be packed together so densely that they might even accept an easy synoptic description (Fig. 4.1).

The number of classes present on a spatial surface is only one indication of its spatial complexity; the relative participation of each one of these classes in the object examined is another. The more different classes participate to cover the surface, the higher the spatial complexity. It is common in the scientific literature of ecology, cartography and geography to define the entropy of a map by using the Shannon formula (Shannon and Weaver 1949; Forman and Godron 1986):

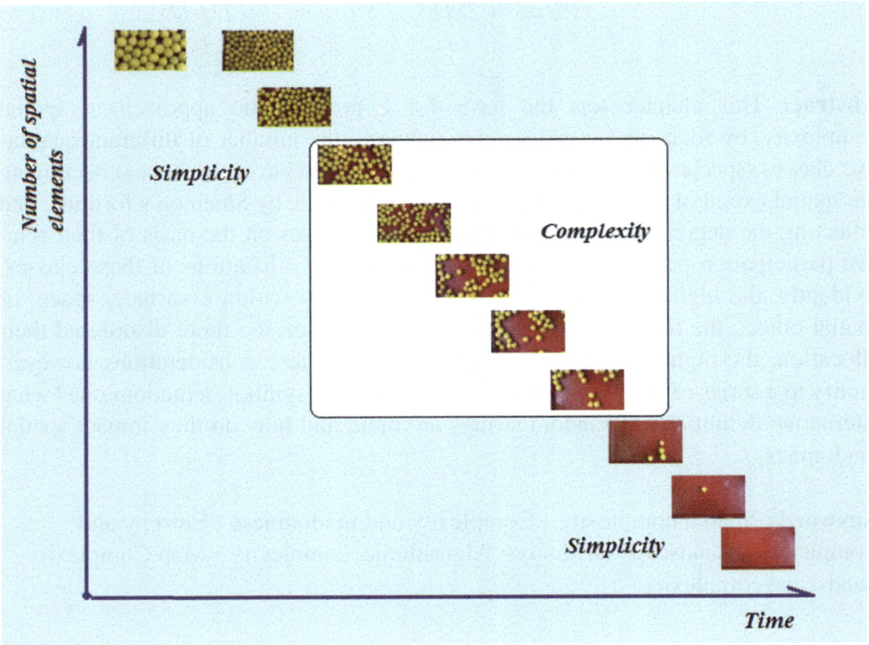

Fig. 4.1 Progressively subtracting fruits from a tray gives a glimpse of spatial complexity: a high number of spatial objects (fruits in this case) is not necessarily an indication of high complexity, as they can be packed together so densely that an overly simplifying description might be attributed to them (i.e. "a fruity area"). The same applies when the spatial objects are too few against the empty backdrop (i.e. "a map with one object in it"). Yet, when both the spatial objects and their backdrop are present at comparable percentages, the increasing difficulty of description becomes obvious. This difficulty is a measure of spatial complexity of the surface which in this case is defined by the number of different covers/categories present on the surface

$$H = -\sum_{i=1}^{V} Q_i \log(Q_i) \tag{4.1}$$

where V is the total number of "colors" present in the map depending on the map examined (i.e. if it is geographic, then V represents land cover classes, land use types, landscape types, population types etc) and Q_i is the percentage of occurrence of each color i in the map's area, with $\sum_{i=1}^{V} Q_i = 1$.

If the entire map is covered by one color only, then H=0. With an increase in the number of colors V, entropy increases, but the entropy formula does not take into account the spatial allocation of these classes, as the maps (*b*) and (*c*) show (Fig. 4.2).

Thus, entropy is only a relative (although highly important) indicator of spatial complexity. Let us see two other examples (Fig. 4.3). The maps of the two upper rows have higher entropy than those of the two bottom rows: H=1 versus H=1.5 respectively. The two upper rows show maps with two colors only, while the two bottom rows show maps of equal size with three colors. The maps with three colours have a higher entropy than the maps with two colours, regardless of the spatial allocation of these colours. These also show why entropy alone is not a sufficient criterion of spatial complexity, since different spatial configurations (and therefore different spatial complexities) can correspond to exactly the same entropy values (Papadimitriou 2012).

In biogeography and ecology, the term "*diversity*" is widely used. In its general form, diversity means the identification of the characteristics of a map, as reflected by the number of different classes of elements in it (i.e. as reflected by the "diversity" of its elements; a term used in spatial ecology) and consists a recurrent theme in ecological research (Clarke and Warwick 1998; Anand and Orloci 2000; Petrovskaya et al. 2006; McShea 1991; Magurran 2004). The connection between diversity and complexity was discussed in ecological context by Zhang et al. (2009), who found that an increase in Shannon diversity appeared concurrently with increasing "landscape complexity". Yet, other studies based on field observations have shown that spatial complexity and entropy (diversity) are not always positively correlated.

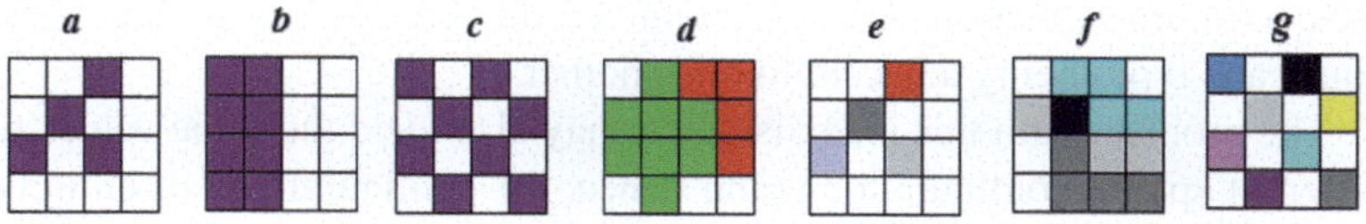

	a	b	c	d	e	f	g
Number of cover types	2	2	2	3	5	6	8
Shannon entropy	0.81875	1	1	1.5	1.875	2.3075	2.5

Fig. 4.2 Increasing spatial differentiation results in increasing Shannon entropy (calculated with a logarithm basis 2). Notice however, that spatial configurations with the same numbers of covers and the same cover percentages may yield the same entropy values for entirely different spatial configurations (i.e. cases *b* and *c*)

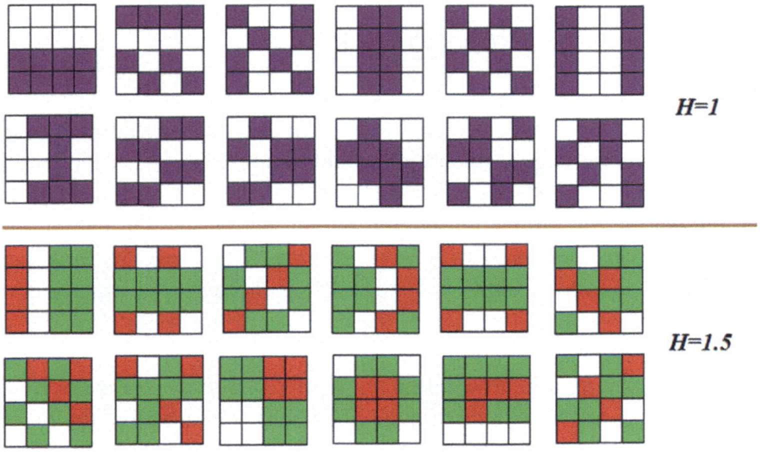

Fig. 4.3 Two sets of maps: the maps of the upper set of two rows are binary and have $H=1$ while the two lower sets of rows have three colors and $H=1.5$ In both cases, the entropy values are the same for each set of maps. Thus, within each set of rows, entropy alone can not help us distinguish between spatial configurations with different spatial complexities each

Species richness and complexity are not always correlated either (Azovsky 2009, p.308).

Besides entropy, another index needs to be parenthetically mentioned, which is *contagion*. The "contagion" index was proposed by O'Neill et al. (1988), Turner (1989; 1990) and Turner and Ruscher (1988) to characterize spatial (landscape) patterns in landscape maps:

$$2n\ln(n) - \sum_{i=1}^{n}\sum_{j=1}^{n} Q_{ij}\ln(Q_{ij}) \quad (4.2)$$

where n is the total number of cover types in the geographical space (or landscape), Q_{ij} is the probability of cover type i being adjacent to cover type j, and $2n\ln(n)$ is the maximum contagion, which is attained if there is an equal probability of any two landscape types being adjacent to one another.

The problem with this index is that it may also yield the same values for entirely different spatial configurations (and hence, for configurations of entirely different topology and spatial complexity). As shown in six example 3×3 binary maps (Fig. 4.4), the 1st binary map has only one black cell, the 2nd has two, while the 3d has 3 and the 4th has four. Yet, all these maps have the same contagion index. The same applies to the binary maps 5 and 6: both have the same contagion, although their spatial configurations are completely different. But different spatial configurations most likely have different spatial complexities (this will be examined in detail in next chapters). Consequently, contagion can not be taken as a measure of spatial complexity.

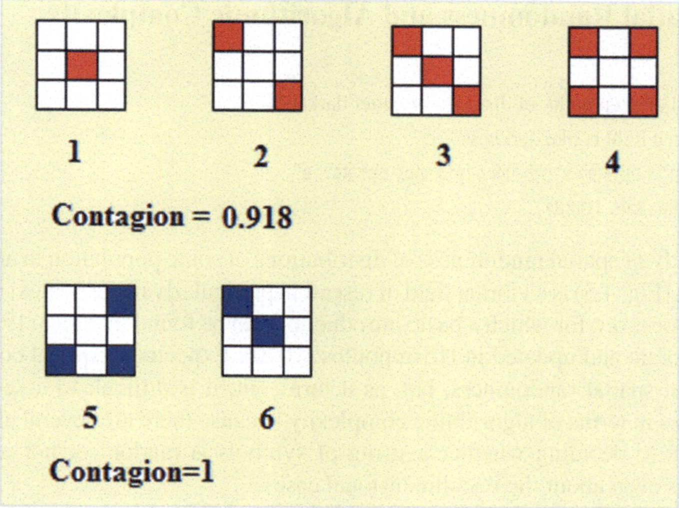

Fig. 4.4 Small binary maps with the same contagion but with different entropy and different complexity (at least, as perceived visually). Binary maps 1 to 4 have the same contagion (equal to 0.918), but their entropies are different (both entropy class r and Shannon entropy H are different in each one of them). The same applies to binary maps 5 and 6

However, it is not only maps that can demonstrate the central role of entropy. As Steinhaus (1954) suggested, there is an interesting association of the concept of entropy to the complexity of a curve: the number of intersections of a plane curve and a random line intersecting that curve is equal to $2L/C$, where L is the curve's length and C is the length of the boundary of the curve's convex hull. Plugging this into Shannon's entropy formula and defining (arbitrarily) as the "temperature" of the curve the quantity

$$\log_2\left(\frac{2L}{2L-C}\right) \tag{4.3}$$

it is possible to derive a thermodynamic analogue of curve complexity. Supposedly, this link between geometry and thermodynamics gives a measure of the "entropy" of a curve (Mendes France 1983; Dupain et al. 1986):

$$H_{curve} = \log_2\left(\frac{2L}{C}\right) + \frac{\log_2\left(\frac{2L}{2L-C}\right)}{e^{\log_2\left(\frac{2L}{2L-C}\right)} - 1} \tag{4.4}$$

The entropy of a curve is also an indicator of its complexity, i.e. if the curve is described by a polynomial of degree d, then its entropy is at most $1+\log_2 d$ (Stewart 1992).

4.2 Spatial Randomness and Algorithmic Complexity

> To a land of deep night, of disorder and utter darkness,
> where even light is like darkness
> ”פ :לְפָא־וּמְכ עַפְתַו םיִרָדְס אֹלְו תֶוָמְלַצ לְפֹא וְמְכ | הָתְפֵיַע ץֶרֶא“
> (The Bible, Job, 10.22)

The study of spatial randomness of distributions of some population in any spatial dimension (Fig. 4.5) is a vibrant field of research, particularly in the context of *spatial random processes*, for which a basic introduction can be found in Adler (1981) and a more elaborate and updated in Hristopoulos (2020). Expectedly, spatial complexity depends on spatial randomness, but, as it turns out, it is difficult to assess spatial randomness in terms of algorithmic complexity because there are several alternative approaches to deciding whether a string of symbols is random or not (diverging approaches even about the one dimensional case).

In an early approach, von Mises (1919) defined an infinite binary string as random, so long as it has as many 0s as 1s at its "limit". Adopting a different approach, Church (1940) defined a random string as every infinite string of which the digits can not be given by a recursive function. Later, Martin-Löf (1966a, b) suggested that random

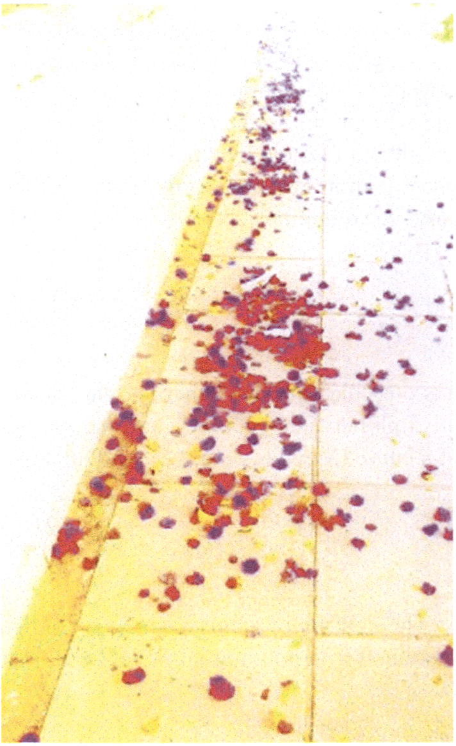

Fig. 4.5 An allocation of flowers on a street created by natural forces acting both deterministically (i.e. gravity) and stochastically (i.e. affected by changing wind speed and direction), thus producing a not entirely random spatial allocation

infinite strings are those that satisfy all statistical tests for randomness. Levin (1973; 1974) and Chaitin (1974; 1975) defined random strings x as those that are endowed with a maximum Chaitin-Levin complexity, meaning there exists a number c, such that for every n, this complexity is higher than the difference n-c. Following the most widely known definition by Kolmogorov (1965) however, an infinite string x is random, if its Kolmogorov complexity $K(x)$ is maximum.

An alternative approach to randomness is Bennett's concept of "logical depth" of an object (Bennett 1973, 1982, 1986, 1988a, b, c, 1990), measuring the time required to compute a minimal program, with its organisation of the studied object. The "logical depth" of a string is the calculation time needed (by a universal machine), in order to produce it from its minimum Kolmogorov description. Bennett's definition has had some applications in physics and biology (Bennett 1986, 1988b) and one of the implication of Bennett's theory is that the possibility to encounter by chance an object with large logical depth is very small. Plausibly, the possibility that Bennett's definition might be used to assess spatial complexity should not be precluded. However, the problem with the definition of randomness becomes explicit even by considering simple cases. Consider, for instance, a binary string composed of 0s and 1s. Let the sum of the string's elements be N. If the string is random in the Church sense, then:

$$\lim_{n \to \infty} \left(\frac{N}{n} \right) = \frac{1}{2} \tag{4.5}$$

But there are strings satisfying this equation at the limit, which are clearly *non-random* (i.e. the string 01010101010101010101...). This simple example illustrates why the riddle of defining string randomness remains unsolvable even for one-dimensional objects. Further, many strings that happen to satisfy the criteria of Von Mises and Church fail to do so for Martin-Löf's criteria. But characterizing a string as random in the sense of Kolmogorov means that it accepts no shorter algorithmic description and therefore it has no regularity at all and passes the statistical tests required by Martin-Löf's definition.

Yet, diverging opinions with respect to strings can be seen in the case of 2d surfaces also. For instance, complexity is often perceived to be a condition "in between order and chaos", in such a way that (i.e. in Fig. 4.6), neither map *a* (ordered) nor map *c* are "complex", because *a* is ordered and *c* is random. But map *b* is perceived as "complex", because it is found in between order and randomness *and* displays distinct patterns (such as clumps of the same colour in B).

Spatial stochasticity can be created by a 2d Brownian stochastic motion. Here, for the purpose of illustration, it is plotted for 400,000 time steps (Fig. 4.7). As the east-west Brownian motion is tantamount to the north-south motion, the joint probability of being at a position u along the horizontal axis and v along the vertical axis is the product of the Gaussian probability densities of the respective motions:

$$P(u, v, t) = \frac{1}{2\pi t} e^{-\frac{(u^2+v^2)}{2t}} \tag{4.6}$$

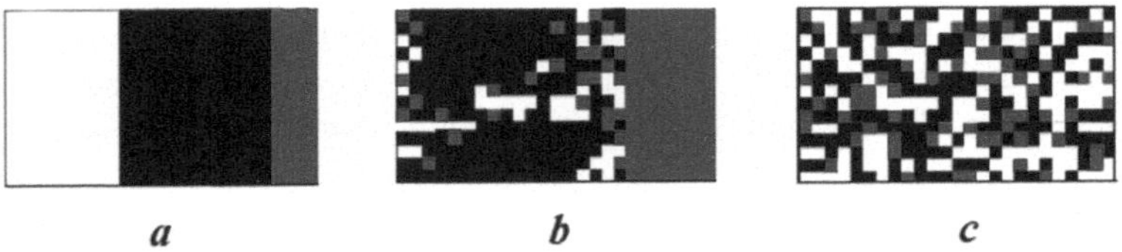

Fig. 4.6 Different perceptions of what is "complex" affects the impression of what a spatially complex surface is. According to some interpretations, map *b* is more complex than both maps *a* and *c*, because it displays both order (patterns of the same colour) and randomness (at the bottom right quadrant). But following the theory of algorithmic complexity, a spatial region is more complex if it is more difficult to describe by an algorithm and if it is closer to randomness: in this way, map *b* is more complex than *a*, while *c* (a random allocation of the three colours) is more complex than *b*

Fig. 4.7 A stochastic Brownian motion over an empty space. When the number of time steps gradually increases (up to 400,000 here), complex patchy, rugged spatial forms emerge

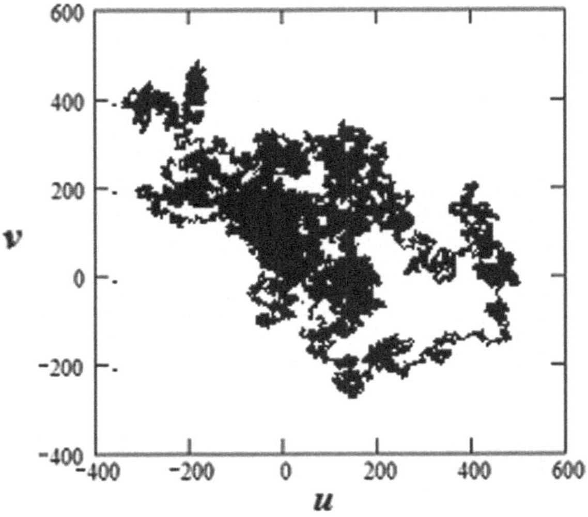

When a rugged surface (noisy, with "ups" and "downs") is examined, its spatial complexity will be even more dependent on the level of spatial resolution at which it is examined. At this point enter issues of choice and technical capabilities: one simply has to experiment with surfaces that can be constructed on the basis of Gaussian-type functions describing a terrain (Fig. 4.8) and follow the general type (a_i,b_i,c_i are constants):

$$F(x, y) = c_1 e^{-(a_1 x^2 + a_2 y^2 + a_3 x + a_4 y + a_5)} + c_2 e^{-(b_1 x^2 + b_2 y^2 + b_3 x + b_4 y + b_5)} \tag{4.7}$$

In a rasterized map, a string of symbols can represent a strip of squares (or parallelograms) of a spatial object. The algorithmic complexity of a string of symbols is equal to the minimum description of this string. In one dimension, if a string of symbols is random, then it has maximum algorithmic complexity; equivalently, if the allocation of colors/covers is random over a map (Fig. 4.9), then it has maximum

Fig. 4.8 The ruggedness of a surface is a parameter indicating its spatial complexity (*a*). The degree of ruggedness essentially reflects the degree of randomness of a surface. The two-dimensional cross-section of a three-dimensional landscape that is created by using Gausian-type distributions for *x* and *y* can be given by a function F(x, y) that describes the "altitude" of the landscape, such as the profile shown in *b*

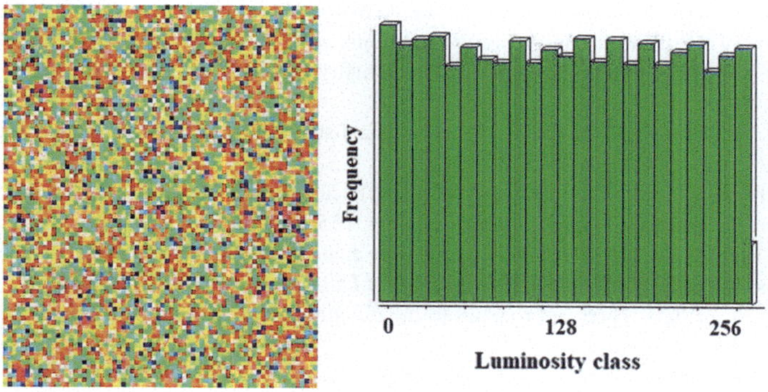

Fig. 4.9 A random image, generated by a computer. It has almost equiprobable allocation of cells per luminosity value (histogram on the right) and zero Moran's autocorrelation index. Such images have maximum algorithmic complexity in the Kolmogorov sense

complexity (with respect to any other such map of the same size and with the same amount of covers/colors). If it has repeating patterns, then its description can be reduced to a simpler one, by taking advantage of these repetitions, and in that case, it has a lower complexity than an incompressible string of symbols.

References

Adler, R. J. (1981). *The Geometry of Random Fields*. New York: Wiley.

Anand, M., & Orloci, L. (2000). On hierarchical partitioning of an ecological complexity function. *Ecological Modelling, 132,* 51–62.

Azovsky, A. (2009). Structural complexity of species assemblages and spatial scale of community organization: A case study of marine benthos. *Ecological Complexity, 6,* 308–315.

Bennett, C.H. (1973). Logical Reversibility of Computation. *IBM Journal of Research and Development,* 525–532.

Bennett, C. H. (1982). The thermodynamics of computation—A review. *International Journal of Theoretical Physics, 21*(12), 905–940.

Bennett, C. H. (1986). On the nature and origin of complexity in discrete, homogeneous locally-interacting systems. *Foundations of Physics, 16*(6), 585–592.

Bennett, C. H. (1988a). *Demons* (pp. 91–97). Janvier: Machines et Thermodynamique. Pour la Science.

Bennett, C. H. (1988b). Logical Depth and Physical Complexity. In R. Herken (Ed.), *The Universal Turing Machine: A half-century survey* (pp. 227–257). Oxford: Oxford University Press.

Bennett, C. H. (1988c). Notes on the History of Reversible Computation. *IBM Journal of Research and Development, 32*(1), 16–23.

Bennett, C.H. (1990). How to define Complexity in Physics, and Why. In W.H. Zurek, (Ed.) *Complexity, Entropy and the Physics of Information* (pp. 137–148). Santa Fe Studies in the Sciences of Complexity, vol. VIII. New York:Addison-Wesley.

Chaitin, G.J. (1974). Information-theoretic limitations of formal systems. *J.A.C.M.*, 403–424.

Chaitin, G.J. (1975). A theory of program size formally identical to information theory. *J.A.C.M. 22,* 329–340.

Clarke, K. R., & Warwick, R. M. (1998). A taxonomic distinctness index and its statistical properties. *Journal of Applied Ecology, 35,* 523–531.

Dupain, Y., Kamae, T., & Mendes France, M. (1986). Can one measure the temperature of a curve? *Archive for Rational Mechanics and Analysis, 94,* 155–163.

Forman, R. T. T., & Godron, M. (1986). *Landscape Ecology*. New York: Wiley.

Hristopoulos, D. (2020). *Random Fields for Spatial Data Modeling*. Cham: Springer.

Kolmogorov, A. (1965). Three approaches to the quantitative definition of information. *Problems of Information Transmission, 1,* 1–17.

Levin, L. A. (1973). On the notion of random sequence. *Soviet Mathematical Doklady, 14,* 1413.

Levin, L. A. (1974). Laws of information conservation and aspects of the foundation of probability theory. *Problems of Information Transmission, 10*(3), 206–210.

Lloyd, S. (2007). *Programming the Universe*. London: Vintage.

Magurran, A. E. (2004). *Measuring Biological Diversity*. Oxford: Blackwell.

Martin-Löf, P. (1966a). On the concept of a random sequence. *Theory Probability Applications, 11,* 177–179.

Martin-Löf, P. (1966b). The definition of random sequences. *Information and Control, 9,* 602–619.

McShea, D. (1991). Complexity and evolution: what everybody knows. *Biology and Philosophy, 6*(3), 303–324.

Mendes France, M. (1983). Les courbes chaotiques. *Courrier du Centre National de la Recerche Scientifique, 51,* 5–9.

O'Neill, R. V. (1988). Indices of landscape pattern. *Landscape Ecology, 1*(3), 153–162.
Papadimitriou, F. (2012). The Algorithmic Complexity of Landscapes. *Landscape Research, 37*(5), 599–611.
Petrovskaya, N. S., Petrovskii, S. V., & Li, B. L. (2006). Biodiversity measures revisited. *Ecological Complexity, 3,* 13–22.
Shannon, C. E., & Weaver, W. (1949). *The Mathematical Theory of Communication.* Urbana: University of Illinois Press.
Steinhaus, U. (1954). Length, shape and area. *Colloquium Mathematicum, 3,* 1–13.
Stewart, I. (1992). *Another fine Math you' ve got me into.* New York: Dover.
Turner, M. G. (1990). Spatial and temporal analysis of landscape patterns. *Landscape Ecology, 4*(1), 21–30.
Turner, M.G., & Ruscher, C.L. (1988). Changes in landscape patterns in Georgia, U.S.A. *Landscape Ecology, 1*(4), 241–255.
Turner, M. G. (1989). Effects of changing spatial scale on the analysis of spatial pattern. *Landscape Ecology, 3*(3/4), 153–162.
Von Mises, R. (1919). Grundlagen der Wahrscheinlich kietsrechnung. *Math.Z.*, 5, 100.
Zhang, F., Tashpolat, T., Ding, J.-L., Wang, B.-C., Wang, F., & Mamat, S. (2009). The change of land use/cover and characteristics of landscape pattern in arid areas oasis: A case study of jinghe county, xinjiang province. *Shengtai Xuebao/Acta Ecologica Sinica, 29*(3), 1251–2126.

Chapter 5
The Topological Basis of Spatial Complexity

1. A point is that of which there is no part.
2. A line is length without breadth.
3. The extremities of a line are points.
"α΄. Σημεῖόν ἐστιν, οὗ μέρος οὐθέν.
β΄. Γραμμὴ δὲ μῆκος ἀπλατές.
γ΄. Γραμμῆς δὲ πέρατα σημεῖα."
(Euclid, 4th century b.C., "Elements", Book A)

Abstract The key topological indicators of spatial complexity are: number of boundaries, genus, dimension, knottingness, braiding, linking/writhing and immersion of the surface or object: i.e. the higher the number of boundaries and/or knots (or braids) and the higher the genus and the dimension of the surface or object, the more spatially complex it is. Some initial experimentations with spatial complexity can be made with easy-to-derive estimates by using the Gage-Hamilton-Grayson theorem, Pick's theorem and intersections of lines with Jordan curves. Then, spatial complexity can be assessed in various ways by using concepts of the complexity of knots and braids, including knot polynomials, linking numbers and Grosberg-Nechaev complexity.

Keywords Spatial complexity · Knot complexity · Braid theory · Map complexity · Topology and Complexity · Knot theory · Pick's theorem

5.1 Boundaries

Ascendance to higher levels of existance
enables even distant entities to develop sympathy to one another...
... because it is easier to encounter disconnected
the elements of the earth than humans.
"ἡ ἐπὶ τὸ κρεῖττον ἐπανάβασις συμπάθειαν

F. Papadimitriou, *Spatial Complexity*, https://doi.org/10.1007/978-3-030-59671-2_5

καὶ ἐν διεστῶσιν εργάσασθαι ἐδύνατο....
...θᾶσσον γοῦν εὕροι τις ἄν γεῶδες τι μηδενὸς γεώδους
προσαπτόμενον ἤπερ ἄνθρωπον ἀνθρώπου ἀπεσχισμένον)"
(Marcus Aurelius, 121–180 a.D., Meditations, 9.9b–9.9c)

The complexity of spatial surfaces and objects depends on their essential topological properties, such as relative position, dimension, genus etc and topological criteria can serve to derive measures of spatial complexity (Papadimitriou 2002, 2013). We will begin with the Gage-Hamilton-Grayson theorem (Gage 1983, 1984; Gage and Hamilton 1986; Grayson 1987, Grayson 1989a, b), according to which, every closed non-self-intersecting curve on the plane can be simplified to a circle, following a "*curve shortening flow*" which moves each point of the curve inwards, proportionately to the curvature at that point. In this way, a spatially complex closed curve without self-intersections, can be reduced to a circle (Fig. 5.1).

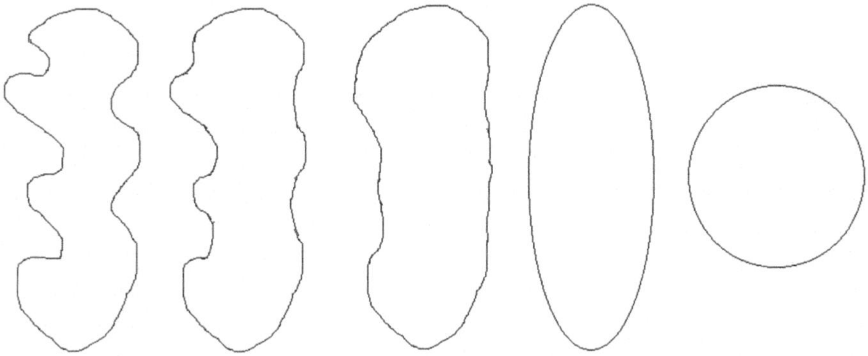

Fig. 5.1 According to the Gage-Hamilton-Grayson theorem, any spatially complex non-self-intersecting curve (on the left) can be reduced to another curve, homeomorphic to a circle, (on the right), following a "curve shortening flow"

Topology alone however, is not a sufficient criterion of spatial complexity; it always needs to be considered in tandem with other criteria (entropy, geometry etc). Take, for instance, a possible topological criterion based on the "*Jordan Curve theorem*" (proven in 1887). As well known from basic topology, this theorem asserts that a simple (non-intersecting) closed curve on the plane separates the plane into two regions, one "inside" the curve, and another "outside" of it. As a consequence of this theorem, if a point is found inside a closed curve on the plane, then there is an odd number of intersections from any point inside it towards the space that is on the outside of the curve.

Would it be plausible to assume that the maximum number of intersections might be an adequate measure of a shape's complexity? Consider for instance, four different closed curves (Fig. 5.2): a circle (*a*) and three other curves homeomorphic to the circle (*b*,*c*). Drawing a line from any point found inside the circle (*a*) to the outside of the circle *c*, there is one intersection point. In the case of the closed curve *b*, there are three intersection points and thus curve *b* is more complex than the circle. Further

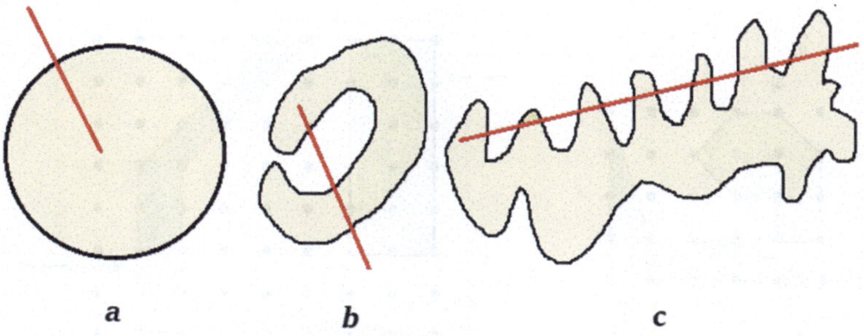

Fig. 5.2 Four different closed curves and straight lines connecting one point inside each curve with one point outside it. Generally, the more complex the shape is, the more intersection points along the lines connecting the shape's inside with its outside

still, from some point found on the inside of curve *c*, there can be as many as thirteen intersection points, so the curve *c* is even more complex than curve *b*. Although it can be objected that an infinity of curves might exist, homeomorphic to any one of those, and they could be spatially more complex and still yield the same numbers of intersections, this theorem may nevertheless be helpful to realize the relevance of boundaries for spatial complexity.

Since closed curves create boundaries that separate the inside regions from those that lie outside the boundary, the role of boundaries in spatial complexity can be further explored by means of Pick's theorem (Pick 1899), which provides a calculation of the area (A_n) of a planar shape, as measured on a square grid, from the number of points the shape has on its boundary (B_p), or enclosed in its interior (I_p), following the simple formula:

$$A_p = \frac{B_p}{2} + I_p - 1 \tag{5.1}$$

Obviously, a more complex shape will have more points (both B_p and I_p) than a simple one. Might then A_p, B_p and I_p be useful to estimate the spatial complexity of spatial objects? A handy answer would be that an estimator of shape complexity to experiment with, might be the sum of the numbers of internal and boundary points. Thus, a "naïve" estimator of spatial complexity to experiment with, might be defined from the shape's boundary and interior points only:

$$C_p = B_p + I_p \tag{5.2}$$

To test its applicability, consider two shapes with the same area (as calculated from Pick's theorem) and then their difference in appearance and in their C_p value (Fig. 5.3). To normalize C_p when comparing more than two shapes, C_p might be divided by the area enclosed. Whether this might explain the complexity of borders may be open to discussion but this does not diminish the fact that the shape of borders constitutes a prominent indicator of spatial complexity (Papadimitriou 2002).

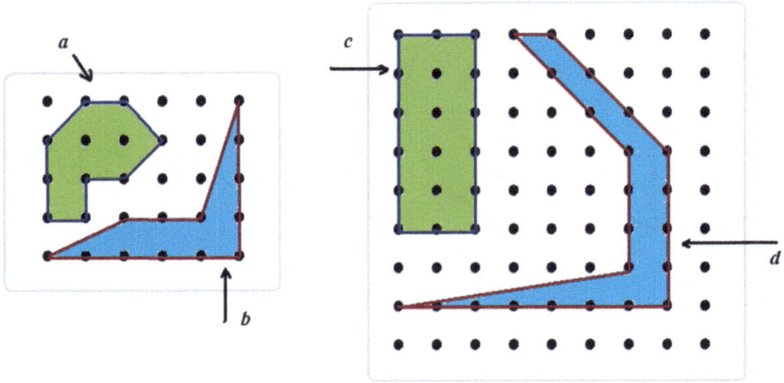

Fig. 5.3 An experiment to link Pick's theorem with spatial complexity can be illustrated by two pairs of shapes (*a–b* on the left and *c–d* on the right), with the same area per pair, but with different spatial complexities. The assumption that spatially more complex shapes should have higher sums of boundary points and internal points is verified here: the area of both shapes *a* and *b* is $A_p(a) = A_p(b) = 5.5$, but *b* is more complex than *a*, as $C_p(b) = 13$, while $C_p(a) = 11$. Similarly, $A_p(c) = A_p(d) = 10$, but $C_p(c) = 18$ while $C_p(d) = 22$

5.2 Knotting

Many riddles must be solved.
Mephistopheles: But many riddles tie also
"Da muß sich manches Rätsel lösen.
Mephistopheles: Doch manches Rätsel knüpft sich auch"
(Johann Wolfgang von Goethe, 1749–1832,"Faust: Der Tragödie erster Teil")

Knots are objects of the 3d space. A main problem of knot theory is to decide whether a knot can be unknotted and how quickly (or, otherwise stated, to decide whether a curve that appears knotted is really knotted). Typically, if a simple closed curve γ is knotted, then there does not exist any subset S of the R^3 that is homeomorphic to the disk $D = \{(x, y) \in R^2, x^2 + y^2 < 1\}$ such that the boundary of S equals the trace of the curve. Two theorems connect the curve's closedness with curvature and knottedness: if γ is closed, then its total curvature is at least 2π (*Fenchel's theorem*), but if it is knotted, then its total curvature is at least 4π (*Fary-Milnor theorem*). Hence, knotted curves have higher curvature than simply closed ones.

In fact, both knots and braids are intricately related to complexity as shown by a significant number of studies, i.e. the complexity of knots (Makowsky and Marino 2003; Diao and Ernst 1998; Kholodenko and Rolfsen 1996; Barenghi et al. 2001; Orlandini et al. 2005) and the complexity of braids (Hamidi-Tehrani 2000; Nechaev et al. 1996; Bangert et al. 2002; Garber et al. 2002; Fiedler and Rocha 1999), to the extent that even new types of complexity have been proposed from the study of knots and braids, such as the "*Matveev complexity*" (Casali and Cristofori 2006) for orientable 3-manifolds. From all these studies, it is important to single out the

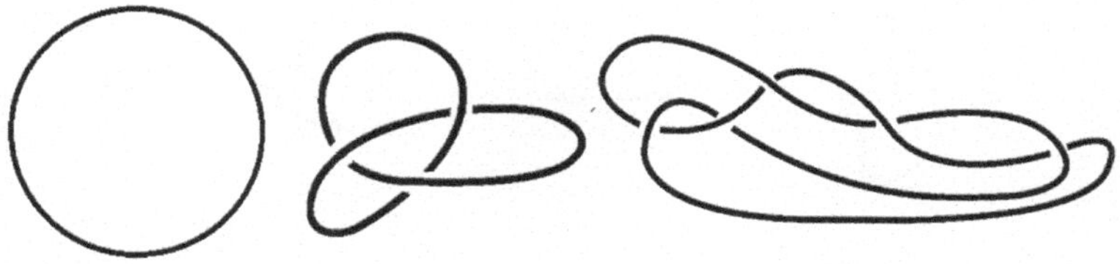

Fig. 5.4 The number of crossings of a knot serves as a measure of knottingness, which, in turn, is an indicator of spatial complexity: a circle (zero crossings) is less complex than a knot with three crossings (middle), which, in turn, is less complex than a knot with five crossings (right)

definition of "knot complexity" as the *number of crossings* of a knot (Fig. 5.4) to be the prominent measure of a knot's complexity (Kholodenko and Rolfsen 1996).

Some knots can thus be reduced to simpler knots. For instance, a complicated knot can be reduced to the simple "trefoil" knot (Fig. 5.5). Their characteristic numberings also differ: the circle is simply the "0_1 knot" (a circle is not a knot), the trefoil knot is the "3_1 knot" , the 10_123 knot is a knot with 10 crossings etc.

Another central problem of knot theory is to decide whether two knotted curves are different forms of the same knot. The first, and indeed the most important factor to help us tackle this problem, is (again) the number of crossings of a knot. The crossing number, called "*beknottedness*" by Tait (1898) is the *minimum* number of changes of sign of crossings which are necessary to eventually simplify the knot to a trivial form. But determining the benottedness of a knot is a difficult problem, even for knots with less than ten crossings (Lickorish 1987). In fact, the same knot may have infinitely many different two-dimensional representations on the *XY*-plane (its "knot diagrams"), and the crossing number of a knot refers always to the diagram which represents the minimal number of crossings. Next to this, but equally significant for spatial complexity, is the problem of "*unknot recognition*" which consists in deciding whether a given knot represents the unknot. Unknotting a knot is made by applying anyone (or combinations) of a simple set of standard moves, the "*Reidemeister moves*" (Fig. 5.6). These consist in a simple set of three basic moves to links and knots in the 3d space (Reidemeister 1932).

Fig. 5.5 Unknotting the complicated knot on the left produces the simple "trefoil" knot on the right

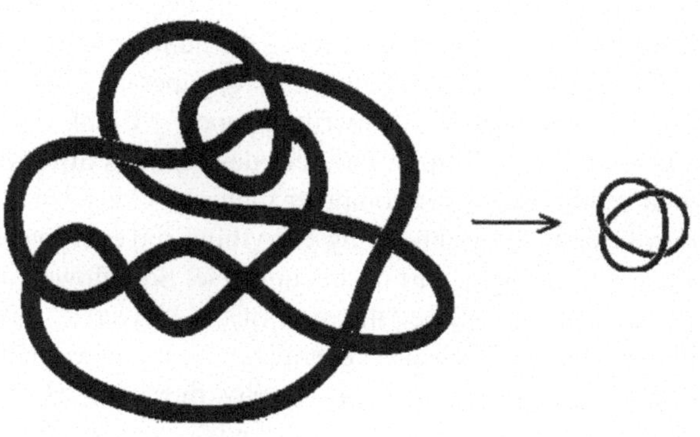

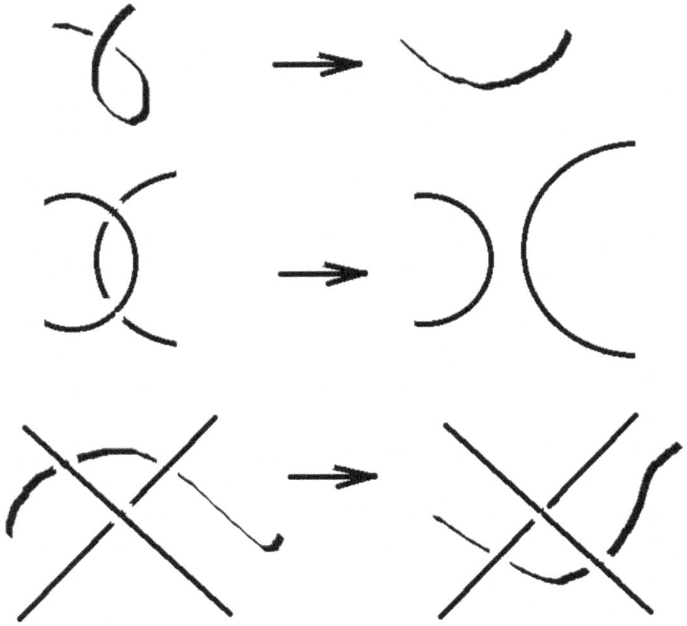

Fig. 5.6 The Reidemeister moves are elementary operations on knots, allowing the unknotting of a knot

In terms of computational complexity, the topological problem of "unknot recognition" belongs to the *NP*-complexity class, because an upper bound of the number of Reidemeister moves (m) necessary to connect a number of crossings (n) of a knot to the unknot is (Hass and Lagarias 2001):

$$m \leq 2^{10^{11} n}, \tag{5.3}$$

This very big number casts doubts whether we might ever be able to solve the problem of whether two knots are the same or not. Later however, it was shown (Burton 2011) that the complexity of deciding whether a knot is the unknot is a low-order exponential function of the number of crossings.

The same piece of a blank white A4 paper may be described as "White paper", "An A4-size white paper", "Unwritten paper", "Blank", or even simply "W" (for white) and so on. But all these different descriptions differ in their sizes and therefore in their requirements for computer memory. Consider the extreme case that someone may be content by annotating everything and everyone as "this part of the Universe", since everything is part of this universe. So a flower, a man, a city, a planet, a solar system, etc will all be "this part of the Universe" alike. Although this may sound like an absolutely absurd generalization, it is nevertheless a valid one. Expressed in terms of geography and cartography, the "peri-urban sprawl", "residential areas", "retail commerce areas", "urban lands", "industrial settings", "city centers" can all be encoded as "built up lands" or "developed areas". It is thus important to consider the spatial-scale dependence of spatial complexity, but also the *encoding-dependence* (the level of semantic generalization at which something is studied). Inevitably, *the level of semantic generalization affects the level of spatial generalization and the reverse.*

Once a spatial object is identified for examination, it can not be classified perceptually without being encoded first. But determining an optimal encoding is theoretically a very difficult (and most likely unsolvable) problem. Thus opens a wide avenue for case-specificity in spatial complexity assessments. Consider, for instance, the following question: What do the following algebraically inequivalent expressions have in common?

$$y_1 = x^4 + 3x^2 + 1 \tag{5.4}$$

$$\begin{aligned} y_2 = x^9y + x^8y^2 + x^7y^3 - x^7y + x^6y^4 - 3x^6y^2 + \\ +2x^6 + x^5y^3 - 2x^5y + x^4y^4 - 4x^4y^2 + 3x^4 \end{aligned} \tag{5.5}$$

$$y_3 = \frac{1}{x^2} + \frac{1}{x^4} + \frac{1}{x^6} - \frac{1}{x^5} - \frac{1}{x^7} \tag{5.6}$$

$$y_4 = x^2 - x + 1 - \frac{1}{x} + \frac{1}{x^2} \tag{5.7}$$

At first, the obvious answer would be: "nothing". Yet, they all describe exactly the same topological figure: the double overhand knot 5_1, or "cinquefoil" (Fig. 5.7).

The first one is the Conway polynomial of this knot, the second is its Kauffmann, the third the Jones and the last one its Alexander polynomial. Therefore, four completely different algebraic expressions describe the same topological reality, the same spatial complexity. But they are so different because they are derived on the basis of different knot description systems. Or, the same topology corresponds to different algebraic encoding schemes. Polynomial invariants have been suggested to describe

Fig. 5.7 The double overhand knot 5_1, or "cinquefoil"

knots, so that a knot is characterized by a polynomial expression. The first such was proposed by Alexander in 1928, another by Conway in 1970, to be followed by Jones's in 1984, the HOMFLY-PT and the Kauffman polynomial in 1987. Computing these polynomials raises questions of complexity, since it is known that this problem is *xP*-hard (Welsh 1993). While recognizing the unknot is an *NP*-hard problem, the computation of the Alexander polynomial is in polynomial time, but that of the Jones polynomial is *#P*-hard. The problem with knot polynomials is that they are not very reliable indicators for distinguishing different knots. For instance, it is difficult for the Jones polynomial to recognize the unknot, while there are different "Kanenobu knots" (Kanenobu 1986), all with the same Jones polynomial. Also some knots have a trivial Alexander polynomial. The problem of different knots with the same polynomials extends to the HOMFLY-PTs (Likorish 1987). Indeed, algebraically "similar" expressions may correspond to completely different spatial objects, but *Khovanov homology* is a generalization of the Jones polynomial of knots and can be used to decide whether a knot is an unknot (Kronheimer and Mrowka 2011), with computational complexity equal to or higher than the problem of computation of the Jones polynomial (a *#P*-hard problem), while computing this homology is in the *EXPTIME* class. Another homology, the *Heegard Floer* (a generalization of the Alexander polynomials) can be used to detect unknots and the genus of knots (Manolescu et al. 2007, 2009).

Yet, *knot complexity* might be directly estimated from knot polynomials: the "*knot complexity*" was defined (Grosberg and Nechaev 1991, 1992) as the maximum degree of the knot's characteristic polynomial. For instance, the Kauffman polynomial of the trefoil knot is:

$$yx^5 + y^2x^4 - x^4 + yx^3 + y^2x^2 - 2x^2 \tag{5.8}$$

and its maximum degree is lower than that of the Kauffman polynomial of a more complex knot (Fig. 5.8):

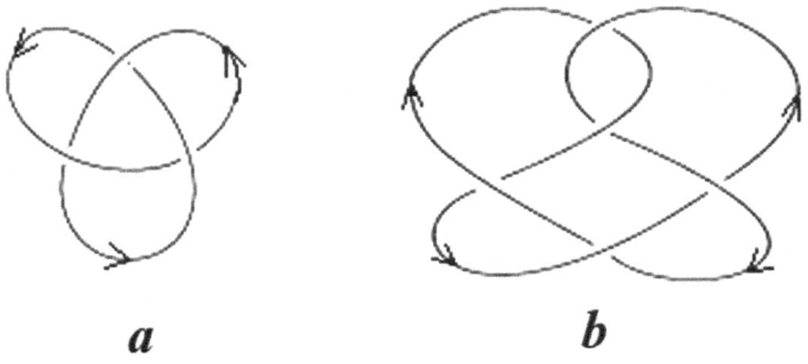

Fig. 5.8 The maximum degree of the Kauffman polynomial of knot *a* (the trefoil knot, 3_1) is higher than that of the knot *b* (the knot 5_2) and therefore knot *b* has higher Grosberg-Nechaev complexity

$$5y^3x^3 - xy^3 + \frac{y^3}{x} + 6y^2x^6 + 7y^2x^4 + \frac{y^2}{x^2} - 3x^5y - 3x^3y - x^6 - x^4 - \frac{1}{x^2}$$
$$- 5x^6y^4 + x^5y^7 + x^3y^7 + x^6y^6 + 2x^4y^6 + x^2y^6 - 5x^5y^5 - 4x^3y^5$$
$$- xy^5 - 8x^4y^4 - 2x^2y^4 + y^4 + 7x^5y^3 \qquad (5.9)$$

5.3 Braiding

With her hands she wove bright braids
"χερσί πλοκάμους ἔπλεξε φαεινούς"
(Homer, Iliad, Ξ. 176)

Much like knots, lines and curved surfaces in 3d space can make interweaving "strands" creating braids (Fig. 5.9).

As with knots, the higher the number of crossings, the more complex the braid. Exploring this concept, Bangert et al. (2002) addressed the topological problem of finding the minimal crossing number of braids, and, likewise with knots, the number of crossings of a braid has been established as a measure of its complexity (Raymer and Smith 2007; Berger 1994; Bangert et al. 2002), to the extent that it is also called "*braid complexity*" (Simsek et al. 2003).

Fig. 5.9 A braid consists in a set of interwoven strands in 3d space

The number of crossings in a braid diagram is equal to the "*Artin word*" of the braid. Concomitant to a braid's complexity is the problem of "Artin words" (which remains unsolved), consisting in finding an algorithm which generates an Artin word of minimal length for a braid with a fixed number of strands. Also, Garber et al. (2002) used the term "braids complexity" for the braids word problem. Aside of the number of crossings however, several other measures of braid complexity have been proposed, such as the Lyapunov exponents of generators of the braid groups (Nechaev et al. 1996), the "braiding exponent" (Thieffeault 2005), the "minimum braid energy" (Bangert et al. 2002), the amount of distortions per braid (Dynnikov and Wiest 2007), as well as measures based on "Dehornoy orderings" (Ito 2011).

5.4 Writhing and Linking

It is impossible to solve a problem while ignoring the link in it
"Λύειν δ' οὐκ ἔστιν ἀγνοοῦντας τὸν δεσμόν"
(Aristotle, 384–322 b.C.,"Metaphysics", 3/995a)

Besides braiding and knotting, writhing is also a significant determinant of complexity and, for linked curves, it is measured by the "*linking number*". The linking number is a positive or negative integer, depending on the orientation of the linked curves and on how many times they cross each other (Fig. 5.10). The higher the number of crossings, the higher the complexity of the setting. Although curves do not really qualify as "spatial objects" within the interpretation of "spatial" that is defined and followed in this book (something is "spatial" if it is at least 2d), they nevertheless define areas in 2d or volumes in 3d space. Thus, the number of times they intersect, cross and link is an indicator of spatial complexity (besides, the same spatial complexity due to writhing would be created if there were bands instead of curves).

A measure of spatial complexity might therefore be based on the absolute value of the linking number (ignoring the orientation of the curves):

$$|L(\gamma_1, \gamma_2)| \tag{5.10}$$

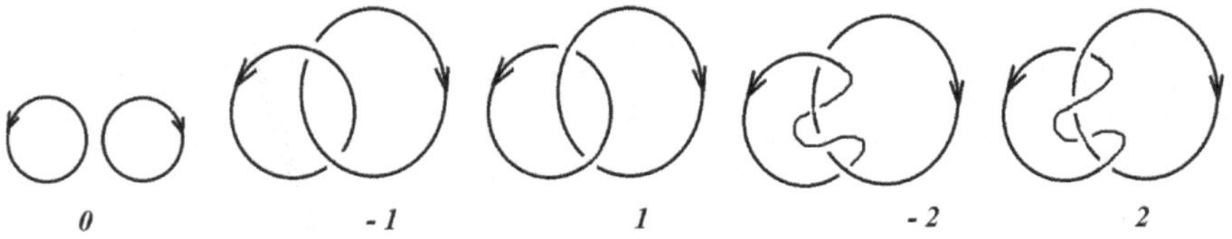

Fig. 5.10 The linking number of two curves is a positive or negative integer, depending on which curve crosses above or below the other: the higher the absolute value of the linking number, the higher the complexity of the curves' relative position

whereas the linking number $L(\gamma_1,\gamma_2)$ of two curves γ_1 and γ_2 is defined in terms of their crossings $\varepsilon(\kappa)$:

$$L(\gamma_1, \gamma_2) = \frac{1}{2} \sum_{\kappa \in (\gamma_1 \cap \gamma_2)} \varepsilon(\kappa) \tag{5.11}$$

Hence, provided that these curves lie both on the same plane, they define areas and the higher the linking number, the higher the number of areas defined by the crossings and therefore the higher the spatial complexity (Fig. 5.11), which might be symbolized by C_{LN}:

$$C_{LN} = 2|L(\gamma_1, \gamma_2)| + 1 \tag{5.12}$$

or, expressed in terms of crossings (for $L \geq 1$):

$$C_{LN} = \left| \sum_{\kappa \in (\gamma_1 \cap \gamma_2)} \varepsilon(\kappa) \right| + 1 \tag{5.13}$$

Notice however, that for smooth, disjoint and closed curves in R^3, the Gauss linking number applies:

$$L(\gamma_1, \gamma_2) = \frac{1}{4\pi} \oint_{\gamma_1} \oint_{\gamma_2} \frac{(\mathbf{r}_1 - \mathbf{r}_2)}{|\mathbf{r}_1 - \mathbf{r}_2|^3} \cdot d\mathbf{r}_1 \times d\mathbf{r}_2 \tag{5.14}$$

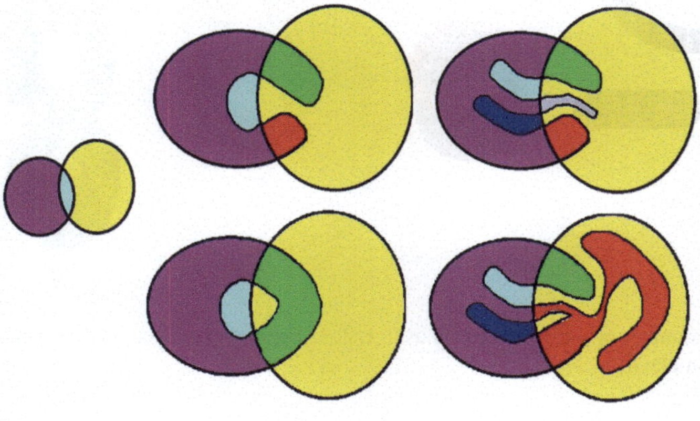

Fig. 5.11 The absolute value of the linking number of two curves, with both of them on the same plane, might be an estimator of the spatial complexity of the areas defined by the curves and their crossings

where $\mathbf{r}_1(t)$ and $\mathbf{r}_2(t)$ are the parametrizations of the curves γ_1 and γ_2. In this case, one would have to define volumes instead of 2d regions. The linking integral applies to non-euclidean geometries also, i.e. in spherical and hyperbolic 3-spaces (DeTurck and Gluck 2004), while Shonkwiler and Vela-Vick (2008) expanded it to submanifolds M and K (in place of curves γ_1, γ_2) with arbitrary dimensions m and k respectively (such that $m+k=n-1$).

5.5 Genus

> "The hidden in men and the hidden in things
> belong in the same topoanalysis"
> (Gaston Bachelard 1994,p. 89)

Expectedly, the higher the genus of a surface, the higher the complexity of its spatial configuration (Fig. 5.12). The genus is a knot invariant also, because transforming a knot by means of Reidemeister moves does not change its genus. The unknot is the only knot with genus zero. The "*Floer homology*" of a knot detects the knot's genus and is computable (Manolescu et al. 2009).

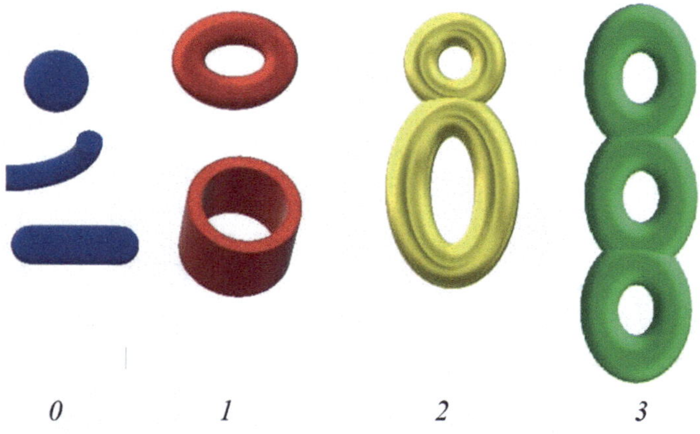

Fig. 5.12 Examples of surfaces with genus 0, 1, 2, 3. The higher the genus of a surface, the higher its spatial complexity

5.6 Dimensions

> "A topological egg progressively differentiating
> into a Euclidean organism"
> (Manuel De Landa 2002, p. 68)

Increasing dimensions inevitably incurs higher computational cost to spatial computations. Two characteristic examples exemplify this: First, translating Pick's theorem to higher dimensions, the volume $V(P)$ of a lattice polyhedron in Z^3 is given by the number of points in its interior and boundary in a similar way as in the case of 2d Pick's theorem, with the exception that it also involves Euler's characteristic of the polyhedron, $\chi(P)$. All these are combined in *Reeve's formula* (Reeve 1957a, b):

$$V(P) = \frac{B_n - nB_1 + 2(I_n - nI_1) + (n-2)[2\chi(P) - \chi(\partial P)]}{2n(n^2-1)} \tag{5.15}$$

which can be generalized to higher (N) dimensions (Macdonald 1963, 1971):

$$V(P) = \frac{\sum_{k=0}^{N-1} (-1)^{k-1} \binom{N-1}{k-1} (B_{N-k} + 2I_{N-k}) + (-1)^{N-1}(2\chi(P) - \chi(\partial P))}{N!(N-1)} \tag{5.16}$$

evidently involving several more calculations than Pick's formula for two dimensions.

Second, recalling Heron's formula for the calculation of the area of a triangle, that is the simplest 2d shape:

$$\sqrt{\tau(\tau-a)(\tau-b)(\tau-c)}, \text{ where } \tau = (a+b+c)/2 \tag{5.17}$$

and increasing spatial dimensions from two to three, the volume of a tetrahedron (the simplest 3d shape) with edges U,V,W, and opposite edges u,v,w (where the edge u is opposite to U, v is opposite to V and w is opposite to W) is given by the formula:

$$Volume = \frac{\sqrt{(a+b-c+h)(b+c+h-a)(a+c+h-b)(a+b+c-h)}}{192uvw} \tag{5.18}$$

where

$$\begin{aligned} a &= \sqrt{xYZ};\ b = \sqrt{yXZ},\ c = \sqrt{zYX},\ h = \sqrt{xyz} \\ X &= (w-U+v)(U+v+x = (U-v+w)(U+v-w)w), \\ Y &= (u-V+w)(V+u+w)\,y = (V-w+u)(V+w-u) \\ Z &= (u-W+v)(u+v+W)\,z \\ &= (W-u+v)(W+u-v) \end{aligned} \tag{5.19}$$

Evidently, a set of calculations requiring a considerably higher number of arithmetic operations (59) than those required for Heron's formula (10): as dimensions increase, spatial complexity of even the simplest spatial elements increases also.

5.7 Orientability and Immersions

Fortunate islands, lands without a place
"São ilhas afortunadas, são terras sem ter lugar"
(Fernando Pessoa 1888–1935, "As Ilhas Afortunadas", v.11–12)

A surface is "closed" if it is "compact", connected and without a boundary. Otherwise stated, a surface is closed if it is a Hausdorff space which is compact and connected, and in which every point has a neighborhood homeomorphic to the plane and "Dehn's classification *theorem*" applies to closed surfaces (1907), assuring that every compact, connected surface is topologically equivalent to a sphere, or to a connected sum of projective planes or to a connected sum of tori. Or, alternatively expressed, any closed surface is homeomorphic either to the sphere, or to the sphere with a finite number of handles added, or to the sphere with a finite number of discs removed from it and replaced by *Möbius bands* added to it (Fig. 5.13). It does not matter for topology where the handles or the Möbius bands are added on the sphere. So long as there are no holes or handles in an object, all closed surfaces on it are homeomorphic to a sphere. Each one of the classes of these surfaces is not homeomorphic to the other. Adding a handle to the sphere is obtained by the removal of the interiors of a pair of disjoint discs first and then by the attachment of a cylinder by gluing its boundary circles to the edges of the two holes in the sphere. Möbius bands are non-orientable surfaces, which highlight one further determinant adding complexity to spatial forms.

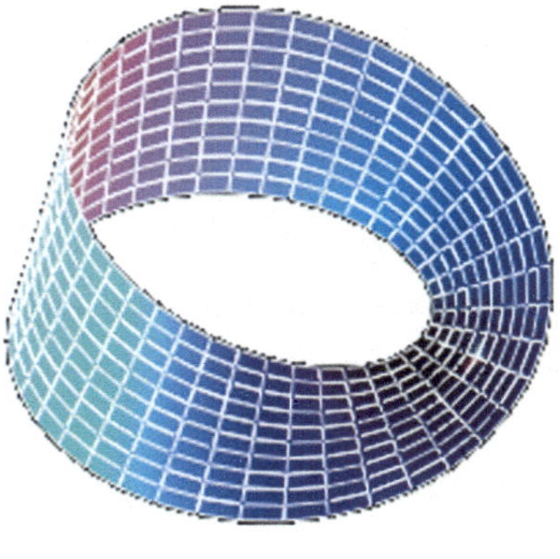

Fig. 5.13 A Möbius band is a non-orientable surface, that is a surface in which its "outside" and "inside" are indistinguishable from one another

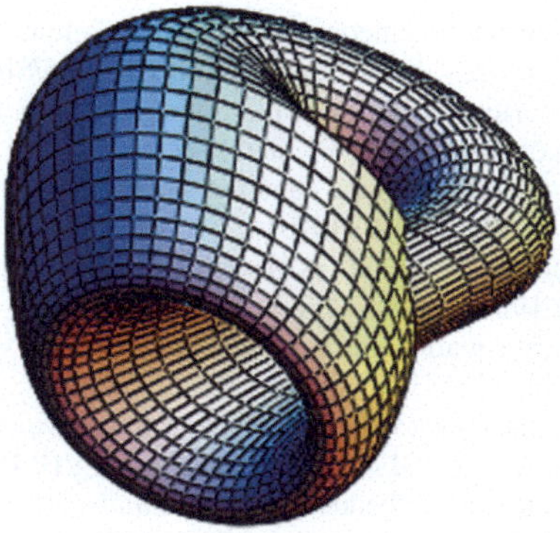

Fig. 5.14 The Klein bottle is a non-orientable surface immersed in R^3

Non-orientability of a surface leads to what seems to be yet another characteristic of high spatial complexity: immersed surfaces. An immersion is a differentiable function between two differentiable manifolds whose derivative is injective everywhere. A common case is mappings from R^2 to R^3. In immersions, the image of the manifold can have transverse intersections. One example is the *Klein bottle* (Fig. 5.14).

There are many non-orientable closed surfaces that can not be embedded in R^3 but they can nevertheless be immersed in it. Immersions of surfaces in 3d space are interesting due to the increased spatial complexity they entail.

References

Bachelard, G. (1994). *The Poetics of Space*. Boston MA: Beacon Press.

Bangert, P. D., Berger, M. A. and Prandi, R. (2002). In search of minimal random braid configurations. *Journal of Physics A: Mathematical and General, 35*(1), 43–59.

Barenghi, C. F., Ricca, R. L. and Samuels, D. C. (2001). How tangled is a tangle? *Physica D: Nonlinear Phenomena, 157*(3), 197–206.

Berger, M. A. (1994). Minimum Crossing Numbers for 3-Braids. *Journal of Physics A: General Physics, 27*(18), 6205–6213.

Burton, B.A. (2011). Maximal admissible faces and asymptotic bounds for the normal surface solution space. *Journal of Combinatorial Theory, A,118* (4),1410–1435.

Casali, M. and Cristofori, P. (2006). Computing Matveev's Complexity via Crystallization Theory: The Orientable Case. *Acta Applicandae Mathematicae, 92*(2), 113–123.

De Landa, M. (2002). *Intensive Science and Virtual Philosophy*. London: Continuum.

DeTurck, D. and Gluck, H. (2005). Electrodynamics and the Gauss linking integral on the 3-sphere and in hyperbolic 3-space. arxiv.org/abs/math.GT/0510388.

Diao, Y. and Ernst, C. (1998). The complexity of lattice knots. *Topology and its Applications, 90*(1), 1–9.

Dynnikov, I. and West, B. (2007). On complexity of Braids. *Journal of the European Mathematical Society, 9*(4), 801–840.

Fiedler, B. and Rocha, C. (1999). Realization of meander permutations by boundary value problems. *Journal of Differential Equations, 156*(2), 282–308.

Gage, M. E. (1983). An isoperimetric inequality with applications to curve shortening. *Duke Mathematical Journal, 50*(4), 1225–1229. https://doi.org/10.1215/S0012-7094-83-05052-4

Gage, M. E. (1984). Curve shortening makes convex curves circular. *Inventiones Mathematicae, 76*(2), 357–364. https://doi.org/10.1007/BF01388602

Gage, M. and Hamilton, R.S. (1986). The heat equation shrinking convex plane curves. *Journal of Differential Geometry, 23*(1), 69-96. MR 840401.

Garber, D., Kaplan, S. and Teicher, M. (2002). A New Algorithm for Solving the Word Problem in Braid Groups. *Advances in Mathematics, 167*(1), 142–159.

Grayson, M.A. (1987). The heat equation shrinks embedded plane curves to round points. *Journal of Differential Geometry 26*(2), 285-314. MR 906392

Grayson, M.A. (1989a). The shape of a figure-eight under the curve shortening flow. *Inventiones Mathematicae, 96*(1), 177-180. MR981740. doi: https://doi.org/10.1007/BF01393973.

Grayson, M.A. (1989b). Shortening embedded curves. *Annals of Mathematics, Second Series,* 129(1), 71-111. MR979601. doi: https://doi.org/10.2307/1971486.

Grosberg, A. Y. and Nechaev, S. K. (1991). Topological constraints in polymer network strong collapse. *Macromolecules, 24*(10), 2789–2793.

Grosberg, A. Y., and Nechaev, S. K. (1992). Averaged Kauffman invariant and Quasi-Knot concept for linear polymers. *Europhysics Letters, 20*(7), 613–619.

Hass, J. and Lagarias, J. C. (2001). The number of Reidemeister moves needed for unknotting. *Journal of the American Mathematical Society, 14*(2), 399–428.

Ito, T. (2011). Braid ordering and knot genus. *Journal of Knot Theory and its Ramifications, 20*(9), 1311–1323.

Kanenobu, T. (1986). Examples on polynomial invariants of knots and links. *Mathematische Annalen, 275*(4), 555–572.

Kholodenko, A. L. and Rolfsen, D. P. (1996). Knot complexity and related observables from path integrals for semiflexible polymers. *Journal of Physics A: Mathematical and General, 29*(17), 5677–5691.

Kronheimer, P. B. and Mrowka, T. S. (2011). Khovanov homology is an unknot-detector. *Publications Mathematiques de l' Institut de Hautes Etudes Scientifiques, 113,* 97–208.

Lickorish, W. B. R. (1987). Linear skein theory and link polynomials. *Topology and its Applications, 27,* 265–274.

Macdonald, I. G. (1963). The volume of a lattice polyhedron. *Proceedings of the Cambridge Philosophical Society, 59,* 719–726.

Macdonald, I. G. (1971). Polynomials associated with finite cell-complexes. *Journal of the London Mathematical Society, 4,* 181–192.

Makowsky, J. A. and Marino, J. P. (2003). The parametrized complexity of knot polynomials. *Journal of Computer and System Sciences, 67*(4), 742–756.

Manolescu, C., Ozsváth, P., Szabo, Z. and Thurston, D. (2007). Coombinatorial link Floer Homology. *Geometry and Topology, 11,* 2339–2412.

Manolescu, C., Ozsváth, P. S. and Sarkar, S. (2009). A combinatorial description of knot Floer homology. *Annals of Mathematics, 169*(2), 633–660.

Nechaev, S. K., Grosberg, A. Y. and Vershik, A. M. (1996). Random walks on braid groups: Brownian bridges, complexity and statistics. *Journal of Physics A: Mathematical and General, 29*(10), 2411–2433.

Orlandini, E., Tesi, M. C. and Whittington, S. G. (2005). Entanglement complexity of semiflexible lattice polygons. *Journal of Physics A: Mathematical and General, 38*(47), L795–L800.

Papadimitriou, F. (2002). Modelling Indicators and indices of landscape complexity: An approach using GIS. *Ecological Indicators, 2,* 17–25.

Papadimitriou, F. (2013). Mathematical modelling of land use/landscape complexity with ultrametric topology. *Journal of Land Use Science, 8*(2), 234–254.

Pick, G. (1899). Geometrisches zur Zahlenlehre. Sitzungsberichte des deutschen naturwissenschaftlich-medicinischen Vereines für Böhmen "Lotos" in Prag. (Neue Folge). 19: 311–319.JFM 33.0216.01. CiteBank:47270

Raymer, D. M. and Smith, D. E. (2007). Spontaneous knotting of an agitated string. *Proceedings of the National Academy of Sciences of the USA, 104*(42), 16432–16437.
Reeve, J. E. (1957a). Le volume des polyedres entiers. *Comptes Rendus de l' Academie de Sciences de Paris, 244,* 1990–1992.
Reeve, J. E. (1957b). On the volume of lattice polyhedra. *Proceedings of the London Mathematical Society, 7,* 378–395.
Reidemeister, K. (1932). *Knottentheorie.* Berlin: Springer.
Shonkwiler, C. and Vela-Vick, D. S. (2008). Higher dimensional linking integrals. *Arxiv, 0801,* 4022.
Simsek, H., Bayram, M. and Can, I. (2003). Automatic calculation of minimum crossing numbers of 3-Braids. *Applied Mathematics and Computation, 144*(2–3), 507–516.
Tait, P. G. (1898). *On knots I, 11, III, Scientific Papers* (Vol. I). London: Cambridge University Press.
Thiffeault, J.-L. (2005). *Measuring Topological Chaos. Physics Review Letters, 94,* 084502.
Welsh, D.J.A. (1993). *Complexity: Knots, Colorings and Counting*. London Mathematical Society Lecture Notes Series, Vol. 186. Cambridge: Cambridge University Press.

Chapter 6
The Algorithmic Basis of Spatial Complexity

-What is the wisest?
-The Number.
-And what is the second wisest?
-Assigning names to things
"Τί τὸ σοφώτατον; Ἀριθμός.
Τί τὸ δεύτερον σοφώτατον;
Τὸ τοῖς πράγμασι τὰ ὀνόματα τιθέμενον"
(Iamblichus, 245–325 b.C., VP82, 47.17–19)

Abstract Building on previous concepts and measures, two metrics of spatial complexity of small 2d maps are proposed here: C_{P1} and C_{P2}. Both are applicable to square maps (binary or not), so long as each square cell of a map can have a one-to-one correspondence to one symbol (taken from an alphabet of symbols). These symbols can be used to represent as many colors, categories etc. as the map requires, on the basis of a conventional "scanning" run-length encoding sequence. The first metric gives an evaluation of spatial complexity on the basis of blocks of cells with repeating patterns that are identified following an iterative process. The second metric is calculated on the basis of vertical and horizontal pairwise comparisons of successive adjacent strips of the map or surface. Both metrics are easy to compute, assume integer values and their magnitudes are comparable for small maps.

Keywords Spatial complexity · Map Complexity · Geography and Complexity · Edit distance · run-length encoding · Algorithmic complexity · Geocomputation

6.1 The Language of Space

lopadotemahoselahogeleokraniolipsanodimypotrimatosilfio
laravomelitokatakehymenokihlepikosyfofatoperisteralektryonopto
kefalliokiglopeliolagosireovafitraganopterygon
"λοπαδοτεμαχοσελαχογαλεοκρανιολειψανοδιμυποτριμματοσιλφιο
λαραβομελιτοκατακεχυμενοκιχλεπικοσσυφοφαττοπεριστεραλεκτρυονοπτο
κεφαλλιοκιγκλοπελειολαγωσιραιοβαφητραγανοπτερύγων"
(Aristophanes, 427-386 b.C., "Ecclesiazusae", line 1163)

F. Papadimitriou, *Spatial Complexity*, https://doi.org/10.1007/978-3-030-59671-2_6

Aristophanes, the ancient greek author of the first comedies in the history of theatre wrote this 171 letters-long word that probably used to be the longest known word throughout the ages, supposedly referring to a name of a dish, a kind of a fricassée containing some 15 ingredients: clearly a "complex" word, describing an equally complex set of ailments recommended for the creation of this dish. But it is not human language alone that can create very long words. Metaphorically, space also "speaks" immensely lengthier "words", so long as one is able to "listen" to them. Geometric, topological and other mathematical aspects of spatial complexity may *indicate* whether a spatial object is endowed with a higher or lower spatial complexity, but they do not constitute measures of spatial complexity themselves. This leads us to a central issue with respect to the measurement of spatial complexity: Since complexity is different than entropy, or any other indicator of complexity, any measure of complexity should be a measure of *complexity* per se, backed by the mathematical literature of complexity. Such a measure should (under certain topological restrictions) be of general applicability to two-(and possibly higher)-dimensional smooth surfaces. A rasterization/digitization significantly facilitates the measurement of spatial complexity, since even maps of points and lines can (under certain conditions) be rasterized and, subsequently, symbols of cover or color can be assigned to each cell of the resulting square grid. The term "raster" that is commonly used in GIS signifies geoprocessing of spatial elements on the basis of squares (pixels). In fact, it is a common practice in geospatial analysis to "rasterize" all spatial elements, that is to attribute to square cells or pixels even linear or point data. The long experience gained from the use of geographic technologies (such as Geographical Information Systems "GIS" and satellite image processing) is particularly useful in this respect. After experimenting with various geo-encoding models, digital arrays of pixels are used to represent either "vector" (0-and-1-dimensional) or "raster" (2-dimensional) spatial entities. The rasterization is easily created by enlarging or reducing the map size as appropriate and assigning a special color to each one of its cells (Fig. 6.1). Whether the term "cell" or "pixel" is used, it

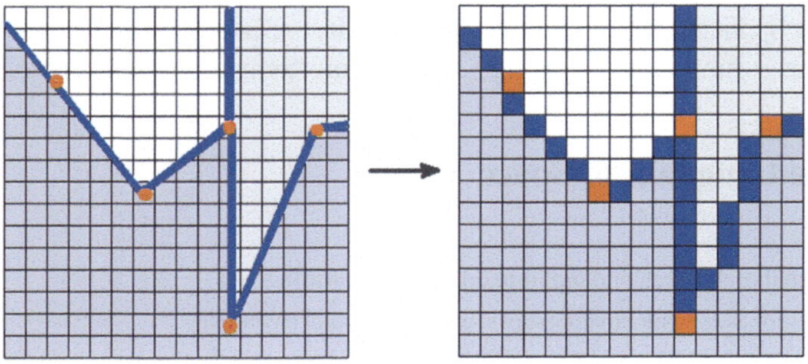

Fig. 6.1 Rasterization" is a widely used process in image analysis, by which points, lines and areas (0,1,2-dimensional spatial elements respectively) are translated to square cells. In this way, all elements of a map, whether points or lines or areas, can be converted to cellular (square) representations, with accuracy depending on the pre-decided resolution of the square map that is to be constructed

invariably concerns discrete square cells on a map of equal square partitions. In the science and practice of image processing, continuous functions are routinely used in image analyses. For instance, the convolution of the map $f(x, y)$ with the map $g(x, y)$ is defined as a function:

$$h(x, y) = \iint f(x_1, y_1)g(x - x_1, y - y_1)dx_1dy_1 \tag{6.1}$$

Thus, analytic expressions of known functions are used in their discrete form, such as, for instance, the discrete Fourier Transform of a 2d array $f(x, y)$ of size mxn:

$$f(u, v) = \frac{1}{\sqrt{mn}} \sum_{x=0}^{m-1} \sum_{y=0}^{n-1} f(x, y)e^{-2i\pi\left(\frac{xu}{m}+\frac{vy}{n}\right)} \tag{6.2}$$

Equivalently, the form of a closed curve on the plain in polar coordinates (with distance D from the barycentre and angle θ with respect to the horizontal axis) is given in terms of a typical Fourier expression:

$$D = \frac{a_0}{2} + \sum_{n=1}^{\infty} [a_n \cos(n\theta) + b_n \sin(n\theta)] \tag{6.3}$$

The analytic approach applies when the curve is "not too rough". When the curve's shape is not so, the analytic approach is of limited applicability, because it renders more than one values for the same angle θ. But when the map is examined in discrete cells, there is always one single value attributed to each cell. Further, the numerical application of such functions practically presupposes discrete spatial domains, so some "discretization" or "rasterization" process is eventually unavoidable for all kinds of computerized image processing.

A *square "map"* is a binary or multicolored planar surface, which has been constructed by any means (satellite or aerial photography, plain photography, field observations, cartographic processing etc.), and can be described mathematically by a function on the discrete plain Z^2 defined as $f{:}Z^2 \rightarrow \{0, 1, 2, ..., V\}$, where V is the number of colors ("covers") appearing on the map. In example, a "binary map" is defined by a function $f{:}Z^2 \rightarrow \{0, 1\}$, where $f(0)$ = white and $f(1)$ = black. In the same way, for higher spatial dimensions, the mapping function f is defined in Z^3 (for three dimensions, in which case, the equivalent of the 2d pixel is a 3d a "voxel"), in Z^4 (for four dimensions, in which case, the equivalent is a 4d a "doxel"), or in Z^n in general.

Obviously, an alphabet of symbols may be chosen so as to adequately represent (under certain conventions) spatial allocations (i.e. colors) on a rasterized map (Fig. 6.2). Yet, we should not lose sight from the fact that both spatial and semantic

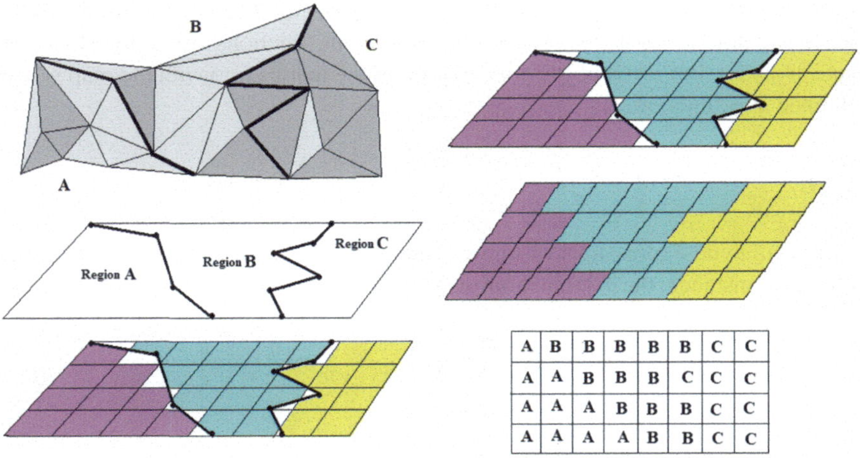

Fig. 6.2 If a three-dimensional surface that is covered by three types of cover (A, B, C) is projected on a 2-dimensional space and then encoded by using an alphabet of three symbols (A, B, C), it eventually yields a map, or, alternatively, a 2d linguistic description of this surface

generalizations made on the initial image, map or surface, affect the accuracy, precision and effectiveness of our assessments of spatial complexity. It is easy to visualize how a map configuration can eventually be converted into a linguistic or numerical description with a "run length encoding", where the spatial sequence of a map's allocation of covers (i.e.landscape or soil types, kinds of land use, geomorphic features) corresponds to symbols of an alphabet. Similarly, three-dimensional spaces can be converted to linguistic descriptions by using some (appropriate for that purpose) alphabet.

Even maps of points can be converted to maps of square cells. This process involves the use of *Voronoi polygons* (or Voronoi cells, or Thiessen polygons) that make a one-to- one correspondence of spatial regions to points. Indeed, areas covered by Voronoi polygons can be converted to a map of square cells (Fig. 6.3). Following similar procedures, lines and curves may be made equivalent to areas, which are eventually displayed as raster maps.

Practically, re-arranging a two-dimensional image as an one-dimensional line has long proven useful in computational analyses, particularly for image classification purposes (Seiffert and Michaelis 1997). It is now natural to ask "how can we measure the complexity of spatial forms of a binary map (or landscape with two land cover types), *without* considering spatial distributions of *particular* physical entities in it?".

An index of spatial complexity (I_S) was initially proposed by Papadimitriou (2002), measuring the number of common boundaries between different kinds of map cover per unit area. Later, a different index was proposed (Papadimitriou 2009), based on Levenshtein distances (Levenshtein 1966) which measure the minimum distance between two strings of symbols as the *minimum* number of the three basic operations necessary to convert one string into another (deletions d, substitutions s,

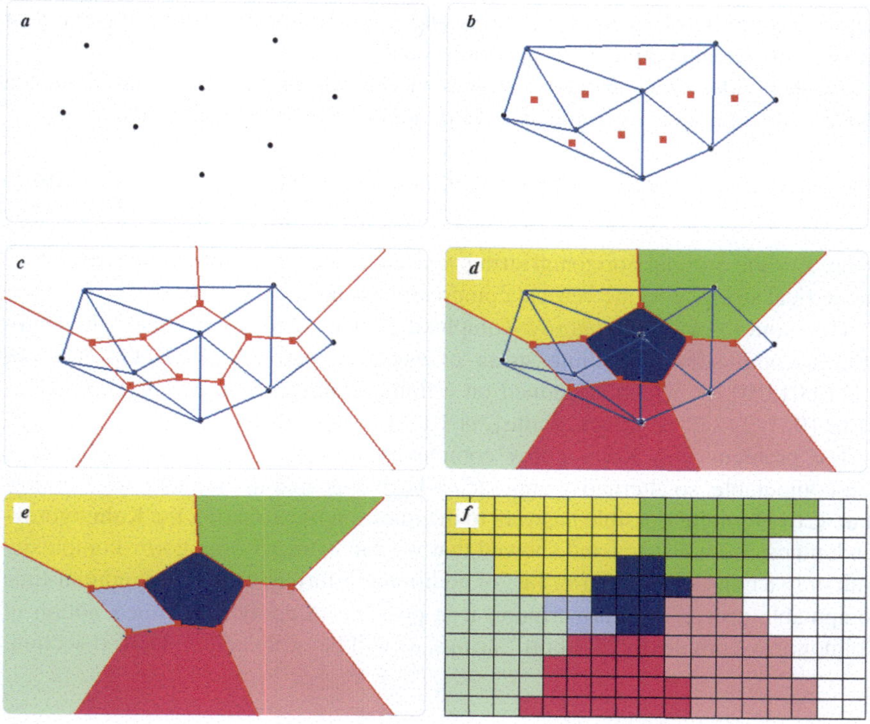

Fig. 6.3 Even spatial distributions of points (**a**) can be converted to areas of square cells (*f*). Voronoi polygons can be used to convert point data to maps of square cells. Beginning with a map (**a**) with point data only, a triangulation of the space is created by connecting the points (**b**). Next, the barycentres of all triangles are defined (**c**). Consequently, the barycentres are joined by lines and these lines are extended as appropriate (**d**) up to the map's boundaries. The number of the resulting Voronoi polygons (**e**) is equal to the number of initial points (8 in this example) and, eventually, the map of Voronoi polygons is rasterized (**f**)

additions t): $a = min\{d + s + t\}$. It was shown (Papadimitriou 2009) that this algorithm can be spatialized and that it can be used to calculate the spatial complexity of a map by means of the formula:

$$C_L = \frac{1}{uv} \sum_{\substack{i=1 \\ j=1}}^{\substack{j=v-1 \\ i=u-1}} \alpha_{ij} \tag{6.4}$$

where u = rows, v = columns of cells of the map, and α_{ij} are the entries of the matrix of the edit distances which are derived by comparing parallel and adjacent strips of landscapes. This formula converts the one-dimensional edit algorithm from a measure of minimum description of string differences to a measure of minimal length description of spatial differences and it is for this reason that it serve as a measure of spatial complexity (Papadimitriou 2009).

Another index (K_S) was later suggested (Papadimitriou 2012), based on the concept of Kolmogorov complexity (Kolmogorov 1965, 1986), defined as

$$K_s = \min\big((U(x),\ U(y), U\big(x'\big), U\big(y'\big)\big) \tag{6.5}$$

where x = the original horizontal string, y = the compressed horizontal string, x' = the vertical string, y' = the vertical compressed string.

The notion of "Kolmogorov complexity" was used to evaluate the algorithmic complexity of a finite string of symbols (e.g. ABDDCBFGAAG... or 0,111,001,010...), which is defined on a finite alphabet, such as {A, B, C, D ... Z} or {0, 1, 2, ...}, or {black, white}, or {0, 1} or any other.

The problem with Kolmogorov complexity is that, in the general case, it is non-computable, so alternative approaches have appeared in the literature. Fortnow et al. (2011) sought possible answers to the question of extracting the Kolmogorov-randomness from a string and showed that we can extract Kolmogorov complexity with only a constant number of bits of additional information in a polynomial-time computable procedure. Furthermore, it is possible to apply a modified notion of Kolmogorov-complexity to "short" strings, as will be explained in the next section.

6.2 Metrics of Spatial Complexity

"Tutor: What is twice three?

Scout: What's a rice tree?

Sub-scout: When is ice free?

Sub-Sub-scout: What's a nice fee?"

(Lewis Carroll, 1832–1898)

With these premises from previous research, two new and more effective metrics of spatial complexity will be defined here. "*Fivos Papadimitriou spatial complexity metric I*" (referred to as C_{PI} hereafter) and the explicit procedure for its computation will be described next.

This metric makes use of an initial concept by Papentin (1973, 1980, 1982, 1983a, b), who suggested that it is possible to evaluate the complexity (not spatial) of a small string by identifying patterns in it and then the string's complexity is defined as the length of the compressed string. However, he did not present any algorithmic procedure for identifying such patterns and his method was not spatial either. Also, the measure previously defined on the basis of Kolmogorov complexity (K_s) that was described by Papadimitriou (2012) did not comprise any algorithmic procedure for finding patterns in a string. And one further difference between the C_{PI} and the procedures developed by both Papentin (1973) and Papadimitriou (2012) mentioned previously, is that C_{PI} is much more restrictive, because it is defined as the minimum algorithmic complexity taking into account *all possible orientation changes* of the square map under consideration.

Consequently C_{PI} differs:

(I) from the Papadimitriou (2012) metric, because: (a) it follows an exact algorithmic procedure for finding patterns, (b) it is based on Papentin complexity rather than on Kolmogorov complexity and (c) it is calculated as the minimum from all possible symmetric transformations (rotations, inversions etc.) of the original map.

(II) from the Papentin (1973) metric, because: (a) it is spatial (Papentin's measure was not) and (b) it is associated with an exact procedure for the identification of patterns.

In computing the string length, the following conventions are used:

(a) If a symbol is repeated twice or more consecutively, then it can be written with a power of the times it appears consecutively, i.e. the string AAAAAAABA can be shortened to A^7BA.

(b) If a block of symbols repeats itself consecutively, then parentheses can be used and powers indicating the number the block of symbols is repeated, i.e. the string ABCABCABCABCABCABC can be represented as $(ABC)^6$.

(c) Lambda (λ) symbolizes a block of consecutive symbols, used to indicate compression. In the compressed string, the lambda block of symbols is separated by a comma (,) from the λ-compressed string. The comma is also counted as a separate symbol in the final string length calculation. For instance, the initial 30 symbols-long string:

BWBBWBBWBBWBBWBBWBBWBBWBBWBWWW

can be compressed by using the compression $\lambda = BWB$ that has 5 symbols to create the following short string: λ^9W^3.

The above notation would be incomplete without citing the compression $\lambda = BWB$, because λ could symbolize just any block of symbols. Hence, the compression of the initial string should be written as: $\lambda = BWB$, λ^9W^3 which consists in 10 symbols (5 symbols for the $\lambda = BWB$ plus one symbol for the comma, and 4 symbols additionally for the short string).

Thus, without any loss of information, it can be said that this compressed 10-symbols long string can fully encode the initial 30-symbols long string. Hence, the formula for the calculation of C_{PI} is simply:

$$C_{PI} = \min_{D_4}\{S_b\} \tag{6.6}$$

where $b = 1, 2, 3, \ldots L$ and S_b are all the compressed strings generated by blocks (b) of symbols, from $b = 1$ symbol up to $b = L$ symbols, and L is the initial length of the string and *for all possible map orientations* derived from the symmetry group of the square (D_4).

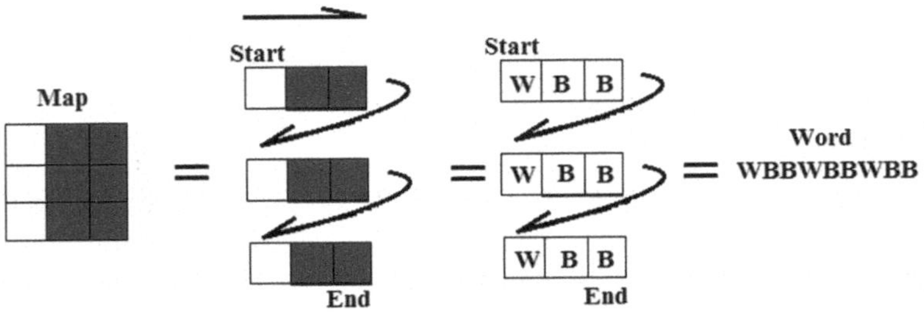

Fig. 6.4 Converting a map to a "word" (a string of symbols) for the calculation of its spatial complexity

Let us now see exactly how the C_{PI}-complexity of a small map can be calculated. Beginning by scanning cells from top to bottom and from left to right (as one reads the lines of a book), consider a 3 × 3 binary map (Fig. 6.4) with initial string (with the position of each symbol numbered below): W B B W B B W B B.

The procedure begins by first trying possible compressions using blocks of two symbols, beginning from the two first symbols on the left of the string and ending with the two last symbols (the abbreviation CSL stands for compressed string length):

(a) Blocks of 2 symbols

Symbols' numbers	λ	Compression	Short string	CSL
1 + 2	WB	λ = WB	λ = WB, λBλBλB	11
2 + 3	BB	λ = BB	λ = BB, WλWλWλ	11
3 + 4	BW	λ = BW	λ = BW, WBλBλB^2	12
4 + 5	WB	λ = WB	λ = WB, λBλBλB	11
5 + 6	BB	λ = BB	λ = BB, WλWλWλ	11
6 + 7	BW	λ = BW	λ = BW, WBλBλB^2	12
7 + 8	WB	λ = WB	λ = WB, λBλBλB	11
8 + 9	BB	λ = BB	λ = BB, WλWλWλ	11

and continues with block sizes increasing by one, up to $L - 1$, as follows:

(b) Blocks of 3 symbols

Symbols'numbers	λ	Compression	Short string	CSL
1 + 2 + 3	WBB	λ = WBB	λ = WBB, $λ^3$	8
2 + 3 + 4	BBW	λ = BBW	λ = BBW, W$λ^2B^2$	11
3 + 4 + 5	BWB	λ = BWB	λ = BWB, WB$λ^2$B	11
4 + 5 + 6	WBB	λ = WBB	λ = WBB, $λ^3$	8

(continued)

(continued)

Symbols'numbers	λ	Compression	Short string	CSL
$5+6+7$	BBW	$\lambda = \text{BBW}$	$\lambda = \text{BBW}, \text{W}\lambda^2\text{B}^2$	11
$6+7+8$	BWB	$\lambda = \text{BWB}$	$\lambda = \text{BWB}, \text{WB}\lambda^2\text{B}$	11
$7+8+9$	WBB	$\lambda = \text{WBB}$	$\lambda = \text{WBB}, \lambda^3$	8

(c) Blocks of 4 symbols

Symbols' numbers	λ	Compression	Short string	CSL
$1+2+3+4$	WBBW	$\lambda = \text{WBBW}$	$\lambda = \text{WBBW}, \lambda\text{B}^2\text{WB}^2$	13
$2+3+4+5$	BBWB	$\lambda = \text{BBWB}$	$\lambda = \text{BBWB}, \text{W}\lambda\text{BWB}^2$	13
$3+4+5+6$	BWBB	$\lambda = \text{BWBB}$	$\lambda = \text{BWBB}, \text{WB}\lambda\text{WB}^2$	13
$4+5+6+7$	WBBW	$\lambda = \text{WBBW}$	$\lambda = \text{WBBW}, \lambda\text{B}^2\text{WB}^2$	13
$5+6+7+8$	BBWB	$\lambda = \text{BBWB}$	$\lambda = \text{BBWB}, \text{W}\lambda\text{BWB}^2$	13
$6+7+8+9$	BWBB	$\lambda = \text{BWBB}$	$\lambda = \text{BWBB}, \text{WB}\lambda\text{WB}^2$	13

(d) Blocks of 5 symbols

Symbols' numbers	λ	Compression	Short string	CSL
$1+2+3+4+5$	WBBWB	$\lambda = \text{WBBWB}$	$\lambda = \text{WBBWB}, \lambda\text{BWB}^2$	13
$2+3+4+5+6$	BBWBB	$\lambda = \text{BBWBB}$	$\lambda = \text{BBWBB}, \text{W}\lambda\text{WB}^2$	13
$3+4+5+6+7$	BWBBW	$\lambda = \text{BWBBW}$	$\lambda = \text{BWBBW}, \text{WB}\lambda\text{B}^2$	13
$4+5+6+7+8$	WBBWB	$\lambda = \text{WBBWB}$	$\lambda = \text{WBBWB}, \lambda\text{BWB}^2$	13
$5+6+7+8+9$	BBWBB	$\lambda = \text{BBWBB}$	$\lambda = \text{BBWBB}, \text{W}\lambda\text{WB}^2$	13

(e) Blocks of 6 symbols

Symbols' numbers	λ	Compression	Short string	CSL
$1+2+3+4+5+6$	WBBWBB	$\lambda = \text{WBBWBB}$	$\lambda = \text{WBBWBB}, \lambda\text{WB}^2$	13
$2+3+4+5+6+7$	BBWBBW	$\lambda = \text{BBWBBW}$	$\lambda = \text{BBWBBW}, \text{W}\lambda\text{B}^2$	13
$3+4+5+6+7+8$	BWBBWB	$\lambda = \text{BWBBWB}$	$\lambda = \text{BWBBWB}, \text{WB}\lambda\text{B}$	13
$4+5+6+7+8+9$	WBBWBB	$\lambda = \text{WBBWBB}$	$\lambda = \text{WBBWBB}, \lambda\text{WB}^2$	13

(f) Blocks of 7 symbols

Symbols' numbers	λ	Compression	Short string	CSL
1 + 2 + 3 + 4 + 5 + 6 + 7	WBBWBBW	λ = WBBWBBW	λ = WBBWBBW, λB^2	13
2 + 3 + 4 + 5 + 6 + 7 + 8	BBWBBWB	λ = BBWBBWB	λ = BBWBBWB, WλB	13
3 + 4 + 5 + 6 + 7 + 8 + 9	BWBBWBB	λ = BWBBWBB	λ = BWBBWBB, WBλ	13

(g) Blocks of 8 symbols

Symbols' numbers	λ	Compression	Short string	CSL
1 + 2 + 3 + 4 + 5 + 6 + 7 + 8	WBBWBBWB	λ = WBBW, BBWB	λ = WBBWBBWB, λB	13
2 + 3 + 4 + 5 + 6 + 7 + 8 + 9	BBWBBWBB	λ = BBWB, BWBB	λ = BBWBBWBB, Wλ	13

From these calculations, it follows that a string of 8 symbols is the shortest one. This shortest string faithfully represents the original string of 9 symbols without loss of information. But 8 is *not* the value of C_{PI}. From all possible rotations of the initial map, a rotation by 90° to the right produces a compressed string shorter than 8 symbols (Fig. 6.5), so the resulting string consists in 4 symbols only: W^3B^6 and hence we eventually conclude that $C_{PI} = 4$ for this map.

Besides C_{PI}, yet another algorithmic metric will be defined here, what can be called the "*Fivos Papadimitriou spatial complexity metric 2*", symbolized by C_{P2}, based on modifications of the formula that was derived by Papadimitriou (2009), so as to make it much easier to calculate and more appropriate for small square maps. The division by the product of rows and columns will be omitted here so it will be only the sum of the total Levenshtein distances; not their average as in Papadimitriou (2009) and is thus calculated as:

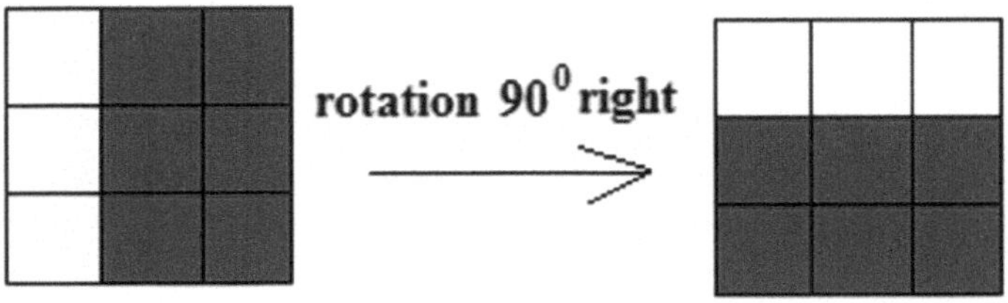

Fig. 6.5 Orientation matters in the calculation of spatial C_{PI} -complexity: rotating the initial binary map (left, with string $WB^2WB^2WB^2$) by 90° to the right results in the string W^3B^6, of which the complexity is readily calculated to be equal to 4 (in contrast to the original map, for which a lengthy calculation process was required, resulting to a complexity value equal to 8)

$$C_{P2} = \sum_{\substack{i=1 \\ j=1}}^{\substack{j=v-1 \\ i=u-1}} \alpha_{ij} \tag{6.7}$$

where a_{ij} are the in-between *i*-rows ($i = 1, 2, \ldots, u$) and in-between *j*-columns ($j = 1, 2, \ldots, v$) Levenshtein distances which result from pairwise comparisons of strips of the surface considered, successively, and for all the strips covering the surface. Thus, the C_{P2} metric differs from the C_L metric in the following:

(a) it requires less calculations to compute compared to C_L for square maps since no averaging for columns and rows is required;
(b) it assumes only positive integers as values;
(c) it can be used to derive values that are comparable to C_{P1} for *small* maps (notice, for instance, that, for 3×3 binary maps the maximum value of both C_{PI} and C_{P2} is 8).

However, instead of seeking a minimum length of string number, C_{P2} seeks a minimum operations number.

Another difference is that while C_{PI} is sensitive to symmetric transformations (rotations, inversions), C_{P2} is not (for an example calculation of C_{P2} see Fig. 6.6).

Complexity calculations of either C_{P2} or C_{PI} may also be carried out (as the case may be) according to the "*entropy encoding*" technique that is widely used in image processing for many electronics applications. This is a procedure of lossless data compression, derived by encoding the image components in a "zigzag" fashion, as shown in Fig. 6.7. It begins from one of the corner cells of the image and then proceeds in a zigzag manner up to the opposite corner of the image, by recording the cell category of each cell encountered in this zigzag process. Yet, there are several other scanning methods that are used in optoelectronics and other technologies.

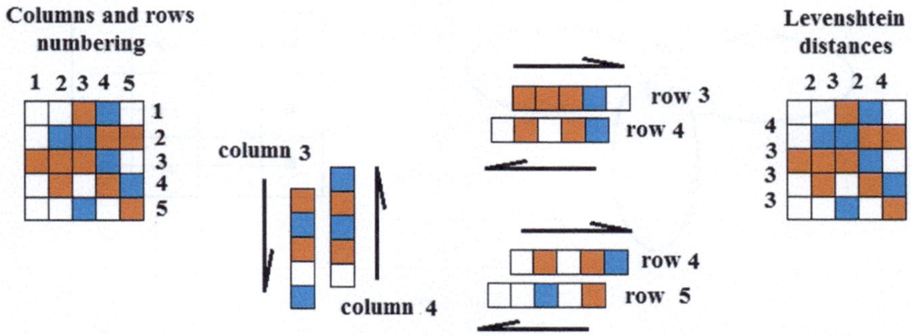

Fig. 6.6 A 5×5 square map with three cover types showing how C_{P2} and C_{PI} values are comparable for small square maps. First, notice that C_{PI}=22 because the map's irreducible string is: $A^2CBA^2B^2C^5$ BA^2CACBA^2BAC. Some characteristic column and row slides are depicted to illustrate the calculations of C_{P2} per pair of rows and columns. Eventually, the calculations lead to C_{P2} = 24 (that is an agreement by more than 90% of the two metrics for this map)

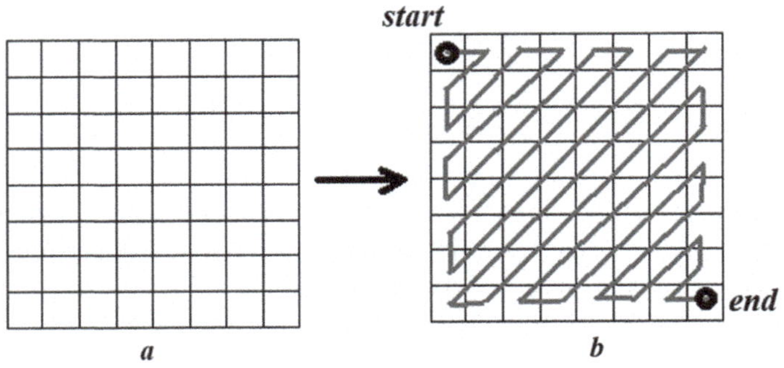

Fig. 6.7 The "entropy encoding" technique applied on a rasterized 8 × 8 initial image (**a**): beginning with the top left corner cell (**b**), continuing with the second upper right cell, then carrying on diagonally until the entire image is "encoded", down to the bottom right cell

Following such other methods will inevitably result in different strings of symbols than the raster scanning (Fig. 6.8).

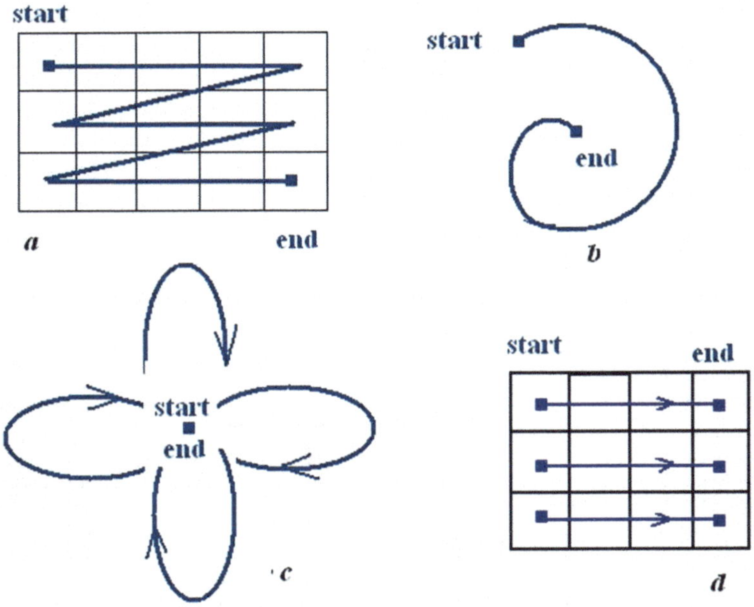

Fig. 6.8 Alternative scanning methods used in various technologies: raster scanning (**a**) that is most commonly used, rosette scanning (**b**), spiral scanning (**c**), linear array scanning (**d**)

6.3 Extrema of Spatial Complexities C_{PI} and C_{P2}

Since things may differ from one another,
to a higher or a lesser degree of difference,
there is also a maximal difference,
and that one I call contrariety
"ἐπεὶ δὲ διαφέρειν ἐνδέχεται ἀλλήλων τὰ διαφέροντα πλεῖον καὶ ἔλαττον,
ἔστι τις καὶ μεγίστη διαφορά, καὶ ταύτην λέγω ἐναντίωσιν"
(Aristotle, 384-322 b.C., "Metaphysics", 10.1055a)

It is easy to verify that the C_{P2} of any multicolored square map of size n ranges in between 2 and $2n - 2\sqrt{n}$. Indeed, the spatial complexity of any multicolored map is lowest when there is only one different cell at any one of the map's corners. In this case, the map's C_{P2} is at least 2, because the corner cell has only two borders with two white cells. So the pairwise comparisons of the 1st row with the 2nd row and the 1st column with the 2nd column will record this difference in C_{P2}. As the size of a multicolored map increases, all cells will have different colors, so the maximum C_{P2} produces $\sqrt{n}$ by pair of rows, as well as by per pair of columns. Hence there are $\sqrt{n} - 1$ comparisons of map rows, plus $\sqrt{n} - 1$ comparisons of map columns, with $\sqrt{n}$ differences per comparison. Consequently, there are $2\sqrt{n}(\sqrt{n} - 1)$ differences between cells, compared by columns and by rows pairwise and thus,

$$2 \leq C_{P2} \leq 2n - 2\sqrt{n} \tag{6.8}$$

Expectedly, the max C_{P2} grows with increasing n.

As concerns C_{PI}, if a square binary map of size n has only one black cell and that particular cell is located at anyone of the map's corners, then its C_{P1} is minimum and equal to:

$$C_{P1\,\min} = 3 + \lfloor \log_{10}(n) \rfloor \tag{6.9}$$

In fact, for any binary map, the lower complexity is attained when there is only one black cell at any one of the map's corners. In this case, the map's C_{PI} is

$$C_{P1}\{\mathrm{BW^u}\} = 3 \tag{6.10}$$

where u is the number of white cells that follow in the string.

If there are up to u = 9 white cells (that is a map with $n = 10$ cells) then the string still has the same minimum:

$$C_{P1}\{\mathrm{BW^9}\} = 3 \tag{6.11}$$

and thus, in this case, the minimum C_{PI} is also $C_{PI} = 3$.

If there are u = 10 white cells following the corner black cell, then the string has $n = 11$ so that:

$$C_{P1}\{BW^{10}\} = 4 \tag{6.12}$$

In fact, C_{PI} will remain equal to 4 for all strings of length up to $n = 99$:

$$C_{P1}\{BW^{99}\} = 4 \tag{6.13}$$

However, with one more cell ($n = 100$), the string will need three digits to be described (of which two characters only to represent the power of W), so C_{PI}-complexity becomes:

$$C_{P1}\{BW^{100}\} = 5 \tag{6.14}$$

Thus, as the map size increases, string codes are determined by a power of 10 (hence by the logarithm of 10) of the exponent u. As $\log_{10}n$ also assumes non-integer values, the floor function $\lfloor\rfloor$ of the logarithm of n applies, so C_{PI} has a bound of its minimum value, depending on the map size n, defined as:

$$C_{P1\,\min} = 3 + \lfloor \log_{10}(n) \rfloor \tag{6.15}$$

or, equivalently,

$$C_{P1} = 2 + \lceil \log_{10}(n) \rceil \tag{6.16}$$

where $\lceil\rceil$ is the ceiling function.

Notice that the floor function is a stepwise function, meaning that the maximum values of the minima of $C_{PI\,min}$ assume integer values only, whose ranges increase slowly with n. To verify, consider the following examples, with increasing mapsize n:

$$\text{For}\, n = 81,\, C_{P1\,\min} = 3 + \lfloor \log_{10}(81) \rfloor = 3 + \lfloor 1.81 \rfloor = 4 \tag{6.17}$$

$$\text{For}\, n = 121,\, C_{P1\,\min} = 3 + \lfloor \log_{10}(121) \rfloor = 3 + 2 = 5 \tag{6.18}$$

$$\text{For}\, n = 625,\, C_{P1\,\min} = 3 + \lfloor \log_{10}(625) \rfloor = 3 + \lfloor 2.25 \rfloor = 5 \tag{6.19}$$

Thus if any one of the four corner cells of a binary square map is black, whatever the map size, the minimum C_{PI} complexity can not be higher than $3 + \lfloor \log_{10}(n) \rfloor$.

More generally, for the entropy class 1, the minimum C_{PI} complexity is attained when there is only one black cell, located at anyone of the map's corners, so for all binary maps

$$C_{P1\ \min} = 3 + \lfloor \log_{10}(n) \rfloor = 2 + \lceil \log_{10}(n) \rceil \tag{6.20}$$

holds, whatever the value of n is.

Given these, it can be proven that if all the black cells of a binary square map appear as a single block of ω-consecutive black cells located after m-consecutive white cells anywhere in the map except for its corners, then the C_{P1} is:

$$C_{P1} = 3 + \lceil \log_{10}(m) \rceil + \lceil \log_{10}(\omega) \rceil + \lceil \log_{10}(n - m - \omega) \rceil \tag{6.21}$$

Indeed, for any black cell located anywhere in the binary map but the corner, the string has the general form:

$$\mathrm{W}^m \mathrm{B} \mathrm{W}^{n-m-1} \tag{6.22}$$

where m is the number of cells preceding the black cell and therefore the C_{P1} of such strings is:

$$C_{P1} = 3 + \lceil \log_{10}(m) \rceil + \lceil \log_{10}(n - m - 1) \rceil \tag{6.23}$$

For three black consecutive cells located anywhere in the map except for the corner, the string is:

$$\mathrm{W}^m \mathrm{B}^3 \mathrm{W}^{n-m-3} \tag{6.24}$$

which has a C_{P1} equal to:

$$C_{P1} = 5 + \lceil \log_{10}(m) \rceil + \lceil \log_{10}(n - m - 3) \rceil \tag{6.25}$$

Consequently, for a block of ω consecutive black cells located anywhere in the map but the corner, the string is:

$$\mathrm{W}^m \mathrm{B}^\omega \mathrm{W}^{n-m-\omega} \tag{6.26}$$

and hence:

$$C_{P1} = 3 + \lceil \log_{10}(m) \rceil + \lceil \log_{10}(\omega) \rceil + \lceil \log_{10}(n - m - \omega) \rceil \tag{6.27}$$

This formula gives the complexity of all binary maps with entropy class 1 with black cells appearing as a single block anywhere in the binary map except for the map's four corner cells.

As an example, consider a binary 10×10 map (Fig. 6.9), with the following incompressible string of 9 symbols:$\mathrm{W}^{36}\mathrm{B}^{25}\mathrm{W}^{39}$. This map has a single block of 25

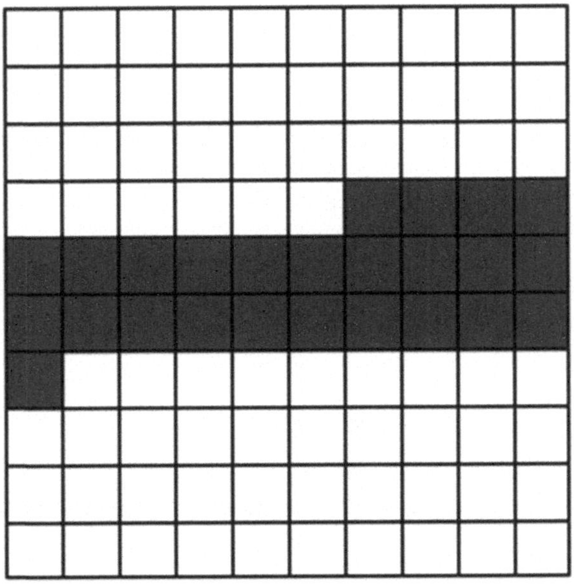

Fig. 6.9 A 10 × 10 binary map, with a single block of 25 consecutive black cells located after 36 initial white cells and 39 white cells that follow immediately after the block of black cells

consecutive black cells located after 36 initial white cells and 39 white cells that follow immediately after the block of black cells.

Applying the previous formula, the C_{PI} complexity of this map is:

$$\begin{aligned} C_{P1} &= 3 + \lceil \log_{10}(36) \rceil + \lceil \log_{10}(25) \rceil + \lceil \log_{10}(39) \rceil \\ &= 3 + 2 + 2 + 2 = 9 \end{aligned} \tag{6.28}$$

Due to to the periodic nature of the floor and ceiling functions, the following proposition is interesting to consider, which holds when the logarithm of n with base 10 is non-integer (as most often the case is).

Also, it if and only if $\lfloor \log_{10}(n) \rfloor \notin Z$, then a lower bound of the minimum C_{PI} of a square binary map is $2 + \log_{10} n$.

This follows immediately by considering the Fourier series expansion (sawtooth function) of the floor function for non-integer real numbers x:

$$\lfloor x \rfloor = x - \frac{1}{2} + \frac{1}{\pi} \sum_{k=1}^{\infty} \left[\frac{\sin(2\pi kx)}{k} \right] \tag{6.29}$$

Substituting the value $x = \log_{10}(n)$ to

$$C_{P1\ \min} = 3 + \lfloor \log_{10}(n) \rfloor \tag{6.30}$$

yields:

$$C_{P1\ \min} = \log_{10} n + \frac{5}{2} + \frac{1}{\pi}\left[\sum_{k=1}^{\infty} \frac{\sin\left[2\pi k(\log_{10}(n))\right]}{k}\right] \tag{6.31}$$

and given the bounded variation of the Fourier series, one lower bound is:

$$2 + \log_{10} n \tag{6.32}$$

In fact, it is easy to verify that

$$\frac{5}{2} + \log_{10} n > 3 \tag{6.33}$$

holds for every $n > 3.16228$ (this is one lower bound; what is the infimum?).

As an example, consider $n = 121$. Then, $\log_{10}(121) = 2.082$, so a lower bound of $C_{PI\ min}$ is $2 + 2.082 = 4.082$. Indeed,

$$C_{P1\ \min} = 3 + \lfloor \log_{10}(121) \rfloor = 5 > 4.082 \tag{6.34}$$

Interestingly, the minimum C_{PI} of binary maps can also be related to the transcendental number π. An alternative expression is derived by expanding the Fourier series by complex numbers. In the previous example, the same result is derived either way:

$$C_{P1\,\min} = \frac{5}{2} + \log_{10}(121) - \frac{i\left[\log(1 - e^{-2i\pi \log_{10}(121)}) - \log(1 - e^{2i\pi \log_{10}(121)})\right]}{2\pi} = 5 \tag{6.35}$$

If and only if all black cells of a binary square map appear as a single block of ω-consecutive black cells located after m-consecutive white cells anywhere in the map except for its corner, and

$$\lfloor \log_{10}(m) \rfloor \wedge \lfloor \log_{10}(\omega) \rfloor \wedge \lfloor \log_{10}(n - m - \omega) \rfloor \notin Z,$$

then a lower bound of the minimumC_{PI} of the map is:

$$3 + \log_{10} m\omega + \log_{10}(n - m - \omega) \tag{6.36}$$

This can be proven by considering that

$$C_{P1\,\min} = 6 + \log_{10} m\omega + \log_{10}(n - m - \omega) - \frac{3}{2} + $$
$$+ \frac{1}{\pi}\left[\sum_{k=1}^{\infty} \frac{1}{k}\left(\sin\left[2\pi k(\log_{10} m)\right] + \sin\left[2\pi k(\log_{10}\omega)\right] + \sin\left[2\pi k(\log_{10} n - m - \omega)\right]\right)\right] \tag{6.37}$$

and as

$$-\frac{3}{2} \leq \frac{1}{\pi}\left[\sum_{k=1}^{\infty} \frac{1}{k}\left(\sin\left[2\pi k(\log_{10} m)\right] + \sin\left[2\pi k(\log_{10}\omega)\right] + \sin\left[2\pi k(\log_{10} n - m - \omega)\right]\right)\right] \leq \frac{3}{2} \tag{6.38}$$

it follows that a lower bound is

$$3 + \log_{10} m\omega + \log_{10}(n - m - \omega) \tag{6.39}$$

Finally, as the map size n of a multicolored square map increases with $n \to \infty$, the ratio of the maximum C_{PI} to the number of the cells' boundaries converges to 1 and the minimum C_{PI} tends to zero.

The proof follows easily by counting the number of boundaries from left to right and converting them into a string of symbols. The first row has $\sqrt{n}$ boundaries, but the last cell of the 1st row is the boundary of the second row. Continuing until the penultimate cell, the number of boundaries in a multicolored square map is $U = n - 1$. All the first column cells have no boundary (they share the same boundary with the last column's right boundaries) and the last cell (down right) has no boundary to any other cell. So, for instance, a 4×4 map has 15 boundaries between cells if the entire map is converted into an one-dimensional map. Consequently, it easily follows that

$$\lim_{n\to\infty}\left(\frac{C_{P1\,\max}}{U}\right) = \lim_{n\to\infty}\frac{n}{n-1} = 1 \tag{6.40}$$

and

$$\lim_{n\to\infty}\left(\frac{C_{P1\,\min}}{U}\right) = \lim_{n\to\infty}\frac{3 + \lfloor \log_{10} n \rfloor}{n-1} = 0 \tag{6.41}$$

References

Fortnow, L., Hitchcock, J. M., Pavan, A., Vinodchandran, N. V., & Wang, F. (2011). Extracting Kolmogorov complexity with applications to dimension zero-one laws. *Information and Computation, 209,* 627–636.

Kolmogorov, A. (1965). Three approaches to the quantitative definition of information. *Problems of Information Transmission, 1,* 1–17.

Kolmogorov, A. (1986). On the notion of infinite pseudorandom sequences. *Theoretical Computer Science, 39,* 9–33.

Levenshtein, V. I. (1966). Binary codes capable of correcting deletions, insertions and reversals. *Soviet Physics Doklady, 10,* 707–710.

Papadimitriou, F. (2002). Modelling indicators and indices of landscape complexity: An approach using GIS. *Ecological Indicators, 2,* 17–25.

Papadimitriou, F. (2009). Modelling spatial landscape complexity using the Levenshtein algorithm. *Ecological Informatics, 4*(1), 51–58.

Papadimitriou, F. (2012). The algorithmic complexity of landscapes. *Landscape Research, 37*(5), 599–611.

Papentin, F. (1973). A Darwinian evolutionary system III. Experiments on the evolution of feeding patterns. *Journal of Theoretical Biology, 39,* 431–445.

Papentin, F. (1980). On order and complexity. I. General considerations. *Journal of Theoretical Biology, 87,* 421–456.

Papentin, F. (1982). On order and complexity. II. Application to chemical and biological structures. *Journal of Theoretical Biology, 95,* 225–245.

Papentin, F. (1983). Binary sequences I. Complexity. *Information Sciences, 31*(1), 1–14.

Papentin, F. (1983a). Binary sequences III. Complexity versus homogeneity and symmetry. *Information Sciences, 31*(1), 33–39.

Seiffert, U., & Michaelis, B. (1997). Growing a 3D-SOMs with 2D-input layer as a classification tool in a motion detection system. *International Journal of Neural Systems, 8*(1), 81–89.

Chapter 7
Exploring Spatial Complexity in 3d

We can only claim that we know something,
when we know its first cause
"Τότε γὰρ εἰδέναι φαμέν ἕκαστον,
ὅταν τὴν πρώτην αἰτίαν οἰώμεθα γνωρίζειν"
(Aristotle, 384-322 b.C., "Metaphysics", A3, 983a, 27–28)

Abstract Calculating the spatial complexity of 3d surfaces and multiply connected spatial objects presents several interesting peculiarities. The author's metrics of spatial complexity C_{P1} and C_{P2} apply to some surfaces such as the Möbius band, cylindrical and toroidal surfaces. But on other surfaces they may not, or they may apply under certain conditions only (i.e. C_{P1} needs some predefined starting point, but C_{P2} does not). Also, Hamiltonian paths can be useful to devise indices of spatial complexity for surfaces and objects in three dimensions. In any case, the genus of the 3d surface plays a key role in assessing its spatial complexity, so an experimental metric C_R is proposed here, based on Reeb graphs.

Keywords Spatial complexity · Map complexity · 3d Complexity · Möbius band · Voxels · Algorithmic Complexity · Reeb graphs

7.1 Simplicial Complexes, Betti Groups and Matveev Complexity

Although objects seem to be solid,
yet they can be porous and formed
from matter mixed with void
"Undique materies quoniam stipata quiesset
praeterea quamvis solidae res esse putentur,
hinc tamen esse licet raro cum corpore cernas"
(Lucretius, 99-55 b.C.,"De Rerum Natura", 1.329)

The problem with describing linear features on the surface of the planet is due to the fact that the sphere is locally diffeomorphic to the plane, but not locally isometric to it and, consequently, a major part of cartographic and geodesic research has had

F. Papadimitriou, *Spatial Complexity*, https://doi.org/10.1007/978-3-030-59671-2_7

to deal with this difficulty. Further still, not all surfaces of 3d spatial objects are differentiable, so the applicability of ordinary topological transformations cannot always be taken for granted. For this reason, for a spatial object that is not composed from smooth and differentiable surfaces, methods of combinatorial topology can be more appropriate for spatial analysis.

One of the first efforts to study the surface of the sphere was by means of differential geometry and geodesics in particular. For many practical applications in geography, cartography and ecology, topological methods are essential and, indeed, the relationship between geography and topology has a long history; for instance, Sen (1976) was one of the first to draw attention to the relationships between topology and geography. Following Egenhofer et al. (1989), the topology of a geographical relief can be broken down to elementary simplices (from 0-simplices up to 3-simplices). In the same way, but with a different terminology, Scholl and Voisard (1989) claimed that maps are essentially sets of "tuples", that is geometric regions to which non-geometric information is associated to. Besides analysing the spatial complexity of 3d objects however, 3d analyses of geographical settings (Papadimitriou 2012) can also be useful to decipher their overall complexity (spatial, functional and qualitative). In fact, simplicial complexes have more intensively been considered in theory and practice of geoinformatics (although not as indicators of spatial complexity). To define simplicial complexes for analysis of 2d and 3d surfaces, it is necessary to make a triangulation first. A triangulation of a surface in 3d space is one of the operations resulting in simplicial complexes. These are "complexes" whose elementary constituents are "simplexes", that is algebraic-topological n-dimensional entities of dimensions 0 (points), 1 (segments), 2 (triangles), 3 (tetrahedra) etc. (Fig. 7.1). The "Betti groups" (or "homology groups") are commutative and determine the invariants of a simplicial complex; for further information, the reader is referred to the classic text of Pontryagin (2015). Here, it suffices to recall that if two polyhedra (corresponding to two different simplicial complexes) are homeomorphic, then the complexes defining them have isomorphic Betti groups in all dimensions and therefore it is possible to characterize the polyhedra by Betti groups. Interestingly, the Betti groups are invariant under barycentric subdivision of the complexes to smaller simplexes.

In a combination of basic structural elements (such as the simplexes) and of the topology of manifolds, a notion of complexity that relates to 3d manifolds is the "*Matveev complexity*" (Matveev 1990) and is based on a combinatorial description

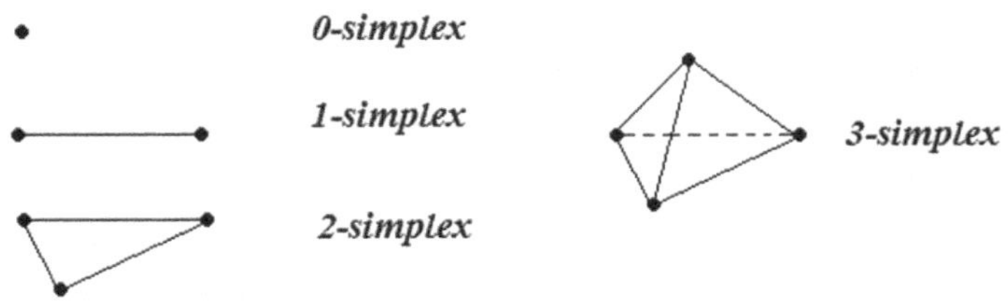

Fig. 7.1 Simplexes of various dimensions

of a manifold's basic structural elements, called "spines" (Matveev 1987; Piergallini 1988); in fact, it is based on the calculation of the minimal amount of vertices of a spine of the manifold. This complexity estimation (which is a non-negative integer) is additive and equal to the minimum number of tetrahedra of the triangulation of a 3-manifold. A related notion is the "*Heegard complexity*" (Cattabriga et al. 2010), but, as research in 3-manifolds has shown (Jaco et al. 2009, 2011), computing the complexity of a manifold by topological criteria is indeed a difficult problem. From the perspective of computational complexity, triangulations on 3-manifolds present *NP*-problems; i.e. to decide (Ivanov 2008, p. 1) "whether a triangulated 3-manifold is homeomorphic to a 3-sphere, or to a 2-sphere bundle over a circle, or to a real projective 3-space, or to a handlebody of genus *g*".

7.2 Evaluation on Möbius Bands, Tori and Multiply-Connected Surfaces

> "Everything that happens, and everything that is said,
> happens or is said at the surface"
> (Gilles Deleuze 2012, p. 150)

As can easily be verified, both C_{P1} and C_{P2} can be applied to closed 2-dimensional surfaces, even if they are of genus 1. The calculation of C_{P2} can be carried out by comparing one strip to its adjacent one, so long as the number of cells in each ring is the same and there is a one-to-one correspondence between them. As concerns C_{P1} however, there are as many possible strings representing each ring, as the number of square cells in each ring suggests, so the calculation of C_{P1} depends on the point of departure for measuring the string: changing the point of departure implies the string will begin from a different symbol and hence its compression may yield different compressed strings (this requirement does not apply to ordinary rectangular maps, since the pre-defined departure point applies to anyone of the four corners). Let us see the calculation of C_{P2} over a 2d ellipsoid surface with genus one, that is embedded in R^3 (Fig. 7.2).

Counting the cells in a counterclockwise direction yields:

Inner ring: W W B W W B W B W B W B B B
Outer ring: W B W B W W B W B B B B W B

and sliding the string of the outer ring one position to the right yields:

Inner ring: W W B W W B W B W B W B B B
Outer ring: W B W B W W B W B B B B W B

from which it follows that $C_{P2} = 6$. The C_{P1} can be evaluated from the concatenation of the string of the inner circle with that of the outer circle. The concatenation is a string of 28 symbols, beginning from the pre-defined starting point:

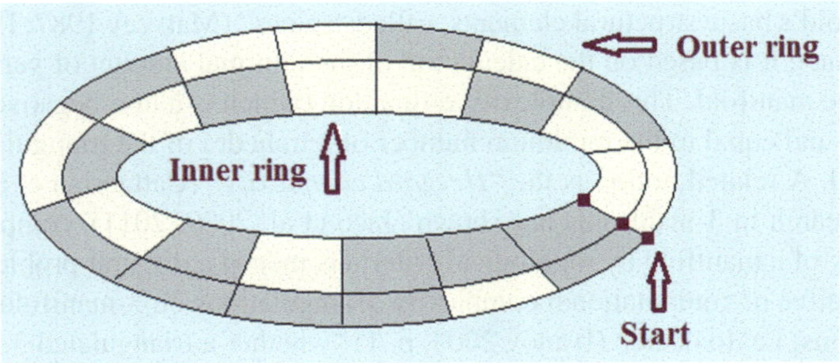

Fig. 7.2 Calculating C_{P1} and C_{P2} on 2-dimensional surfaces with curvature: C_{P1} needs a predefined starting point, although C_{P2} does not

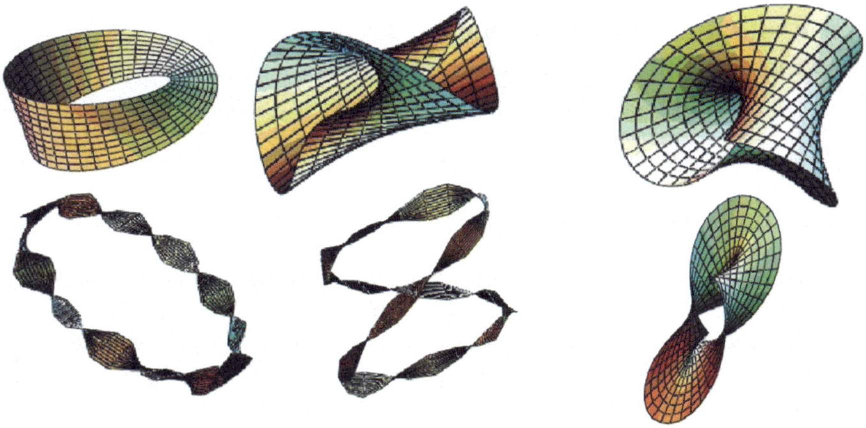

Fig. 7.3 Both C_{P1} and C_{P2} are applicable for calculations on surfaces spreading over various kinds of Möbius bands

WBWBWWBWBBBBWBWWBWWBWBWBWBBB

of which the compressed string is a string of 24 symbols:

$$\lambda = \text{BWBWWBW},\ \text{W}\lambda\text{B}^3\lambda\text{WBWBWBWB}^3$$

Also, the calculation of C_{P1} and C_{P2} applies to Möbius bands, as well as to other surfaces that result from Möbius bands twisted several times (Fig. 7.3).

Similar procedures follow on maps of squares that cover in a chessboard-like manner a cylindrical surface (Fig. 7.4), so spatial complexity may also be calculated by means of both C_{P1} and C_{P2} complexities, strap by strap, until the entire cylinder's surface is completely covered (and similarly, on conchoids, pseudospheres and other surfaces).

Fig. 7.4 C_{P1} and C_{P2} complexities both apply to cylindrical surfaces

Allowing for calculations that bypass edges of solid bodies, C_{P1} and C_{P2} complexities, can also be calculated on many other surfaces, such as *bipyramidal astroids*

$$\begin{aligned} x(u, v) &= a\cos^2 u \cos^3 v \\ y(u, v) &= a\sin^3 u \cos^3 v \\ z(u, v) &= a\sin^3 v \end{aligned} \tag{7.1}$$

and even on holey surfaces such as the *Lissajous curves*

$$\begin{aligned} x &= a\sin t \\ y &= b\sin(nt + \varphi) \\ z &= c\sin(mt + \psi) \end{aligned} \tag{7.2}$$

As in the case of Möbius bands, the process for the calculation of C_{P1} and C_{P2} complexities along strips that imperceptibly turn from outside to inside applies over areas of other surfaces also, such as the surface created by *Weierstrass' elliptic function:*

$$\wp(z) = \frac{1}{z^2} + \sum\left(\frac{1}{(z-w)^2} - \frac{1}{w^2}\right) \tag{7.3}$$

But, on some surfaces, two different calculations are required (one for the inside and another for the outside). Examples of such surfaces are "finite minimal Riemann surfaces" (where $w = u + iv$), e.g. of the type:

$$\begin{aligned} x &= a\mathrm{Re}\left[w\left(1-k^2\right) + \frac{8wk^2}{w^2-1}\right] \\ y &= b\mathrm{Re}\left[iw\left(k^2+1\right) - \frac{8wk^2}{w^2-1}\right] \\ z &= a\mathrm{Re}\left[k\left(w + 2\ln\left(\frac{w-1}{w+1}\right)\right)\right] \end{aligned} \tag{7.4}$$

Fig. 7.5 Binary square maps on torus and double torus surfaces

Yet, in other cases, more than two calculations may be necessary. One such example is the *sphero-cylindrical curve* created from a sphere of radius a, a cylinder of radius b and axis c, with $x^2 + y^2 + z^2 = a^2$ and $x^2 + (z{-}c)^2 = b^2$ with the parametrization:

$$
\begin{aligned}
x &= b\cos t \\
y &= \pm\sqrt{a^2 - b^2 - c^2 - 2bc\sin t} \\
z &= c + b\sin t
\end{aligned}
\tag{7.5}
$$

for $b + c$ equal to or less than a.

While C_{P2} applies without any complication on a map that is spread over a torus surface (Fig. 7.5), the C_{PI} is applicable only with the provision of "stripping off" the square cells of the torus, band by band. This essentially means that there should necessarily exist a "starting" column (or row) on this surface. Identifying the optimal such starting point on the surface so as to minimize the C_{PI} of all possible strings is obviously a hard problem, even for small such surfaces. The same does not necessarily apply to maps on surfaces of higher genus (Fig. 7.6).

With these considerations, an obvious question to ask is "what is the difference between C_{PI} and C_{P2} complexities of square planar maps with holes?". This question prompts to conjecture that C_{P2} would still apply to planar maps with holes, so long as the corresponding columns and rows of squares are comparable in terms of Levenshtein distances.

Fig. 7.6 Do C_{PI} and C_{P2} metrics apply when the genus of a surface increases?

7.3 Spatial Complexity of Simple 3d Solids and Voxels

"For whom are these serpents?"
(Georges Bataille 1991, p. 162)

The metrics C_{P1} and C_{P2} are not appropriate for calculations over surfaces on lattices such as the internal faces of successive cubes but they can be applied in case of voxelized surfaces, if the calculation proceeds even when tipping over an edge (Fig. 7.7).

Voxelization is useful in many applications, from geology and MRI (Windreicha et al. 2003; Dickie et al. 2015) to estimations of poses of the human body (Schick and Stiefelhagen 2015). Inferring changes in values of distributions per voxel from images is significant for some scientific domains, such as CT-angiography (computed tomography angiographic images) that is used for discovering problems in the arteries (Chen and Molloi 2003). Also, in computer tomography, binary images need to be reconstructed from their projections and this procedure has significant applications in nanotechnology and electron microscopy (Batenburg et al. 2009; Balazs 2013). Voxels can also be combined in aggregates so as to approximate the shape of a sphere (the so-called "naïve sphere").

Evidently, any 3d object can be approximated in a similar way, by "naïve" 3d-objects consisting of voxels. Voxelization algorithms have been developed for 3D lines, 3D circles, etc. and have appeared in various forms, depending on whether they are applicable on solids, polygons, polyhedra, quadric objects etc. (Laine 2013; Kaufman and Shimony 1986; Kaufman 1987a), as well as curved objects such as spheres and cylinders (Kaufman 1987b).

For a suitable introduction to the theory of voxelization, the reader is referred to the relevant literature (Cohen-Or and Kaufman 1995; Herman 1998; Rosenfeld 1998).

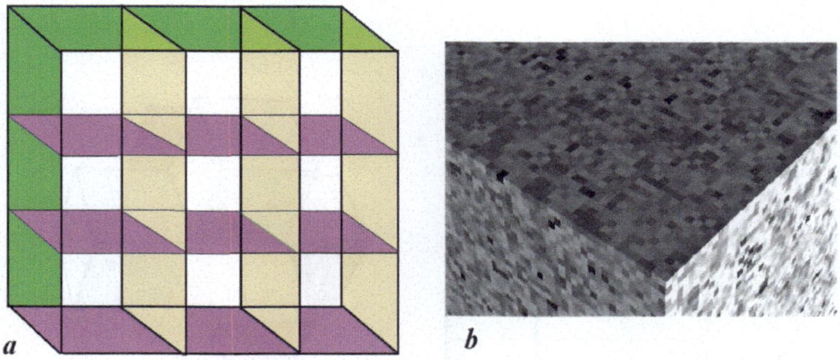

a *b*

Fig. 7.7 While neither C_{P1} nor C_{P2} are appropriate for calculations over surfaces on lattices in R^3 such as the internal faces of successive cubes (**a**), they can nevertheless be useful when the calculation is allowed to apply even on successive faces of cubic surfaces as if they were covered by a tapestry (**b**)

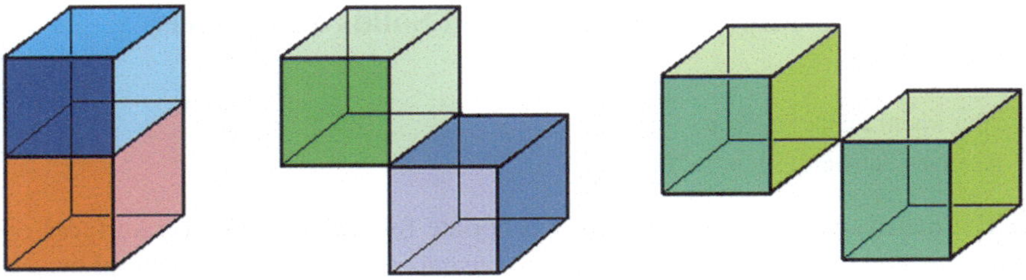

Fig. 7.8 The three possible cases of adjacency of two cubes: 2-adjacency (when the two cubes share a common face), 1-adjacency (when they share a common edge) and 0-adjacency (when they share a common vertex)

Voxelization begins by idenifying the three possible $(n - 1)$-dimensional adjacency types for voxels in $n = 3$ dimensions (Fig. 7.8) in the discrete Euclidean space Z^3: (a) six voxels that are 6-adjacent to the central voxel, (b) eighteen voxels that are 18-adjacent to the central voxel, and c) twenty six voxels that are 26-adjacent to the voxel (Fig. 7.9).

The spatial complexity of the colored faces of a cube might be evaluated by adopting an "entropy encoding" approach, essentially mimicking a typical "*Hamiltonian path*", that is a path from vertex to vertex, so that no vertex is visited twice. Given that there are three "visible" faces of the cube each time, if we need to record the colorings of all 6 faces whereas each one of them is covered by a 3×3 map, then we have 54 cells to account for. Half of them appear on each view of the cube at each time step (Fig. 7.10). The first view displays face F_1 (front), followed by F_2 (top) and face F_3 (left rear) and the second view the faces F_4 (back), F_5 (right rear) and F_6 (bottom).

Thus, beginning with the topmost cell of the front face and following an "entropy encoding" path, it is possible to cover all the 54 voxels in a spiral-like and continuous manner, down to the last 54th, that is the back lowest voxel at the far rear end of the cube's diagonal (Fig. 7.11), as a snake making rounds with its tail.

Fig. 7.9 Two adjacency types for voxels in 3d space: 26 adjacent voxels surrounding the central voxel and 6 adjacent voxels

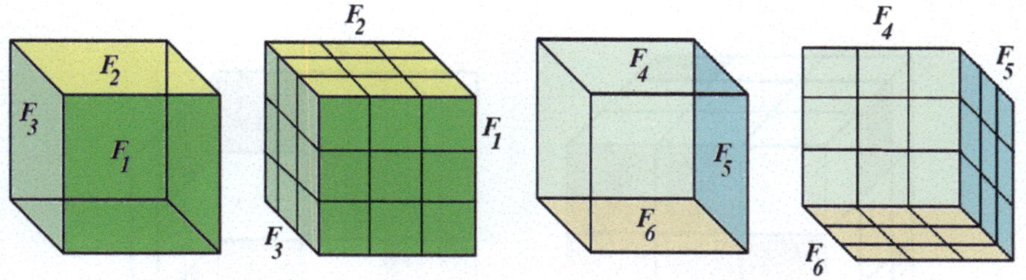

Fig. 7.10 Numbering of cube faces: F_1 = front, F_2 = top, F_3 = rear left, F_4 = back, F_5 = rear right, F_6 = bottom

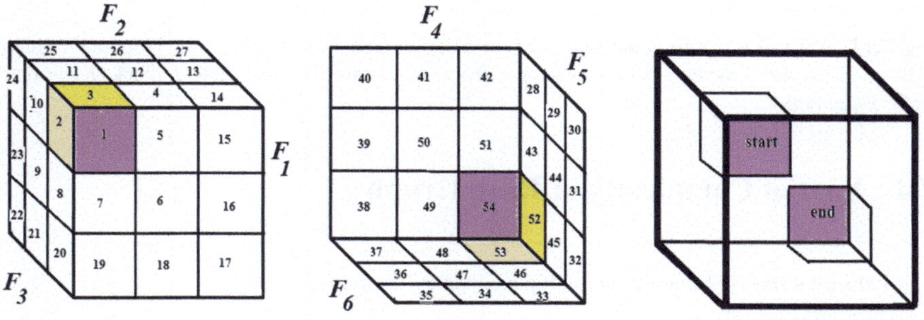

Fig. 7.11 Calculation of spatial complexity following a 3-dimensional analogue of the "entropy encoding" method. Registration of faces begins from top left and, following a spiral curve, ends up at the exactly opposite face at the bottom right of the $9 \times 9 \times 9$ cube (F_1 = front, F_2 = top, F_3 = rear left, F_4 = back, F_5 = rear right, F_6 = bottom)

As an example, consider a cube with each one of its faces covered by a different binary 3×3 map (Fig. 7.12).

The calculation of C_{PI}-complexity along the entropy encoding path begins with the initial 54-symbols long string, which, after compression with the replacement $\lambda = W^4B^3W^2B$, yields the final compressed string:

$$\{\lambda = W^4B^3W^2B,\ \lambda^2WB\lambda^2W^3B^2W^3BWB\}$$

and therefore 25 symbols are required for the surface of this cube if this particular encoding scheme is adopted, but, evidently, there are several alternative entropy encoding schemes.

Thus, besides the possibility to use C_{PI}-complexity as a metric of the spatial complexity of the cube's surfaces, there are ample margins for alternative encodings and complexity evaluations. Plausibly, the 3d equivalent of C_{PI} should be the *minimum* of all compressed strings measured over *all* possible entropy encoding itineraries over the cube's surface (unless a universal agreement is made on the exact direction of the entropy encoding itinerary).

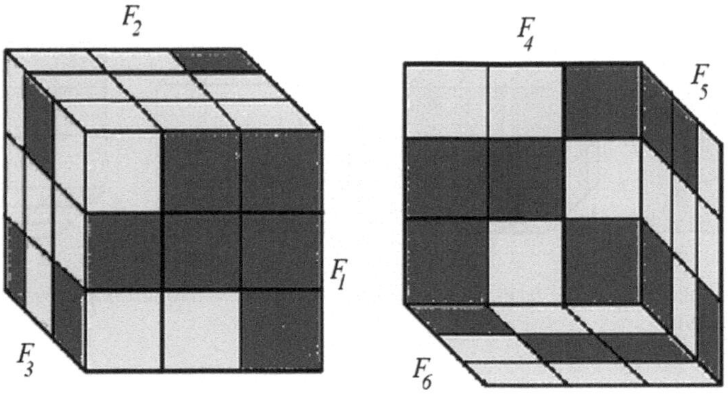

Fig. 7.12 A 9 × 9 × 9 cube consisting of 81 voxels, of which the outer faces are 54 in total (9 map cells on each one of the cube's 6 faces). F_1 = front, F_2 = top, F_3 = rear left, F_4back, F_5 = rear right, F_6 = bottom

7.4 Spatial Complexity of Reeb Graphs

"With what fantasy he conferred multiple curvature on space!"
(Gaston Bachelard 1994, p. 157)

Recalling that the Euler characteristic $\chi(s)$ is an invariant of a surface s and is given by the simple formula $\chi(s)$ = nodes-vertices + polygons, it might be a good guess that a combination of areas (polygons), lines (vertices) and points (nodes) should be useful to provide us with an index of spatial complexity of 3d surfaces; it has already been used to analyse the complexity of geographical settings (Papadimitriou 2013). Further to Euler's formula, Lhuilier's formula takes into account cavities and tunnels and relates to Euler's formula, although, as was proved later, that formula needed modifications, so the extent to which the determinants of $\chi(s)$ might serve as estimators of the spatial complexity of 3d objects is open for future research. Applying a simple "additive" approach to *Reeb graphs* gives a glimpse of how adding elementary constituents (sources, sinks and saddles in this case) of a 3d spatial object might provide an metric of its spatial complexity (Fig. 7.13). The Euler characteristic is calculated as

$$\chi = \text{Source points} + \text{Sink points} - \text{Saddle points} \tag{7.6}$$

so an "additive" complexity metric C_R applicable to Reeb graphs might be the following:

$$C_R = \text{Source points} + \text{Sink points} + \text{Saddle points} \tag{7.7}$$

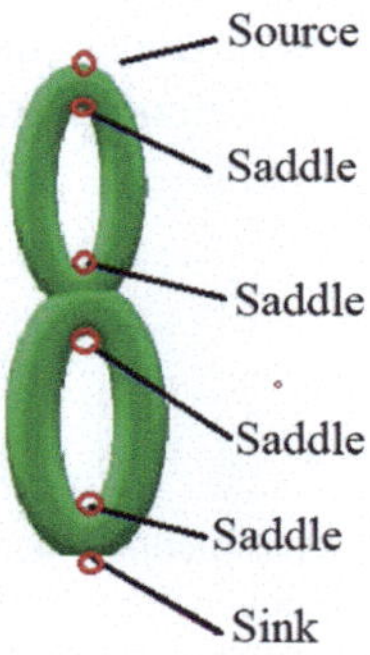

Fig. 7.13 Identifying source, sink and saddle points on a surface of genus 2. The Euler number of this object is: χ = Source points + Sink points − Saddle points = $1 + 1 - 4 = -2$. A complexity metric is then: C_R = Source points + Sink points + Saddle points = $1 + 1 + 4 = 6$

Indicatively, the values for eight shapes of different genus each (Fig. 7.14) are calculated (Table 7.1). Notice how shapes *d* and *g* have the same genus (3) and the

Fig. 7.14 Different 3d objects, with varying numbers of source points, sink points and saddle points and with different genus each

Table 7.1 Example calculations of C_R for the shapes of Fig. 7.14

Shape code	Euler's χ	Source points	Sink points	Saddle points	Complexity C_R
a	2	1	1	0	2
b	0	1	1	2	4
c	− 2	1	1	4	6
d	− 4	1	1	6	8
e	− 6	1	1	8	10
f	− 8	1	1	10	12
g	− 4	4	1	9	14

same Euler number (−4) but shape g is obviously more complex than shape d and this difference is reflected their respective complexity values: $C_R(d) = 8$ against $C_R(g) = 14$.

References

Balazs, P. (2013). Complexity results for reconstructing binary images with disjoint components from horizontal and vertical projections. *Discrete Applied Mathematics, 161,* 2224–2235.
Bataille, G. (1991). *The impossible.* San Francisco, CA: City Lights Books.
Batenburg, K. J., Bals, S., Sijbers, J., Kuebel, C., Midgley, P. A., Hernandez, J. C., et al. (2009). 3D imaging of nanomaterials by discrete tomography. *Ultramicroscopy, 109*(6), 730–740.
Cattabriga, A., Mulazzani, M., & Vesnin, A. (2010). Complexity, Heegaard diagrams and generalized Dunwoody manifolds. *Journal of the Korean Mathematical Society, 47,* 585–599.
Chen, Z., & Molloi, S. (2003). Automatic 3D Vascular Tree construction in CT angiography. *Computerized Medical Imaging and Graphics, 27,* 469–479.
Cohen-Or, D., & Kaufman, A. (1995). Fundamentals of surface voxelization. *Graphical Models and Image Processing, 57*(6), 453–461.
Deleuze, G. (2012). *The logic of sense.* London: Continuum.
Dickie, D. A., Mikhael, S., Job, D. E., Wardlaw, J. M., Laidlaw, D. H., & Bastin, M. E. (2015). Permutation and parametric tests for effect sizes in voxel-based morphometry of gray matter volume in brain structural MRI. *Magnetic Resonance Imaging, 33,* 1299–1305.
Egenhofer, M. J., Frank, A. U., & Jackson, J. P. (1989). A topological data model for spatial databases. In *Springer Lecture Notes in Computer Science 409 "Design and Implementation of Large Spatial Databases"* (pp. 271–286).
Herman, G. T. (1998). *Geometry of digital spaces.* Boston: Birkhauser.
Ivanov, S. V. (2008). The computational complexity of basic decision problems in 3-dimensional topology. *Geometriae Dedicata, 131*(1), 1–26.
Jaco, W., Rubinstein, H., & Tillmann, S. (2009). Minimal triangulations for an infinite family of lens spaces. *Journal of Topology, 2,* 157–180.
Jaco, W., Rubinstein, H., & Tillmann, S. (2011). Coverings and minimal triangulations of 3-manifolds. *Algebra Geometry Topology, 11,* 1257–1265.
Kaufman, A. & Shimony, E. (1986). 3D Scan-Conversion Algorithms for Voxel-Based Graphics. Proceedings of the ACM Workshop on Interactive 3D Graphics, Chapel Hill, NC, October 1986, pp.45–76.
Kaufman, A. (1987a). An algorithm for 3D scan-conversion of polygons. In: *Proceedings of Eurographics'87,* Amsterdam, Netherlands, pp. 197–208.
Kaufman, A. (1987b). Efficient Algorithms for 3D Scan-Conversion of Parametric Curves, Surfaces, and Volumes. *Computer Graphics* 21(4), 171–179
Laine, S. (2013). A topological approach to voxelization. *Computer Graphics Forum, 32*(4), 77–86.
Matveev, S. (1987). Transformations of special spines, and the Zeeman's conjecture. In Izv. Akad. Nauk SSSR Ser. Mat. 51, 1104–1116 (Math. USSR Izv. 31, pp. 423–434, English Trans.).
Matveev, S. (1990). Complexity theory of three-dimensional manifolds. *Acta Applicandae Mathematicae, 19,* 101–130.
Papadimitriou, F. (2012). Modelling landscape complexity for land use management in Rio de Janeiro. *Brazil. Land Use Policy, 29*(4), 855–861.
Papadimitriou, F. (2013). Mathematical modelling of land use and landscape complexity with ultrametric topology. *Journal of Land Use Science, 8*(2), 234–254.
Piergallini, R. (1988). Standard moves for standard polyhedra and spines. *Rendiconti Del Circolo Matematico Di Palermo, 18,* 391–414.

Pontryagin, L. S. (2015). *Foundations of combinatorial topology*. New York: Dover (republication of the original text that was published in 1952).

Rosenfeld, A. (1998). Digital geometry: Introduction and bibliography. In R. Klette, A. Rosenfeld, & F. Sloboda (Eds.), *Advances in digital and computational geometry* (pp. 1–54). Singapore: Springer-Verlag.

Schick, A., & Stiefelhagen, R. (2015). 3D pictorial structures for human pose estimation with supervoxels. In *Proceedings 2015 IEEE Winter Conference on Applications of Computer Vision, WACV 2015*, 7045880, pp. 140–147.

Scholl, M. & Voisard, A. (1989). Thematic map modelling: Design and implementation of large spatial databases. In *Springer Lecture Notes in Computer Science 409, "Design and Implementation of Large Spatial Databases"* (pp. 167–190).

Sen, A. (1976). On a class of map transformations. *Geographical Analysis, VIII,* 23–37.

Windreich, G., Kiryati, N., & Lohmann, G. (2003). Voxel-based surface area estimation: From theory to practice. *Pattern Recognition, 36,* 2531–2541.

Chapter 8
Spatial Complexity in 4-and-Higher-Dimensional Spaces

> *"There is an analysis situs of more than three dimensions.*
> *I do not say that it is an easy science""*
> *(Henri Poincaré, 1854–1912, "Dernieres Pensées", 1913, p. 43)*

Abstract One of the greatest challenges in the exploration of spatial complexity consists in measuring it on 4d surfaces and objects. Some interesting results are already available for hypercubes and, in this respect, Hamiltonian cycles and cubical complexes appear promising for solving some problems of 4d spatial complexity on hypercubes. Topologically surprising objects, such as the exotic spheres, abound in 4d. Yet, in some cases, higher-dimensional topologies can be easier to work out calculations on manifolds, thus leaving 3d and 4d objects as likely more difficult to examine topologically, and, by consequence, with respect to the spatial complexity of surfaces and objects that are in them. We need to explore what algorithmic measures of spatial complexity might apply to objects in 4d spaces and, more ambitiously perhaps, decide whether the 3d-and-4d-spaces are the ones capable of sustaining the highest spatial complexity among all *n*-dimensional spaces.

Keywords Spatial complexity · Fourth dimension · Hypercube · Tesseract · Exotic spheres · Doxel · 4d Complexity

8.1 Hypercubes, Tesseracts, Doxels, Clifford Tori

> "Boredom, neat parallels, oh how neat parallels
> are beneath God's perpendicular"
> (André Breton, 1896–1966, "Soluble Fish", 1924)

Ever since Bernhard Riemann delivered his "Habilitationsschrift" titled "Über die Hypothesen welche der Geometrie zu Grunde liegen" in 1854, and until today, the 4d space presents innumerable surprises. As will be explained later, the word "innumerable" can also be taken literally. It thus goes without saying that spatial complexity measurements and computations in 4d can be overwhelmingly more

F. Papadimitriou, *Spatial Complexity*, https://doi.org/10.1007/978-3-030-59671-2_8

complex than those in the 3d space. To complicate things even further, research in differential topology in the later half of the 20th century has revealed that 3d and 4d spaces are far more difficult to understand than other spaces (Hinton 1980; Kaku 1995; Rucker 1996). The difficulty to prove the famous "Poincaréconjecture" bears testimony to this.

The notion of the "*hypecube*" (Coxeter 1974; Bowen 1982) is indispensable to any research in four and higher dimensions. The hypercube is a generalization of the 3-cube to n-dimensions and is a special case of the hyper-rectangle. It is composed by $(n(n-1)(n-2)2^{n-4})/3$ cubes, $n(n-1)2^{n-3}$ squares, $n2^{n-1}$ edges and $2n$ nodes. In four dimensions, the hypercube is also called "tesseract", a geometric object consisting in 8 cubes, 24 squares, 32 edges, 16 vertices. One of the most widely known representations of the shape of a hypercube is Salvador Dali's famous painting "Corpus Hypercubus" presenting the Crucifixion on a hypercube, while the tesseract has also been suggested as a potentially useful shape for house building (Capanna 2013). The hypercube can be a metaphor to represent all three kinds of landscape complexity (spatial, functional, qualitative) with the additional fourth dimension of time (Papadimitriou 2010). In higher than four dimensions, exactly as the 4-cube is the tesseract or "octachoron", the 5-cube is the penteract, the 6-cube is the hexeract, and with increasing dimensions by one, the hepteract, octeract, enneract, dekeract in the 10th dimension, followed by hendekeract, dodekeract and so on. Hypercubes have served as models for possible applications in interconnection networks and parallel computation. Visualizing a tesseract is probably the simplest visual encounter with an object in the 4th dimension (Fig. 8.1). It may be conjectured that the evaluation of spatial complexity on such a surface might follow a similar procedure to that followed in the case of the ordinary 3d cube.

This is because the concept of "*doxel*" (dynamic voxel) in 4d space has been suggested to keep track of changes of voxels or their movements in time; at this point, a distinction is needed to avoid confusion: following changes in spatial complexity with time is different than documenting changes in time on the same map or surface

Fig. 8.1 A sketch depicting Dali's painting "corpus hypercubus"

(Papadimitriou 2009, 2012). The use of doxels is instrumental in several applications: i.e. in informatics (Pacheco and Real 2011), in 3d printing (Tanaka 2015) and in geographic and geo-environmental representations of 3d data at different time instants (Brovelli and Zamboni 2012). Evidently, the requirements in terms of computational resources for such 4d analyses of spatial objects are exuberantly high and we still lack efficient algorithms for handling doxels. For some scientific and technical analyses applicable on restricted spatial sizes, it seems that handling doxel data is nevertheless feasible. In terms of theory however, we are still short of assessments of computational complexity of problems in 4d. Yet, in the case of vertical decompositions of *n hyperplanes* of arrangements of linear surfaces in 4d, a bound was found (Koltun 2001) for *n* 3-simplices in 4d to be of the order of $O(n^4 \log n)$. L-trominoes in 4d euclidean spaces were studied by Befumo and Lenchner (2018) and an interesting problem was posed, that is to identify "a polyomino of dimension *k* that cannot be used to tile any rectangular board in dimension *k* or higher" (Befumo and Lenchner 2019, p. 40).

No doubt, it is very difficult for the human mind to figure out how a 4d object may look like, so a framework is needed to help us to keep track of such an object's spatial elements. One method might be found in "*Hamiltonian paths*", that can be applied to hypercubes (Fig. 8.2). They pass through each vertex once and thus the possibility to assess spatial complexity on the surface of a hypercube derived on the basis of Hamiltonian paths might be considered as a possibility. Alternatively, if the path leads back to the same vertex which it started from, then the assessment can be made on the basis of a "*Hamiltonian cycle*".

Deriving estimates of spatial complexity by means of Hamiltonian cycles might follow an "entropy encoding" procedure, as was the case in 3d. It should be noticed however, that, in its general form, the problem of finding a Hamiltonian cycle is *NP*-complete (Karp 1972; Garey and Johnson 1983) and, indeed, from the perspective of computational complexity, there are several interesting results concerning

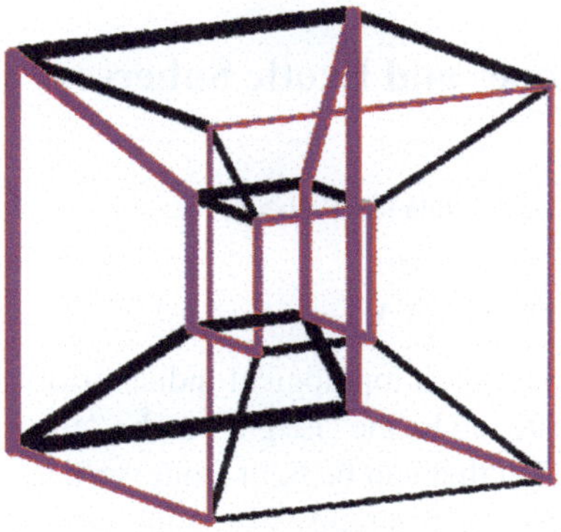

Fig. 8.2 A Hamiltonian path of a graph passes only once through each vertex of the graph and can be defined on hypercubes also. As such, it might serve as a guide for spatial complexity calculations on 4d hypercubes, i.e. by adopting "entropy encoding" procedures, as in the case Hamiltonian cycles on (3d) cubes

the computational complexity of problems emerging from hypercubes, which have turned out to be *NP*-complete (Afrati et al. 1985) or *NP*-hard (Dvořák and Koubek 2010; Baldi 2012).

As in 3d objects, a "*cubical complex*" is a set of elementary cubes and can be of any dimension strictly higher than 2. The Euler-Poincaré characteristic of a 4d cubical complex C is given by an alternating sum of its constituent cubes, where C_n is the number of cubes of dimension n in C:

$$\chi(C) = C_0 - C_1 + C_2 - C_3 + C_4 \tag{8.1}$$

The boundary of a n-dimensional cube is the collection of its $(n-1)$-dimensional faces, so the number of simplices of a n-dimensional triangulation of a manifold (tetrahedrization etc) might serve as a criterion of a manifold's complexity. In fact, any manifold of dimension other than 4 has a handlebody decomposition. This basic idea in studying the complexity of 4d manifolds is, in a sense, not very different from Matveev's complexity for 3d manifolds (Constantino 2011). Yet, in order to classify 4-dimensional shapes by means of their topological properties, the cohomology and homology groups of cubical complexes need to be defined, since they are topological invariants. Towards this end, the concepts of "*effective homology*" (Sergeraert 1994) and "*Steenrod squares*" may be useful for distinguishing spaces of isomorphic homology rings; this field is open for further research.

Despite the apparent difficulties in calculating complexity in 4d, and aside of graphs and hypercubes, it is easier to unknot knots in 4d space than in 3d, but 2d surfaces can form more complex non self-intersecting knots in 4d space, since 2d-surfaces can form knots of higher complexity than strings in 3d space. The "*Clifford torus*" (a 4d torus) can be flattened to a plane, and hence, it can be conjectured that measures of spatial complexity that apply to surfaces in 2d might also apply to a 4d Clifford torus as well, but this is yet another issue to be explored.

8.2 Manifolds, Gropes and Exotic Spheres

A monster that is more complex than the Typhoon
"Θηρίου Τυφῶνος πολυπλοκώτερον"
(Plato, 428-348 b.C., "Phaidros", 230A)

From a different perspective, a topological indicator of complexity of objects in 4d spaces has been suggested to be the "height" or the "class" of a *grope*. Gropes are 2-complexes (Cannon 1978) that can be built from surfaces spreading out branches in space like octopus tentacles, eventually extending to 4-manifolds and it has been suggested (Freedman and Quinn 1990) that the complexity of a grope is measured by the number of its layers.

Aside of gropes, manifolds of any dimension can be joined together to create composite manifolds, the "*connected sums*" (Fig. 8.3). An obvious question to ask

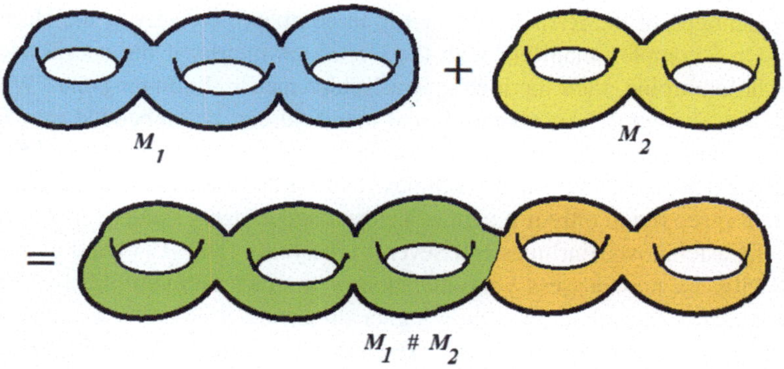

Fig. 8.3 The topological operation of creating a "connected sum" of two manifolds. Manifolds M_1 and M_2 are connected together so as to create the connected sum of manifolds $M_1\#M_2$

is how much more complex such a composite may be compared to a non-connected manifold. A partially satisfactory assessment of complexity in this respect can be sought in the concept of "*prime manifolds*". These are *n*-manifolds that can not be expressed as non-trivial connected sums of two manifolds (non-trivial means they are not *n*-spheres). Hence, there are no homeomorphisms by which prime manifolds can be simplified topologically and they are regarded as equivalent to prime numbers. Whether the spatial complexity of the connected sum might be calculated from the addition of the spatial complexities of each manifold is an open problem.

Manifolds are said to "admit" a "*differential structure*" if the differential calculus can be applied to them. In 1923, Kerékjártó proved that all 2d or 3d Riemannian manifolds, as well as Euclidean spaces higher than 4d admit a single differential structure each. Later, it was proven (Milnor 1956, 1959a, b, 2000) that some higher-dimensional manifolds may accept several differential structures: a 7d sphere for instance, accepts 28 differential structures. Even more amazingly, the number of differential structures varies very strangely with increasing dimension: i.e. while 1,2,3,5-dimensional spaces have only one differential structure each, the 8d has 2, the 11d has 992, the 12d has only one, the 15d has 16256 etc., while 10d manifolds do not accept any differential structure at all (Kervaire and Milnor 1963). Hence, applying calculus to any spatial dimension is a tricky subject since it is still unclear whether all Riemannian manifolds can accept differential structures or not, and if they can, how many. Calculating spatial complexity on such differential structures of higher-dimensional spheres is yet another open problem.

And even trickier problems emerge if the representation space changes from real to complex. The form omega (ω) of *symplectic geometry* measures areas of surfaces of even-dimensional spaces. It can be applied to R^4 and the advantage is that it is 2d, as complex numbers are used and therefore symplectic forms omega of C^2 can be used to measure areas of R^4. The extent to which algorithmic complexity measures can be applied to assess spatial complexity in C^2 (and, consequently in R^4) is also open for future research.

The analogue of the Riemannian metric in complex geometry is the Hermitian metric. On a complex manifold with the Hermitian metric, if the coordinates are normal holomorphic, then the metric is a Kähler metric. Following the "Whitney Embedding Theorem", every smooth real manifold can be embedded into R^n for some n. Extending this theorem to the complex space, and according to the "Kodaira Embedding Theorem", a compact complex manifold is embeddable in the complex projective space if and only if it admits a closed and integral Kähler form. With the works of Castelnuovo, Enriques and Severi in the years 1891–1949 and by Kodaira in the 1950s, we now possess a classification of complex 2d manifolds by virtue of the Enriques-Kodaira theorem, while higher dimensional complex manifolds (3d-complex) were classified by Hironaka, Yau and Mori in the 1970s and 1980s. Back in 1950, Pontryagin discovered the terms under which a n-dimensional surface is a boundary of a $n+1$ dimensional surface. Shortly afterwards (in 1954), his theory was expanded by Thom with his "*Cobordism Theory*" and, based on Thom's theory, Milnor discovered "*exotic spheres*" in 7d-spaces, which are not boundaries of balls. Also, it was the expansion of cobordism theory to k-cobordism ("H_o homotopy") that led Novikov (in 1962) to a classification of differential manifolds with dimension 5 and higher. The classification of simply-connected and compact 5-manifolds (Barden 1965) is a lot simpler than that of 3d and 4d manifolds. But non-simply connected 5-manifolds are hard to classify.

Poincaré's famous problem was to prove whether the hypersphere is the only closed and orientable 3d surface whose fundamental group is trivial (the fundamental group of the sphere is trivial because any closed path on the sphere can be contracted to a point). If the sphere has one handle, it is non-trivial, because there is (at least) one path which can not be contracted to a point. So the fundamental group of a surface differentiates the surfaces according to their homotopy.

Three spherical coordinates are required x_1, x_2, x_3 to identify a point on the sphere S^2 with radius r, inclination θ, azimuth φ, where θ is in between 0 and π and φ is in between 0 and 2π:

$$\begin{aligned} x_1 &= r \sin\theta \cos\varphi \\ x_2 &= r \sin\theta \sin\varphi \\ x_3 &= r \cos\theta \end{aligned} \tag{8.2}$$

But a location of a point of the four dimensional hypersphere has hyperspherical coordinates on S^3 that are calculated from four coordinates:

$$\begin{aligned} x_1 &= r \cos\psi \\ x_2 &= r \sin\psi \cos\theta \\ x_3 &= r \cos\psi \sin\theta \cos\varphi \\ x_4 &= r \sin\psi \sin\theta \sin\varphi \end{aligned} \tag{8.3}$$

In fact, there are closed and bounded regions of 4d space that can not be parts of a hypersphere. In terms of spatial complexity, we take notice of two issues here: one quantitative and one qualitative. The first is that, as the dimension increases only by one, not only one additional coordinate is necessary to identify the position of a point from its corresponding angles, but also that (although the trigonometric equations are analogous to those that apply to S^2) the number of arithmetic operations increases significantly, from 10 to 18. The second, and more important, is that it is very difficult for humans to visualize the glome (the four-dimensional sphere).

Poincaré conjectured in 1904 that the topological characterization of the hypersphere is equivalent to that of a sphere. As happened with other manifolds, so in this conjecture also, with increasing spatial dimension, it is often easier to prove equivalent theorems, but the problem persists in the third dimension. For instance, it was shown (Smale 1959, 1962) that it is possible to turn a left glove inside out and convert it into a right-hand glove in a 4d space, but this is impossible in the 3d space. The existence of "exotic" smooth structures in 4d space (R^4) however, presents a peculiar anomaly of this space. These "*exotic spheres*" are smooth manifolds that are homeomorphic but not diffeomorphic to R^4 (Gompf 1983). Furthermore, there is an uncountable quantity of such smooth 4-manifolds, whereas each one of them is homeomorphic to R^4 but endowed with its own smooth structure which makes it non-diffeomorphic to R^4. Milnor discovered "exotic spheres" in 7d-spaces, that are not boundaries of balls. In a remarkable discovery, it was proven (Gompf 1985) that the number of "exotic" structures in 4d space is not only infinite, but, in fact, it has the cardinality of the continuum. Might the implication of this result, along with other analogous (Donaldson 1983), indicate that the 3d space is our "upper bound" to understanding spaces of any dimension? Fortunately however, experimental evidence from virtual reality experiments (Aflalo and Graziano 2008; Ambinder et al. 2009), suggests that (if appropriately trained to do so) humans *are* able (up to a certin extent) to understand 4d environments.

Thus, emerge (at least) two hitherto unanswered (or unanswerable?) basic questions: (a) What measures of spatial complexity should apply to spaces with dimensions higher than 3d? (b) Is there sufficient evidence on the basis of which it can be conjectured that *3d-and-4d-spaces should be singled out as the ones capable of sustaining the highest spatial complexity among all n-dimensional spaces*?

References

Aflalo, T. N., & Graziano, M. S. A. (2008). Four dimensional spatial reasoning in humans. *Journal of Experimental Psychology: Human Perception and Performance, 34*(5), 1066–1077.

Afrati, F., Papadimitriou, Ch. C., & Papageorgiou, G. (1985). The complexity of cubical graphs. *Information and Control, 66,* 53–60.

Ambinder, M. S., Wang, R. F., Crowell, J. A., Francis, G. K., & Brinkmann, P. (2009). Human four-dimensional spatial intuition in virtual reality. *Psychonomic Bulletin & Review, 16*(5), 818–823.

Baldi, P. (2012). Boolean autoencoders and hypercube clustering complexity. *Designs, Codes and Cryptography, 65*(3), 383–403.

Barden, D. (1965). Simply connected five-manifolds. *Annals of Mathematics*, Ser. *2*, *82*(3), 365–385. https://www.jstor.org/stable/1970702.
Befumo, A., & Lenchner, J. (2018). Tiling one-deficient rectangular solids with trominoes in three and higher dimensions. *Mathematical Magazine, 91*, 62–69.
Befumo, A., & Lenchner, J. (2019). More fun with L-trominoes in three and higher dimensions. *Mathematics Magazine, 92*(1), 32–40.
Bowen, J. P. (1982). Hypercubes. *Practical Computing, 5*(4), 97–99.
Brovelli, M. A., & Zamboni, G. (2012). Virtual globes for 4d environmental analysis. *Applied Geomatics, 4*(3), 163–172.
Cannon, J. (1978). The recognition problem: What is a topological manifold? *Bulletin of the American Mathematical Society, 84*, 832–866.
Capanna, A. (2013). Tesseract houses. In *Proceedings 12th Conference on Applied Mathematics*, Aplimat.
Constantino, F. (2011). Complexity of 4-manifolds. *Experimental Mathematics, 15*(2), 237–249.
Coxeter, H.S.M. (1974). *Regular complex polytopes*. London & New York: Cambridge University Press.
Donaldson, S. K. (1983). Self-dual connections and the topology of smooth 4-manifold. *Bulletin of the American Mathematical Society, 8*, 81–83.
Dvořák, T., & Koubek, V. (2010). Computational complexity of long paths and cycles in faulty hypercubes. *Theoretical Computer Science, 411*(40–42), 3774–3786.
Freedman, M.H. & F. Quinn,F. (1990). *The topology of 4-manifolds*. Princeton Mathematical series 39, Princeton NJ: Princeton University Press.
Garey, M. R., & Johnson, D. S. (1983). *Computers and Intractability: A guide to the theory of NP-completeness*. New York: W. H. Freeman.
Gompf, R. E. (1983). Three exotic R4's and other anomalies. *Journal of Differential Geometry, 18*, 317–328.
Gompf, R. E. (1985). An infinite set of exotic R^4s. *Journal of Differential Geometry, 21*, 283–300.
Hinton, C. H. (1980). *Speculations on the fourth dimension: Selected writings of Charles H. Hinton*. New York: Dover.
Kaku, M. (1995). *Hyperspace: A scientific odyssey through parallel universes, time warps, and the tenth dimension*. Oxford: Oxford University Press.
Karp, R. M. (1972). Reducibility among combinatorial problems. In R. E. Miller & J. W. Thatcher (Eds), *Complexity of computer computations* (pp. 85–103). New York: Plenum Press.
Kervaire, M., & Milnor, J. W. (1963). Groups of homotopy spheres:I. *Annals of Mathematics, 77*(3), 504–537.
Koltun, V. (2001). Complexity bounds for vertical decomposistions of linear arrangements in four dimensions. *Lecture Notes in Computer Science, 2125*, 99–110.
Milnor, J. W. (1956). On manifolds homeomorphic to the 7-sphere. *Annals of Mathematics, 64*(2), 399–405.
Milnor, J. W. (1959a). Sommes de variétes différentiables et structures différentiables des sphères. *Bulletin de la Société Mathématique de France, 87*, 439–444.
Milnor, J. W. (1959b). Differentiable structures on spheres. *American Journal of Mathematics, 81*(4), 962–972.
Milnor, J. W. (2000). Classification of (n−1)-connected 2n-dimensional manifolds and the discovery of exotic spheres. Surveys on surgery theory. *Annals of Mathematics Studies, 145*, 25–30.
Pacheco, A., & Real, P. (2011). Associating cell complexes to four dimensional digital objects. *Lecture Notes in Computer Science, 6607*, 104–115.
Papadimitriou, F. (2009). Modelling spatial landscape complexity using the Levenshtein algorithm. *Ecological Informatics, 4*(1), 51–58.
Papadimitriou, F. (2010). Conceptual modelling of landscape complexity. *Landscape Research, 35*(5), 563–570.
Papadimitriou, F. (2012). The algorithmic complexity of landscapes. *Landscape Research, 37*(5), 599–611.

Poincaré, H. (1913). *Dernieres Pensées.* (Paris: Flammarion). Transl. Mathematics and Science: Last essays, New York: Dover.

Rucker, R. (1996). *The fourth dimension: A guided tour of the higher Universe.* Boston: Houghton Mifflin.

Sergeraert, F. (1994). The computability problem in Algebraic Topology. *Advances in Mathematics, 104*(1), 1–29.

Smale, S. (1959). Diffeomorphisms of the 2-sphere. *Proceedings of the American Mathematical Society, 10,* 621–626.

Smale, S. (1962). On the structure of manifolds. *American Journal of Mathematics, 84,* 387–399.

Tanaka, H. (2015). New file format for 3d printing, its extensions and applications. In *NIP& Digital Fabrication Conference* (pp. 3–6).

Part III
Numbers Behind Spatial Complexity

Chapter 9
Squares, Cats and Mazes: The Art and Magic of Spatial Complexity

Intelligence in labyrinths
(Michel De Certeau 1984, p. 90)

Abstract Spatial complexity can be playful, surprising and artful: some of its pleasant facets are highlighted in this chapter, as they emerge from the examination of spatial partitions and by means of games playable on boards of squares: chess, go, tic-tac-toe, checkers (among many other spatial games) revealing how charming the spatial complexity of square arrangements can be. Indeed, square maps are scientifically interesting as well as a source of inspiration throughout the ages, from Latin squares and famous modern painters to the mysterious Arnold Cat Maps and video games. Square maps can be both symbols of minimalism in art as well as genitors of highly complex mazes and labyrinths. With innumerable algorithmic challenges pertaining to them, they are a source of entertainment and endowed with a geometric shape perfectly suited for displaying and exploring the puzzling, mystic and aesthetic aspects of spatial complexity.

Keywords Spatial complexity · Spatial Computing · Arnold Cat Map · Spatial games · Mazes · Complexity and Art · Game complexity

9.1 Square Partitions

> "Our physical world not only is described by mathematics, but it is mathematics: a mathematical structure, to be precise"
>
> (Max Tegmark 2014, p. 6)

Creating square grids by intersecting horizontals and verticals at equal lengths is not the only way to partition a given spatial region. Although they are the commonest and by far the easiest to handle numerically, hexagons, triangles, and other shapes can as well be used to tile the plane, or even combinations of shapes (Fig. 9.1). In fact, any rectangular space may also be partitioned by a fixed ratio, such as the golden

F. Papadimitriou, *Spatial Complexity*, https://doi.org/10.1007/978-3-030-59671-2_9

Fig. 9.1 Various spatial partitions, based on simple geometric shapes: triangles, hexagons and combinations of geometric shapes (hexagons, squares and triangles, dodecagons and triangles, octagons and squares etc.)

section, in which any larger rectangle can be made proportional to its adjacent and smaller rectangle, by a factor equal to the *golden section* (Fig. 9.2). Thus, the ratio of the length of the larger rectangle over that of the smaller one is given by

$$\frac{1}{2}(1+\sqrt{5}) = 1 + \frac{1}{1+\frac{1}{1+\frac{1}{1+\cdots}}} \tag{9.1}$$

and the ratio of any pair of consecutive numbers is given by the Fibonacci sequence (these numbers representing length and width of the rectangle) yielding the golden section, i.e.

Fig. 9.2 Partitioning rectangular spaces can be made by a fixed ratio. In this case, the ratio is the "golden section"

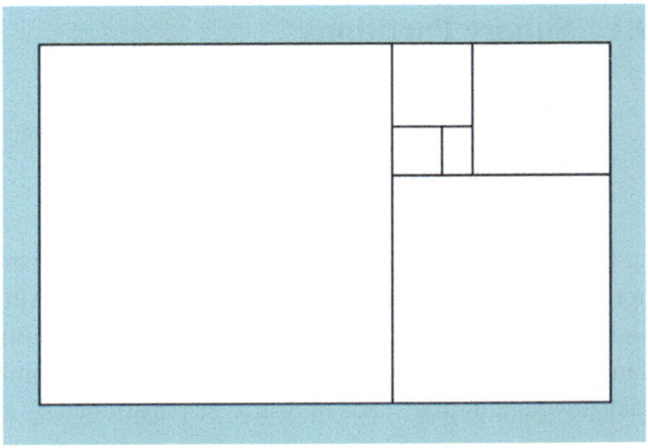

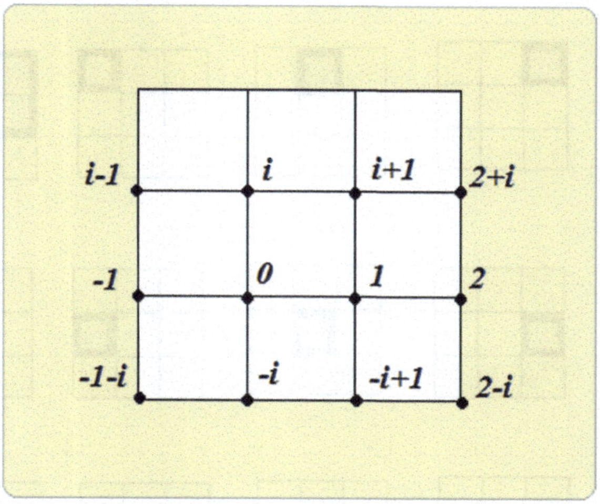

Fig. 9.3 The squares on this 2-dimensional space correspond to points defined by complex numbers with integer real parts, thus partitioning a 2d surface according to the structuring induced by the algebraic integers R_d

$$\frac{89}{55} = 1 + \frac{1}{1 + \frac{1}{1+\frac{1}{1+\cdots}}} \tag{9.2}$$

Space partitioning may as well be the result of the application of algebraic structures, such as the ring of algebraic integers, R_d. This ring contains all the numbers of the form $a + bs$, where a, b are ordinary integers. When $d = -1$, the ring R_{-1} is the "*ring of Gaussian integers*", that is the ring of complex numbers defined on the complex plain, by a and b integers (Fig. 9.3). When $d = -3$, the ring R_{-3} can represent vertices of equilateral triangles. Notice that four numbers $(1, -1, i, -i)$ suffice to define a square in the ring R_{-1} and six numbers in R_{-3} (the numbers 1, -2, $(1 + 3i)/2$, $-(1 + 3i)/2$, $-(1 - 3i)/2$, $(1 - 3i)/2$).

Besides, there also exist iterative schemes for partitioning, making use of pyramidal numbers or fractal patterns. The former are based on the rule that the total number of squares contained in a grid of $m \times m$ unit square is the square "pyramidal number":

$$\frac{m(m+1)(2m+1)}{6} \tag{9.3}$$

In the case of 3×3 maps for instance ($m = 3$), the pyramidal number is 14 (Fig. 9.4). These numbers correspond to alternative coverings of the same map, by varying squares, either non-overlapping (1×1 squares) or overlapping if more than 1×1 cells are used, up to $m \times m$. The overlapping ones are of no apparent usefulness for spatial analyses, so the use of pyramidal numbers is unsuitable for spatial complexity assessments. It is however interesting from the point of view of computational complexity, since identifying squares which correspond to pyramidal numbers is a computationally-hard problem, because it eventually leads to unsolvable diophantine equations (Ma 1985; Anglin 1990).

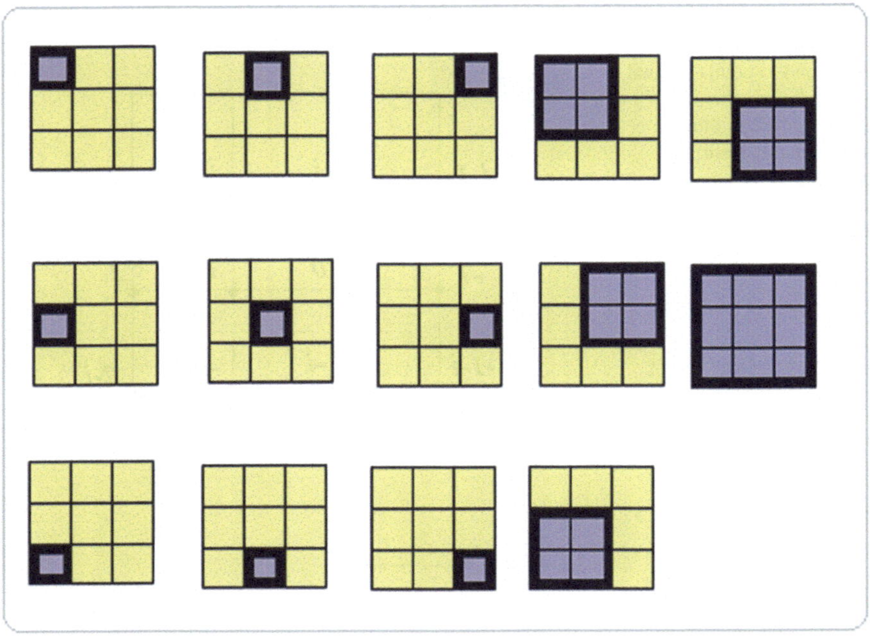

Fig. 9.4 Squares corresponding to the pyramidal number 14 for 3 × 3 binary maps

Another way to divide a square space is to use fractal methods, such as the Sierpinski square (or Cantor gasket) (Fig. 9.5) which, at each step, yields 8 squares of side length 1/3 and therefore has a fractal dimension equal to log8/log3 = 1.89…

Aside of being conceptually closer to the human perception of space however, square partitions have the additional benefit that they can easily emerge by appropriately converting triangular, hexagonal and other symmetric partitions of space to square grids, although this does not preclude deriving parallelogram lattices instead of squares (Fig. 9.6).

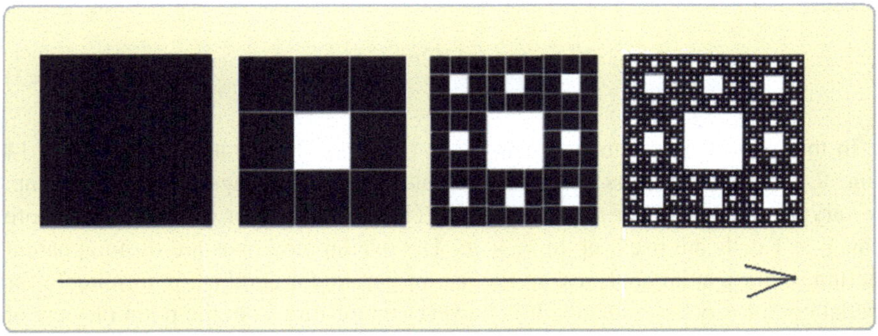

Fig. 9.5 The Sierpinski square is a fractal object dividing the square at every step in more squares, eventually ending up with a "dust" of isolated points around the central square

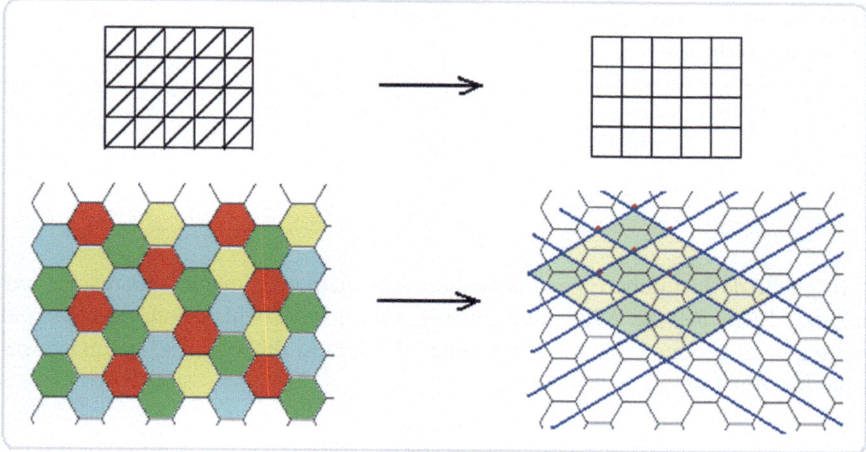

Fig. 9.6 Hexagonal, triangular (and other) symmetric partitions of the plane can easily be transformed to correspond to parallelogram or square grids

Given these, it has become perhaps more clear why our digital technologies rely so much on square arrays. It is thus understandable why maps of square areas are ideally suited for the analysis of spatial complexity (Papadimitriou 2002, 2009, 2012, 2013). There has never been a period of human history in which squares ruled everyday life more than they do now: pixels are squares, and so are digital screens of mobile devices, televisions, computers, and many other essential electronic devises; and all these are outlets displaying spatial complexity. But we are not the first ones to be fascinated by the power of square arrays.

9.2 Squares, Minimalism and Art

Between two words, you have to choose the lesser
"Entre deux mots, il faut choisir le moindre"
(Paul Valéry, 1871–1945, "Tel quel", 1929)

The power of squares in understanding spatial extents has been widely recognized across cultures and civilizations (the square as a sacred form is encountered in the four arms of Vishnu or Shiva, the Tibetan mandalas, the Kaaba cube of Mecca, etc.) and square arrangements have long been sources of inspiration and puzzlement for artists and thinkers. Perhaps nowhere is this more explicit than in the case of "magic squares", of which an example is the 4×4 magic square depicted in Albrecht Durer's famous "Melancholia" gravure (1514). The sum of rows of this magic square is 34 that is as much as the sum of columns, as the sum of diagonals and the sum of its four corner cells too:

16	3	2	13
5	10	11	8
9	6	7	12
4	15	14	1

In 1693, de la Loubere gave a method for calculating magic squares for any odd size. Likewise, the "diabolical squares" are those of which both negative and positive diagonals produce the same sums. The oldest diabolic magic square was found inscribed in India (12th b.C.):

15	10	3	6
4	5	16	9
14	11	2	7
1	8	13	12

But there is more to art than puzzles and "magic" tricks. Spatial complexity means, signifies, creates meanings, or diffuses meanings. For this reason, it poses as an ideal ground for matching mathematics with art. Besides, as Hilbert said (in 1922) "In the beginning was the sign" in his "The new grounding of Mathematics: First Report" (as reported by Ewald 2001).

In the nineteenth century, the mathematician-writer Abbott (1838–1926) begun his celebrated story "Flatland" (written in 1884) by exclaiming "How franticly I square my talk!". In this famous fictitious two-dimensional story, Abbott (1991) wrote an (unreal) correspondence between human beings and (essentially)...spatial complexity. In the class-sensitive period that this novel was written, the various inhabitants of "Flatland" were probably imagined by the author to correspond to increasing spatial complexity (although he did not specify this), according to his own personal criteria: straight lines would "correspond" to women, triangles to soldiers and "lowest classes of workmen", equilateral triangles or equal-sided triangles to middle class and squares to "professional men and gentlemen". Hexagons are reserved for the nobility and circles for priests. Observing the attribution of shapes to social classes, it easily follows that the higher the social class, the higher the spatial complexity.

Besides, square divisions of space constitute a recurrent and classic theme in visual arts, encountered within various artistic currents. One of the founders of the De Stijl movement for instance, Theo Van Doesburg (18831931), presented four black and progressively enlarging squares in his "Arithmetic Composition" (1929–1930) (Bridgeman Art Library, Switzerland). In 1906, Henri Matisse created his famous painting "Luxury" (Gallery Orsai, Paris), depicting a calm and pleasant space, giving the impression of being composed from small pixels.

The devotion to squares however, is characteristic of *minimalism* in art. Within the context of the Russian constructivism, Alexander Rodchenko (1891–1956) painted some completely monochrome square paintings, Piet Mondrian (1872–1944), with his 1935 picture titled "Composition C: Yellow, Red and Blue" presented three squares (a red, a blue and a yellow) in a black grid of white colours. In another painting, he presented a plain red square titled "Pure red color" (1921), Supposedly, he was affected by the dutch theosophic school of "plastic mathematics", which contrasted horizontal and vertical lines to curves.

But probably the most exquisite representative of the links between square maps and art was Kazimir Malevich (1878–1935). His famous "Black Square in a White Font" (1915) is, as its title suggests, nothing but a big black square with a white border around it, apparently inviting the viewer to reflect on the mystery of binary square arrangements. The painting "Quadrilateral" (1914) or "Black Square" is explained by the painter himself in his "*Suprematism*" manifesto. His "Suprematist Elements: Squares" (1923) consisted of two black squares with a beige backdrop. Further, in his "Suprematist Composition" titled "Red Square and Black Square", Malevich presented a black square and a tilted red square. In interpreting this painting, Altieri (2001) contended that the red square's tilt posed a geometric challenge to the system of coordinates established by the black square. In the context of Malevich's "Suprematism", squares signify feelings and white domains the void.

Similarly, Joseph Albers (1888–1976) presented two red squares in beige ground in his "Homage to the Square" (1961). Perhaps even more characteristically, Piet Mondrian's works display sets of lines intersecting orthogonally forming square arrangements. In these remarkable cases, spatial complexity was intentionally kept to a minimum in two ways: not only there was one color only (or two), but the spatial shape was also the simplest convex shape to describe algorithmically: the square.

Square grids consisting of squares painted with different colors are representative of "concrete art", i.e. the "Polychrome of pure colors" (1956) by Karl Gerstner (1930–2017) who used painted cubes of plexiglas to print various multicolored square maps. Conceptually very similar was the "arte programmata", in which binary orthogonal geometric features and patterns are used with non-repetitive patterning, i.e. with the works of Gianfranco Chiavatti (1936–2011). But, the charm of squares is not confined to art only.

9.3 Mazes, Labyrinths and Spatial Games

> (Ariadne) gave Theseus a string of which the one end he attached to the labyrinth's gate and when he found the Minotaur at the labyrinth's end he killed him by smiting him with his fists, then made his way out of the labyrinth by following the string again (back to the gate)
>
> "λίνον εἰσιόντι Θησεῖ δίδωσι: τοῦτο ἐξάψας Θησεὺς τῆς θύρας
> ἐφελκόμενος εἰσῄει. καταλαβὼν δὲ Μινώταυρον
> ἐν ἐσχάτῳ μέρει τοῦ λαβυρίνθου παίων πυγμαῖς ἀπέκτεινεν,
> ἐφελκόμενος δὲ τὸ λίνον πάλιν ἐξῄει"
>
> (Apollodorus, "Epitome", 1.9)

Mazes probably qualify for the title of the "temples of spatial complexity". Deeply impressing humans throughout the ages, they constitute the most characteristic example of how spatial complexity can be useful for the creation of games. From king Minos' famous "labyrinth" in ancient Greece, to the floor of the Chartres cathedral, mazes have been created in gardens, palaces and public areas, all over the world. Some are famous, such as the maze of the gardens of Schönbrunn Palace in Austria, some are particularly large, as the Gardens Shopping Mall in Dubai (currently the world's largest indoor maze) and the Samsø Labyrinten in Denmark (the world's largest maze, with an area of 60,000 m^2).

Nowadays, several algorithms have been devised for *generating mazes* (i.e. Prim's, Kruskal's, Sidewinder, Aldous-Broder, Binary Tree, Eller's Recursive Backtracker, Wilson's, Growing tree, Hunt and Kill, Growing Forest). Equivalently, there are algorithms for *solving mazes* (Pledge algorithm, Recursive backtracker, Chain algorithm, Dead and Filler, Tremaux's algorithm, Wall follower, Cul-de-sac filler, Blind alley filler, Blind Eye Sealer, Shortest Path Finder etc.). The reader may find a rich literature documenting these algorithms, but presenting them here analytically is beyond the scope of this book.

From a computational complexity perspective however, it is interesting to notice that two maze problems, the "rolling block" and "Alice" mazes have been shown to be *PSPACE*-complete (Holzer and Jakobi 2012). But many other *spatial games* (and video-games) with maze-like forms (such as the games Lemmings, Loder Runner, Mindbender, Skweek, Starcraft, Tron and the famous Pac-Man) are all *NP*-hard (Viglietta 2013).

Spatial games fascinated people since the early antiquity. The game "Go", invented in China two millennia b.C., based on a 19 × 19 square board, can host as many as 10^{768} possible games and "future conflicts may resemble the oriental game of Go more than the western game of chess" (Arquilla and Ronfeldt 2001, p. 2). Recreations with spatial complexity involve a wide range of spatial games that are notoriously difficult to play, precisely due to their very large number of combinations, i.e. there are 6,670,903,752,021,072,936,960 possible configurations of sudoku (Stewart 2008), while "Eternity-II puzzle" (a game invented in 2007 and played on a 16 × 16 grid) has 1.115×10^{557} possible configurations (Pickover 2009).

Besides these, there are old games challenging the player to discover possible square allocations of numbers, complying to certain rules. Latin squares is one such, Sudoku is another, in which the player is expected to assign positive integers to cells of a big 9 × 9 square composed of 3 × 3 squares, so that in no column or row of the big square appears anyone of the numbers 1–9 twice. A 9 × 9 map needs few steps only to check whether a solution is valid (exactly 81 steps), but the number of steps required to search for a solution if it is not known beforehand is impossibly high: 6.6×10^{27} (Aron 2011). Most spatial games essentially draw their complexity from the breadth of possible spatial combinations of cells on a board. Chess and checkers are only two such games (and most well known), among many others: Laska, Lanrik, Kriegspiel, Zetan, Chancellor's chess, Satrange, Japanese chess, Marseille chess, Alice, Kamikaze, Ming Mang, Hazami Sogi, etc. The size of game boards for spatial games varies depending on the game, but it is usually a square (i.e. 8 × 8 in chess,

10×10 for Snakes and Ladders, 15×15 for Scrabble, 18×18 for Go), although it can also extend in higher than two dimensions.

The computational complexity classes also vary: chess is *EXPTIME*-complete (Fraenkel and Lichtenstein 1981), as are checkers (Robson 1984) and "Go" (Robson 1983). These games can have a time duration that is exponential with respect to the size of their playing board. The game "Reversi" (or "Othello") playable on a square board is *PSPACE*-complete (Iwata and Kasai 1994). In fact, even the simplest of all spatial games, the "tic-tac-toe" with its 9 cells, is *PSPACE*-complete (Reisch 1980), making it an excellent example of how high spatial complexity may emerge from very simple spatial arrangements. Similarly, the game "Tetris" has been shown (Demaine et al. 2002) to be "intractable" for the human mind ("*NP*-complete") and sudoku is *NP*-hard (verifiable in polynomial number of steps, but solvable in exponentially high number of steps).

Several famous spatial problems have been examined in chess. For instance, Euler's "Knight's Tour Problem" (1759) asks for the tour of a knight over the board passing once through all the chessboard's squares. It has a solution for the 8×8 chessboard but not for the 4×4 chessboard. "*Schwenk's theorem*" characterizes the rectangular boards that can support a knight's tours and defines that (Stewart 2010) a $m \times k$ parallelogram chessboard supports a knight's tour unless either (a) m and k are both odd, (b) m equals 1, 2, 3, 4, or (c) $m = 3$ and $k = 4$ or $k = 6$ or $k = 8$.

Chess on Klein surfaces is spatially more complex than common planar 2d chess (Fig. 9.7), so calculations of movements of chess pieces on this surface presented by Watkins (2004) are interesting to see how more complex formulas emerge depending on whether the piece "king" moves on a Klein surface.

The number of kings required to cover a Klein $m \times m$ chessboard, depends on two calculations Watkins (2004):

Fig. 9.7 A chessboard on a Klein surface

$$\left\lceil \frac{m}{6} \right\rceil \left\lceil \frac{2m}{3} \right\rceil \tag{9.4}$$

But if the chessboard surface is asymmetric (non-square, that is $m \times k$), then the complexity of the previous calculations increases to (Watkins 2004):

$$\left\{ \begin{matrix} \left\lceil \frac{m}{6} \right\rceil \left\lceil \frac{2k}{3} \right\rceil - \left\lceil \frac{k-1}{3} \right\rceil & m = 1, 2, 3 \bmod 6 \\ \left\lceil \frac{m}{6} \right\rceil \left\lceil \frac{2k}{3} \right\rceil & m = 4, 5, 6 \bmod 6 \end{matrix} \right\} \tag{9.5}$$

Again, spatial asymmetry induces increases in spatial complexity. By far the most important problem in chess mathematics however, is the "*Covering Problem*", consisting in the determination of the number of pieces of a particular type of movement (i.e. kings, queens, knights, rooks etc.) required to cover a square chessboard. Nine kings are necessary to cover the 8×8 chessboard and the same can be done with 8 bishops or 8 rooks. For queens (whose movement is the most far-reaching over the chessboard), the "*Spencer-Welch theorem*" defines the number of queens required to "cover" the chessboard. Some mathematical chess problems have also been been studied over 3d and 4d chessboards (Gibbins 1944; Jelliss and Marlow 1987; DeMaio 2007; Kumar 2008).

Another chess-like spatial game is John Conway's "*Game of Life*" that can be played on square boards and provides useful insights into how self-organisation can emerge in space. One of its variants, the "Garden of Eden", of size $5k \times 5k$, produces a large number of configurations (Berelkamp et al. 2004):

$$(2^{25} - 1)^{k^2} \tag{9.6}$$

and, given adequate time for self-replication, spatial patterns eventually emerge. The "Game of Life" begins with 3×3 sub-squares, to which simple rules apply, depending on whether the cell is occupied by a digital entity or not. The rules define how many entities are required in the 3×3 sub-square in order for that entity to survive or reproduce at the next time step and in this way, artificial ecosystems can be created *in silica*, which led to the exciting research field of "*Artificial Life*" that aims to simulate life-like behaviors and processes by using computer-made (artificial) animals and plants.

9.4 The Arnold Cat Map

God is sufficiently wise and powerful to mix the many into one and to dissolve again the one into many. But there is no man, nor will ever be, who will be able to do this

"ὅτι θεὸς μὲν τὰ πολλὰ εἰς ἓν συγκεραννύναι καὶ πάλιν

ἐξ ἑνὸς εἰς πολλὰ διαλύειν ἱκανῶς ἐπιστάμενος ἅμα καὶ δυνατός,

ἀνθρώπων δὲ οὐδεὶς οὐδέτερα τούτων ἱκανὸς οὔτε ἔστι νῦν οὔτε εἰς αὖθίς ποτε ἔσται"

(Plato, 428–348 b.C., "Timaeus", 68d)

Or, may be, not? Entropy implies irreversibility: whatever is will never be the same again. This is what Physics says. Physics exploits Mathematics but has no much room for magic. But there is plenty of room in Mathematics for unexpected truths and bewildering results.

Occasionally, mathematics may give the impression of a touch of "magic": the "Arnold Cat Map" (ACM) is one such a case and it is interesting to examine it here in the context of spatial complexity, because it shows unexpected properties of 2d maps (although it can be expected to extend to 3d volumes also). The ACM is a discrete map transformation of an image converting it into another, and iteratively into another, so that after successive iterations, the final image that eventually appears is completely identical to the original. It was invented (or discovered?) by Vladimir Arnold and as he used a cat's face to show the power of the mapping, it was since called Arnold's Cat Map (Arnold and Avez 1968).

This simple yet almost magical transformation rearranges the position of each map cell and repositions it elsewhere on the image, according to a predefined (and unchanging) rule. After a number of iterations, the cell returns to the same position as it initially was and it therefore contributes (along with all other cells, which have been transformed according to the same rule) to a reproduction of the original image again after all the iterations have been performed (Fig. 9.8).

For their amazing behaviours, ACMs have applications in cryptography and steganography (data encoding in images). They have positive Kolmogorov-Sinai

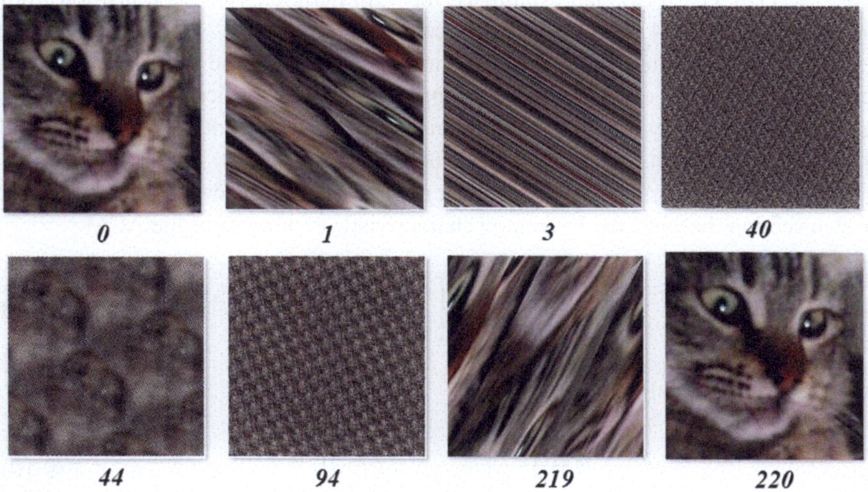

Fig. 9.8 The Arnold cat map transformations of the image of Liuliuta. Numbers beneath each image show the iteration number of the Arnold Cat Map transformation. Soon after the second iteration, the image has lost all its resemblance to the original cat's image and it looks chaotic at the 40th. Oddly, at the 44th iteration, ghost-like features of the original image reappear but do not last. Eventually, after 219 iterations, the 220th suddenly produces exactly the original again: "Cats have nine lives"

entropy (Lichtenberg and Lieberman 1992) and lie at the heart of classical dynamical chaos (Chirikov 1979; Kornfeld et al. 1982).

The general formula transforming the position of a cell located at (x, y) to another position on the map is:

$$\begin{bmatrix} x_{k+1} \\ y_{k+1} \end{bmatrix} = \begin{bmatrix} 1 & p \\ q & pq+1 \end{bmatrix} \begin{bmatrix} x_k \\ y_k \end{bmatrix} \bmod \left(\sqrt{n}\right) \tag{9.7}$$

where n is the size of the square map (thus the root is a positive integer), k is the number of iteration (a positive integer), p and q are the parameters of the ACM (some positive integers).

Since the determinant of the transformation matrix equals to 1, the map is area-preserving and the final image is identical to the initial.

As an example, consider the case of an 124×124 map, with parameters $p = q = 1$. The ACM thus is:

$$\begin{bmatrix} x_{k+1} \\ y_{k+1} \end{bmatrix} = \begin{bmatrix} x_k + y_k \\ x_k + 2y_k \end{bmatrix} \bmod (124) \tag{9.8}$$

The first iteration of a cell described by coordinates $(x, y) = (8, 6)$ yields $(x, y) = (14, 20)$. The second, $(x, y) = (34, 54)$. In this way, after visiting the positions (88, 18) (106, 0), (106, 106), (88, 70), (34, 104), (14, 118), (8, 2), (10, 12), (22, 34), (56, 90), (22, 112), (10, 122), and eventually, the 15th iteration yields a transition from the position (132, 254) to the original place of the cell: (8, 6).

This shows how the Arnold Cat Map circulates a cell around the image and then returns it back to its original position. Apparently, as this process is valid for one cell, it simultaneously applies to all the image's cells. Hence, after some iterations, all cells have returned back to their original positions.

Noticeably, the simplest ACM is when $p = q = 1$ and this eventually entails the golden section, because the Lyapunov characteristic exponents of the ACM

$$\begin{bmatrix} x_{k+1} \\ y_{k+1} \end{bmatrix} = \begin{bmatrix} 1 & 1 \\ 1 & 2 \end{bmatrix} \begin{bmatrix} x_k \\ y_k \end{bmatrix} \tag{9.9}$$

are given by the equation

$$\begin{bmatrix} x_{k+1} \\ y_{k+1} \end{bmatrix} = \begin{bmatrix} 1-u & 1 \\ 1 & 2-u \end{bmatrix} \begin{bmatrix} x_k \\ y_k \end{bmatrix} \tag{9.10}$$

which leads to

$$u^2 - 3u + 1 = 0. \tag{9.11}$$

Hence

$$u = \frac{3 \pm \sqrt{5}}{2} \tag{9.12}$$

which, if plugged into

$$\begin{bmatrix} 1-u & 1 \\ 1 & 2-u \end{bmatrix} \begin{bmatrix} x \\ y \end{bmatrix} = \begin{bmatrix} 0 \\ 0 \end{bmatrix} \tag{9.13}$$

yields

$$y = \left(\frac{1+\sqrt{5}}{2} \right) x = \varphi x \tag{9.14}$$

A somewhat similar behavior results from the *chaotic* "*Chebyshev Map*", described by the equation

$$x(n+1) = \cos\left(\frac{k}{\cos(x(n))} \right) \tag{9.15}$$

where $k(n)$ is the modulo of

$$\left\lfloor \frac{x(n)+1}{2} \right\rfloor$$

The Arnold Cat Map can be applied to 3d objects also (Chen et al. 2004):

$$\begin{bmatrix} x_{k+1} \\ y_{k+1} \\ z_{k+1} \end{bmatrix} = \mathbf{A} \begin{bmatrix} x_k \\ y_k \\ z_k \end{bmatrix} \bmod (n)$$

where

$$\mathbf{A} = \begin{bmatrix} 1 + a_x a_z b_y & a_z & a_y + a_x a_z + a_x a_y a_z b_y \\ b_z + a_x b_y + a_x a_z b_y b_z & 1 + a_z b_z & a_x + a_y b_z + a_x a_y a_z b_y b_z + a_x a_z b_z + a_x a_y b_y \\ a_x b_x b_y + b_y & b_x & 1 + a_x b_x + a_y b_y + a_x a_y b_x b_y \end{bmatrix} \tag{9.16}$$

with all parameters alpha and beta being positive integers (it can be verified that **A** has determinant equal to 1).

Despite its random and chaotic appearance, ACM is an invertible, ergodic and structurally stable type of Anosov diffeomorphisms, essentially a homeomorphism of a closed surface preserving the two-dimensional Lebesgue measure and has the "*Poincaré Recurrence Theorem*" inbuilt into it. This theorem guarantees ergodicity for all dynamical systems (under the condition that the system is Hamiltonian and preserves its volume in the phase space).

But the completely accurate reproduction of the image after successive iterations (despite the fact that each and all cells seem randomly transposed) is not the only enigmatic behavior of ACMs. There is yet another, perhaps even more intriguing phenomenon, and this has to do with the still poorly understood relationship between map size and number of iterations. For instance, for $p = q = 1$, while the 100×100 map needs as many as 150 iterations to bring back any cell at its original position, the slightly larger 101×101 map needs only 25 iterations, the 124×124 only 15, but the 150×150 needs 300. So simple map transformations acting on 2d square maps display complex associations with the map size. This inevitably leads us to hypothesize that some map sizes might be endowed with some peculiar properties, but we don't know which ones these map sizes are.

References

Abbott, E. A. (1991). *Flatland: A romance of many dimensions*. Princeton, N.J.: Princeton University Press.

Altieri, C. (2001). Representation, representativeness and non-representational Art. In C. Shukla (Ed.), *Art and representation: Contributions to contemporary aesthetics* (pp. 243–261). Westport, CT: Praeger.

Anglin, W. S. (1990). The square pyramid puzzle. *The American Mathematical Monthly, 97*(2), 120–124.

Arnold, V., & Avez, A. (1968). *Ergodic problems in classical mechanics*. New York: Benjamin.

Aron, J. (2011, June 4). The world's hardest problem: There's a solution that will change everything. *New Scientist, 2815*, 37–41.

Arquilla, J., & Ronfeldt, D. (2001). The advent of netwars (revisited). In J. Arquilla, & D. Ronfeldt (Eds.), *Networks and Netwars*. Santa Monica, CA: Rand.

Berelkamp, E. R., Conway, J. H., & Guy, R. K. (2004). *Winning ways for your Mathematical Plays* (Vol. 4). Wellsley, MA: A.K. Peters.

Chen, G., Mao, Y., & Chui, C. (2004). A symmetric image encryption scheme based on 3d Chaotic Cat Maps. *Chaos, Solitons and Fractals, 21,* 749–761.

Chirikov, B. V. (1979). A universal instability of many-dimensional oscillator systems. *Physics Reports, 52,* 263.

De Certeau, M. (1984). *The practice of everyday life*. Berkeley, CA: University of California Press.

Demaine, E. D., Hohenberger, S., Liben-Nowell, D. (2002, October 21). *Tetris is hard, even to approximate*. arXiv:cs/0210020v1.

DeMaio, J. (2007). Which chessboards have a closed knight's tour within the cube? *The Electronic Journal of Combinatorics, 14,* R32.

Ewald, K. C. (2001). The neglect of aesthetics in landscape planning in Switzerland. *Landscape and Urban Planning, 54*(1), 255–266.

Fraenkel, A., & Lichtenstein, D. (1981). Computing a perfect strategy for $n \times n$ chess requires time exponential in n. *Journal of Combinatorial Theory A, 31,* 199–214.

Gibbins, N. M. (1944, May). Chess in 3 and 4 dimensions, *Mathematical Gazette*, 46–50.

Holzer, M., & Jakobi, S. (2012). On the complexity of rolling block and Alice mazes. *Lecture Notes in Computer Science, 7288,* 210–222.

Iwata, S., & Kasai, T. (1994). The Othello game on an $n \times n$ board is PSPACE-complete. *Theoretical Computer Science, 123*(2), 329–340.

Jelliss, G. P., & Marlow, T. W. (1987). 3d tours. *Chessics, 2*(29/30), 162.

Kornfeld, I. P., Fomin, S. V., & Sinai, Ya. G. (1982). *Ergodic theory*. New York: Springer.

Kumar, A. (2008). Magic knight's tours for chess in three dimensions. *Mathematical Gazette, 92,* 111–115.

Lichtenberg, A., & Lieberman, M. (1992). *Regular and chaotic dynamics*. New York: Springer.

Ma, D. G. (1985). An elementary proof of the solution to the diophantine equation $6y^2 = x(x + 1)(2x + 1)$. *Sichuan Daxue Xuebao, 4,* 107–116.

Papadimitriou, F. (2002). Modelling Indicators and indices of landscape complexity: An approach using GIS. *Ecological Indicators, 2,* 17–25.

Papadimitriou, F. (2009). Modelling spatial landscape complexity using the Levenshtein algorithm. *Ecological Informatics, 4*(1), 51–58.

Papadimitriou, F. (2012). The algorithmic complexity of landscapes. *Landscape Research, 37*(5), 599–611.

Papadimitriou, F. (2013). Mathematical modelling of land use and landscape complexity with ultrametric topology. *Journal of Land Use Science, 8*(2), 234–254.

Pickover, C. (2009). *The Maths book*. London: Sterling.

Reisch, S. (1980). Gobang ist PSPACE-vollstandig (Gomoku is PSPACE-complete). *Acta Informatica, 13,* 5966.

Robson, J. M. (1983). The complexity of go. In *Information Processing; Proceedings of IFIP Congress* (pp. 413–417).

Robson, J. M. (1984). N by N checkers is Exptime complete. *SIAM Journal on Computing, 13*(2), 252–267.

Stewart, I. (2008). *Professor Stewart's cabinet of mathematical curiosities*. London: Profile Books.

Stewart, I. (2010). *Cows in the Maze and other Mathematical explorations*. Oxford: Oxford University Press.

Tegmark, M. (2014). *Our Mathematical universe*. New York: Vintage.

Viglietta, G. (2013). *Gaming is a hard job, but someone has to do it*. arXiv:1201.4995v5.

Watkins, J. J. (2004). *Across the board: The Mathematics of chessboard problems*. Princeton, NJ: Princeton University Press.

Chapter 10
Entering the "Spatium Numerorum": Creating Spatial Complexity from Numbers

White queen:- Can you do addition? What's one and one and one and one and one and one and one and one and one and one? Alice:- I don't know. I lost count. Red Queen:- She can't do addition.
(Lewis Carroll, 1832–1898, "Through the Looking Glass", 1871)

Abstract Spatial complexity can be created from simple square maps. By partitioning space according to a partitions formula, the total number of possible spatial partitions can be derived and then, applying the Burnside lemma gives the total number of symmetric maps allowed by combinatorics. The number of possible map configurations quickly "explodes" and this poses restrictions to spatial analysis. Beginning with a restricted and manageable number of generic maps and subjecting them to symmetric transformations of the symmetry group of the square, it is possible to create big numbers of possible spatial configurations. Thus a space of numbers (a "Spatium Numerorum") is created, beginning with partitions of numbers which are calculated by partitions formulas (i.e. the Hardy-Ramanujan, Rademacher and Bruinier & Ono).

Keywords Spatial complexity · Spatium Numerorum · Number theory and Complexity · Burnside Lemma · Map Complexity · Geocomputation · Partitions function

10.1 Calculating Spatial Partitions

"Though leaves are many, the root is one"
(W.B. Yeats, 1865–1939, "The coming of wisdom with time")

Evidently, there are n^n possible map configurations of n-colored squares over a square map with size n. Consider for instance a 2×2 map. This map has $n = 4$ square cells and therefore the number of all possible color map configurations that can be created from it is $n^n = 4^4 = 256$. How to identify them? A starting point is to determine the possible classes of these configurations and can be found in

F. Papadimitriou, *Spatial Complexity*, https://doi.org/10.1007/978-3-030-59671-2_10

number theory, by making use of *partition functions*. A partition function is a function returning for every positive integer n the number of possible forms by which this n can be "partitioned", that is how many possible sums add up to n, or otherwise stated, how many entropy classes are possible. For instance, the partition function for $n = 5$ gives the following P(5)=7 possible partitions:

n=5,	P(5)=7:
	5
	4+1
	3+2
	3+1+1
	2+2+1
	2+1+1+1
	1+1+1+1+1

Essentially, the partition function gives the number of possible ways that a number can be "decomposed" (not to be confused with factoring). As a further example, consider the partitions of $n = 4$:

n=4,	P(4)=5:
	4
	3+1
	2+2
	2+1+1
	1+1+1+1

Translating these figures to maps of square cells, it is easy to see how a sum of possible map configurations corresponds to each one of these partitions of a square map of 4 cells. To identify these configurations, we need to allow as many as n colors on the map, so the maximum number of colors is $n = 4$. All the possible map configurations per partition of 4 for the partitions 4, $3 + 1$, $2 + 2$ and $1 + 1 + 1 + 1$ are given in Figs. 10.1, 10.2, 10.3 and 10.4 and in summary in Table 10.1.

It was the Fibonacci sequence that was first used to determine the number of partitions $P(n)$ of a number n:

$$P(n) = \frac{1}{\sqrt{5}}\left(\left[\frac{1+\sqrt{5}}{2}\right]^{n+1} - \left[\frac{1-\sqrt{5}}{2}\right]^{n+1}\right) = \sum_{k=0}^{n/2}\binom{n-k}{k} \tag{10.1}$$

This "old" formula encapsulates the golden ratio, since

$$\lim_{n\to\infty}\frac{P(n)}{P(n-1)} = \left[\frac{1+\sqrt{5}}{2}\right] = 1.61803... \tag{10.2}$$

Most commonly however, the Hardy and Ramanujan formula is used to calculate the partitions of any positive integer (Hardy and Ramanujan 1918) which provides an asymptotic solution of P with respect to n:

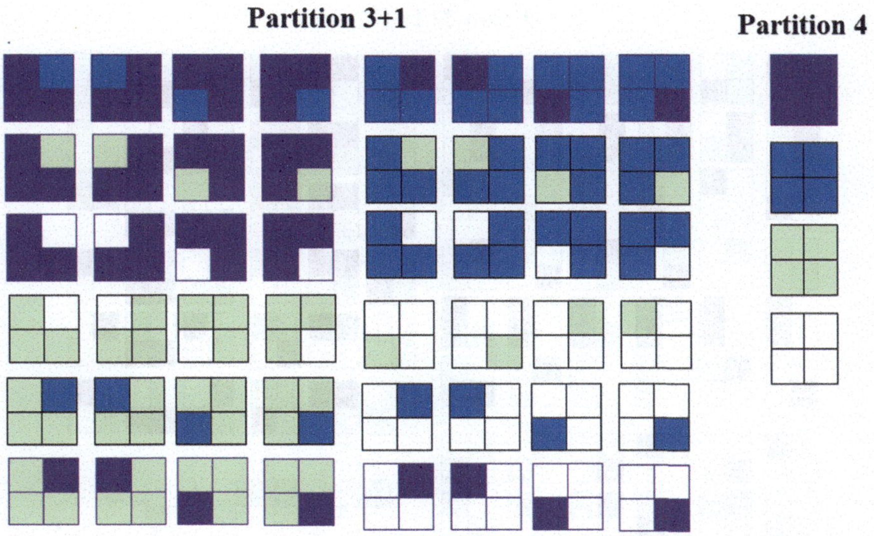

Fig. 10.1 The possible map configurations of a 4-colored 2×2 map for the partitions $4 = 4$ and $4 = 3 + 1$ (4 and 48 partitions respectively)

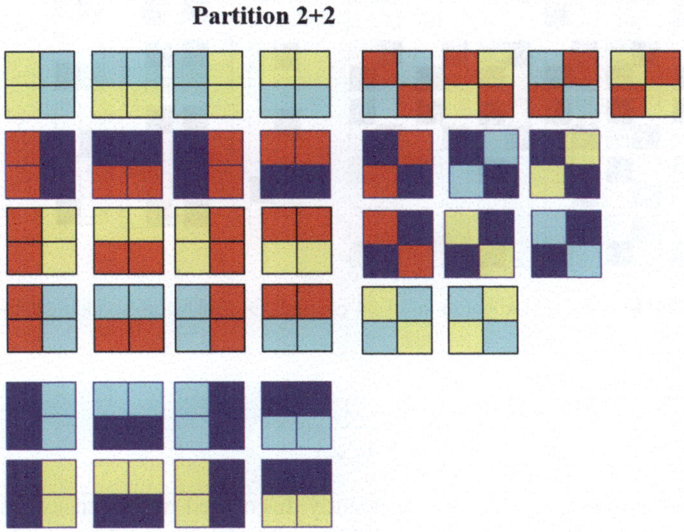

Fig. 10.2 The possible map configurations of a 4-colored 2×2 map corresponding to the partition $4 = 2 + 2$

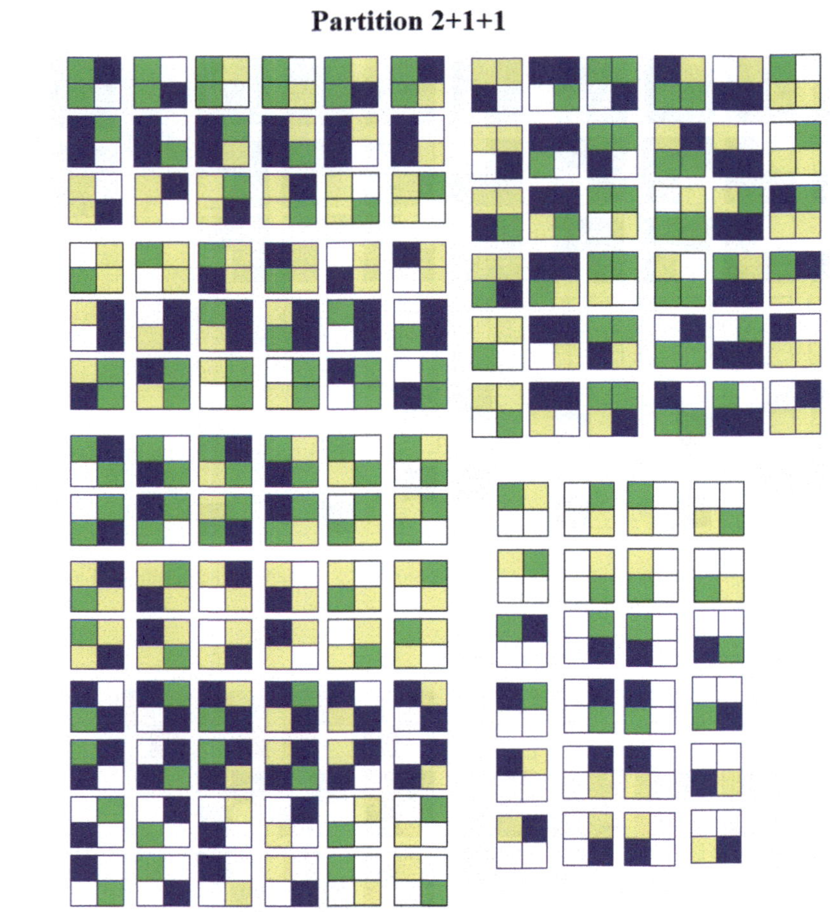

Fig. 10.3 The 144 possible configurations of a 4-colored 2×2 map corresponding to the partition $4 = 2 + 1 + 1$

$$P(n) \approx \frac{1}{4n\sqrt{3}} e^{\pi\sqrt{2n/3}} \quad (10.3)$$

The same formula was also independently discovered by Uspensky (1920); the reader may refer to Hardy and Wright (1979) and to Hardy (1999).

Some values of the partition function for some small quadratic maps, per map size, are given in Table 10.2: the ratio $P(n)/N(n)$ diminishes close to zero, even for small n.

Later, Rademacher (1937) obtained an exact convergent series solution which includes the Hardy-Ramanujan formula (Rademacher 1932, 1937, 1943):

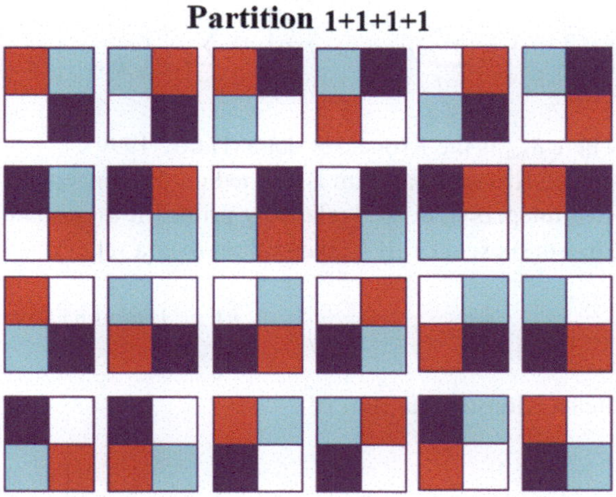

Fig. 10.4 The 24 possible configurations of a 4-colored 2×2 map corresponding to the partition $4 = 1 + 1 + 1 + 1$

Table 10.1 Number of possible map configurations $N(n)$ per partition class for $n = 4$ square multi-colored maps. The sum total of all possible configurations is $n^n = 4^4 = 256$

Partition class	Possible configurations $N(n)$
4	4
$3 + 1$	48
$2 + 2$	36
$2 + 1 + 1$	144
$1 + 1 + 1 + 1$	24

Table 10.2 Values of the partition function $P(n)$ for small map sizes, numbers of possible map configurations $N(n)$ and $P(n)/N(n)$ ratios respectively. Notice how quickly the ratio attains extremely small values

Map size n	Partitions $P(n)$	Possible map configurations $N(n)$	Ratio $P(n)/N(n)$
4	5	$4^4 = 256$	19.5×10^{-3}
9	30	$9^9 = 387{,}420{,}489$	7.743×10^{-8}
16	297	$16^{16} = 1844 \times 10^{19}$	1.61×10^{-17}
25	2436	$25^{25} = 8882 \times 10^{34}$	2.742×10^{-32}
36	21637	$36^{36} = 1064 \times 10^{56}$	2.033×10^{-52}

$$P(n) = \frac{1}{\pi\sqrt{2}} \sum_{k=1}^{\infty} A_k(n)\sqrt{k}\frac{d}{dn}\left[\frac{\sinh\left(\frac{\pi}{k}\sqrt{\frac{2}{3}}\left(n - \frac{1}{24}\right)\right)}{n - \frac{1}{24}}\right] \tag{10.4}$$

with the sequence $A_k(n)$ expressed as a Kloosterman sum (an exponential sum involving natural numbers):

$$A_k(n) = \sum_{h=1}^{k} \delta_{GCD(h,k)} \exp\left[\pi i \sum_{j=1}^{k-1} \frac{j}{k}\left(\frac{hj}{k} - \left\lfloor \frac{hj}{k} \right\rfloor - \frac{1}{2}\right) - \frac{2\pi i h n}{k}\right] \tag{10.5}$$

where δ_{mn} is the Kronecker delta (Hardy 1999).

The Kloosterman sum is defined on the concept of "relative primes" of integers: Two integers n, m are "relatively prime" if they share no common positive factors (divisors) except 1. If h takes values over a set of residues relative to prime to n and

$$\hat{h}h = 1(\mathrm{mod}\, n), \tag{10.6}$$

then a Kloosterman sum is:

$$S(u, v, n) \equiv \sum_{h} \exp\left[\frac{2\pi i\left(uh + v\hat{h}\right)}{n}\right]. \tag{10.7}$$

For further information the reader may consider the relevant literature of number theory (Kloosterman 1926, 1946; Hardy and Wright 1979; Katz 1987; Apostol 1976; Hardy 1999).

A more recent formula for partitions is given by Bruinier and Ono (2011):

$$P(n) = 2\pi(24n - 1)^{-\frac{3}{4}} \sum_{k=1}^{\infty} \frac{A_k(n)}{k} I_{\frac{3}{2}}\left[\frac{\pi\sqrt{24n-1}}{6k}\right] \tag{10.8}$$

where $I_{3/2}$ stands for the modified Bessel function of the first kind and $A_k(n)$ is the Kloosterman sum.

The modified Bessel function of the first kind $I_{\mathrm{n}}(z)$ is defined as an integral

$$I_n(z) = \frac{1}{2i\pi} \oint e^{\frac{z}{2}\left(1+\frac{1}{t}\right)} t^{-1-n} \mathrm{d}t \tag{10.9}$$

or, in terms of the gamma function

$$I_n(z) = \left(\frac{z}{2}\right)^n \sum_{k=0}^{\infty} \frac{\left(\frac{z^2}{4}\right)^k}{k!\Gamma(n+k+1)} \tag{10.10}$$

and appears as a solution of the second-order modified Bessel differential equation:

$$x^2 \frac{\mathrm{d}^2 y}{\mathrm{d}x^2} + x\frac{\mathrm{d}y}{\mathrm{d}x} - \left(x^2 + n^2\right)y = 0 \tag{10.11}$$

This formula for $P(n)$ is given as a finite sum of algebraic numbers:

$$P(n) = \frac{Tr(n)}{24n - 1} \tag{10.12}$$

where the trace $Tr(n)$ is defined as

$$Tr(n) = \sum_{Q \in Q_n} R(\alpha_Q). \tag{10.13}$$

$R(z)$ stands for a function:

$$R(z) = -\left(\frac{1}{2\pi i}\frac{\mathrm{d}}{\mathrm{d}z} + \frac{1}{2\pi y}\right) f(z), \tag{10.14}$$

where $z = x + iy$ and Q_n is any set of representatives of the equivalence classes of the integral binary quadratic form

$$Q(x, y) = ax^2 + bxy + cy^2 \tag{10.15}$$

with $a > 0$ and $b = 1$ mod(12), with the property that for each $Q(x, y)$, we let a_Q be "CM point" in the upper half-plane, for which $Q(a_Q, 1) = 0$, recalling that a point is CM if its corresponding elliptic curve has complex multiplication.

The function $f(z)$ is the weight-2 meromorphic modular form entailing Eisenstein series and Dedekind eta functions:

$$\begin{aligned} F(z) &= \frac{1}{2}\frac{E_2(z) - 2E_2(2z) - 3E_3(3z) + 6E_2(6z)}{\eta^2(z)\eta^2(2z)\eta^2(3z)\eta^3(6z)} \\ &= q^{-1} - 10 - 29q - 104q^3 - 273q^3 \ldots \end{aligned} \tag{10.16}$$

where $q = \mathrm{e}^{2\pi i z}$ is the nome, $E_2(q)$ are Eisenstein series, and $\eta(q)$ are Dedekind eta functions.

The Eisenstein series is defined as:

$$G_r(\tau) = \sum_{m=-\infty}^{\infty} \sum_{m=-\infty}^{\infty} \frac{1}{(m + n\tau)^r} \tag{10.17}$$

where $r > 2$ is an integer, $\tau > 0$ and the sums exclude $m = n = 0$, while satisfying the following relationship in terms of Riemann zeta functions:

$$G_{2k}(\tau) = 2\zeta(2k) + \left[\frac{\left(\sum_{n=1}^{\infty} \sigma_{2k-1}(n) e^{2\pi i n \tau} \right) \left(2(2\pi i)^{2k} \right)}{(2k-1)!} \right] \tag{10.18}$$

for $n > 1$, with $\zeta(s)$ the Riemann zeta function and $\sigma_k(n)$ the divisor function.

With an elliptic modulus k and a nome $q = e^{i\pi\tau}$, the first values of the Eisenstein series $E_{2n}(q)$ are (Apostol 1976):

$$E_2(q) = 1 - 24 \sum_{k=1}^{\infty} \sigma_1(k) q^{2k} \tag{10.19}$$

$$E_4(q) = 1 + 240 \sum_{k=1}^{\infty} \sigma_3(k) q^{2k} \tag{10.20}$$

$$E_6(q) = 1 - 504 \sum_{k=1}^{\infty} \sigma_5(k) q^{2k} \tag{10.21}$$

Also, the Eisenstein series is defined as:

$$\sum_{k=-\infty}^{\infty} \sum_{j=-\infty}^{\infty} \begin{bmatrix} (j+k\tau)^{-2n} & j^2 + k^2 \neq 0 \\ 0 & otherwise \end{bmatrix} = 2\zeta(2n) + \frac{\left(\sum_{k=1}^{\infty} \sigma_{2n-1}(k) e^{2\pi i k \tau} \right) \left(2(2\pi i)^{2n} \right)}{(2n-1)!} \tag{10.22}$$

for $n > 1$, with $\zeta(s)$ the Riemann zeta function and $\sigma_k(n)$ the divisor function.

The Dedekind eta function is a modular form defined over the upper half-plane $\{I(\tau) > 0\}$ by the formula:

$$\eta(\tau) = (\overline{q})^{\frac{1}{24}} \sum_{n=-\infty}^{\infty} (-1)^n (\overline{q})^{-n\frac{(3n-1)}{2}} = (\overline{q})^{\frac{1}{24}} \left(1 - (\overline{q})^2 + (\overline{q})^5 + (\overline{q})^7 - (\overline{q})^{12} - \cdots \right) \tag{10.23}$$

where $(\overline{q}) = e^{2\pi i \tau}$ is the square “nome” q and τ is the half-period ratio (Atkin and Morain 1993; Berndt 1994) eventually transforming the Dedekind eta function to the form:

$$\eta(\tau) = e^{\frac{\pi i \tau}{12}} \prod_{k=1}^{\infty} \left(1 - e^{2\pi i k \tau} \right) \tag{10.24}$$

For some values, it relates to other known functions such as the Jacobi theta functions, the gamma function etc. For instance, for theta functions with zero argument:

$$\vartheta_{\cdot 3}\left(0, \mathrm{e}^{i\pi\tau}\right) = \frac{\eta^2\left(\frac{\tau+1}{2}\right)}{\eta(\tau+1)} \tag{10.25}$$

and with the gamma function

$$\eta(i) = \frac{\Gamma\left(\frac{1}{4}\right)}{2\pi^{\frac{3}{4}}} \tag{10.26}$$

The Eisenstein series E_2 is related to partitions $P(n)$ as follows:

$$E_2(z) = 1 - 24\sum_{n=1}^{\infty}\sum_{d/n} d(P(n))^n. \tag{10.27}$$

The nome q is defined on Jacobi theta functions as (Borwein and Borwein 1987):

$$q = \mathrm{e}^{\frac{i\pi K\sqrt{1-k^2}}{K(k)}} \tag{10.28}$$

with τ the half-period ratio, $K(k_e)$ the complete elliptic integral of the first kind, k the elliptic modulus and the elliptic integral of the first kind has the general form

$$F(\varphi, k) = \int_0^{\tan\varphi} \frac{dv}{\sqrt{(1+v^2)((1+(1-k)^2v^2)}} \tag{10.29}$$

where $v = \tan\theta$ (Abramowitz and Stegun 1972).

Hence, the divisor function and the Jacobi theta functions enter in the calculation of the Dedekind eta function and for the Eisenstein series.

The divisor function of an integer n is the sum of k-th powers of the positive integer divisors of n:

$$\sigma_k(n) = \sum_{d/n} d^k \tag{10.30}$$

and relates to the Riemann zeta function, by means of Ramanujan's formula (Wilson 1923):

$$\sum_{n=1}^{\infty}\left(\frac{\sigma_a(n)\sigma_b(n)}{n^s}\right)=\frac{\zeta(s)\zeta(s-a)\zeta(\sigma-b)\zeta(s-a-b)}{\zeta(2s-1-b)} \tag{10.31}$$

while also satisfying

$$\lim_{n\to\infty}\left(\frac{\sigma(n)}{n\ln\ln n}\right)=e^{\gamma} \tag{10.32}$$

where γ is the Euler-Mascheroni constant.

The Jacobi theta functions are quasi-periodic, expressed in terms of the nome q, given in the form $\vartheta_{\cdot n}(z,q)$ where q is defined in terms of a quasi-period τ as:

$$q=\mathrm{e}^{2\pi i\tau}. \tag{10.33}$$

Setting thus the nome, leads to different Jacobi forms for successively higher n, i.e.:

$$\vartheta_{\cdot 1}(z,q)=\sum_{n=-\infty}^{n=\infty}(-1)^{n-\frac{1}{2}}q^{\left(n+\frac{1}{2}\right)^2}\mathrm{e}^{(2n+1)iz} \tag{10.34}$$

$$\vartheta_{\cdot 2}(z,q)=\sum_{n=-\infty}^{n=\infty}q^{\left(n+\frac{1}{2}\right)^2}\mathrm{e}^{(2n+1)iz} \tag{10.35}$$

$$\vartheta_{\cdot 3}(z,q)=\sum_{n=-\infty}^{n=\infty}q^{(n)^2}\mathrm{e}^{2niz} \tag{10.36}$$

Elliptic integrals of the 1st kind are expressed in terms of Jacobi functions and an elliptic modulus, which can be expressed in terms of Jacobi theta functions:

$$k=\frac{\vartheta_2^2(0,q)}{\vartheta_3^2(0,q)}. \tag{10.37}$$

The Jacobi theta functions with $z=0$ relate to the gamma function for some values of the nome, i.e.

$$\vartheta_{\cdot 3}\left(0,\mathrm{e}^{-\pi}\right)=\frac{\sqrt[4]{\pi}}{\Gamma\left(\frac{3}{4}\right)} \tag{10.38}$$

10.2 Entropy Class

...and either the white becomes black, or the black becomes white...
"...καὶ γὰρ εἰ λευκὸν ὑπάρχον μελαίνοιτο καὶ εἰ μέλαν λευκαίνοιτο"
(Galen, 129–199 A.D., "On the Natural Faculties", 1.2)

The possible spatial partitions define entropy. In place of Shannon's formula for entropy H, a simpler measure will be used here instead and will hereafter be named "entropy class" (r). This is the number of colored cells in a map (Fig. 10.5) and assumes only integer values. In a binary map, the colored cells are the "black" ones and are considered to be those that constitute the second largest population after the population bearing the dominant color (the whites in the case of binary maps). If, for instance, a binary 3×3 map is dominantly white, then the black cells can not be more than 4, that is $r = (n - 1)/2$, where n is the total number of cells ($n = 9$ in this case). If the binary map is even-numbered, then the number of colored cells cannot be higher than $r = n/2$.

Obviously, all binary maps of the same entropy class also have the same Shannon entropy. The higher the number of black cells, the higher the entropy class and this applies up to the maximum entropy class of the square binary map. Notice however, that that Shannon entropy H reflects the percentage of the relative participation of each map type on the map and it is independent of observation scale. Contrary to this, the entropy class r is *scale-dependent*, representing the number of colored cells in a binary map for a precise size and resolution (Fig. 10.6). This difference makes r more advantageous to Shannon entropy for the analysis of the spatial complexity of square maps, also due to the fact that it assumes only integer values.

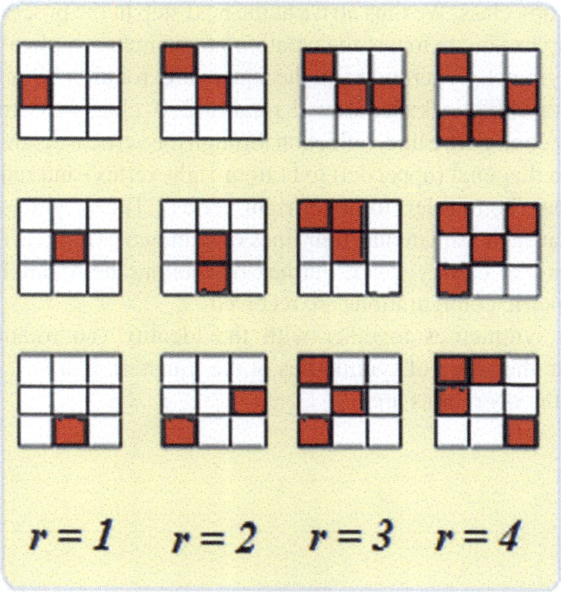

Fig. 10.5 Entropy class r is defined here as the total number of colored cells (red in this case) in the binary map. Some 3×3 binary maps with "entropy classes" $r = 1, 2, 3$ and 4 are shown

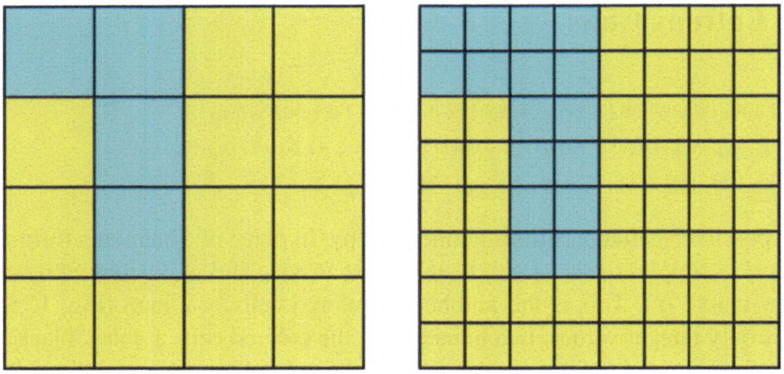

Fig. 10.6 The difference between Shannon entropy H and entropy class r consists in the fact that r is scale-dependent. These two maps have both the same entropy H, but their entropy classes differ: the map on the left has $r = 4$, while the map on the right has $r = 16$. Thus, by using r instead of H, map analysis becomes scale-dependent, which is essential for the assessment of spatial complexity

10.3 Generic Maps and Symmetry

> "Et plutard un Ange, entr' ouvrant les portes,
> viendra ranimer, fidèle et joyeux les mirroirs ternis"
> (Charles Baudelaire, 1821–1867, "La mort des amants")

Partitioning a space is equivalent to the identification of possible "entropy classes" in it. But partitioning alone is inadequate to determine all possible map configurations for a certain entropy class. We thus arrive at the next step in the process of generation of spatially complex square maps: the creation of symmetric replications.

The typical symmetry operations on the square are: rotation 90° clockwise about the centre, rotation 180° clockwise around, rotation 270° clockwise around, reflection through the horizontal centre line, reflection through the vertical centre line, reflection through the main diagonal (upper-left to bottom-right vertex) and reflection through the other diagonal (bottom-left to upper-right vertex). These symmetric operations are possible because the square has four lines of symmetry (Fig. 10.7): the two axes and the two lines $y = x$ and $y = -x$. Further, by rotating the square by 90°, 180° or 270°, new symmetric configurations are received.

These seven symmetries together with the identity (no rotation, or "trivial symmetry) create the group of symmetries of the square (8 in total).

Specifically, the operations are:

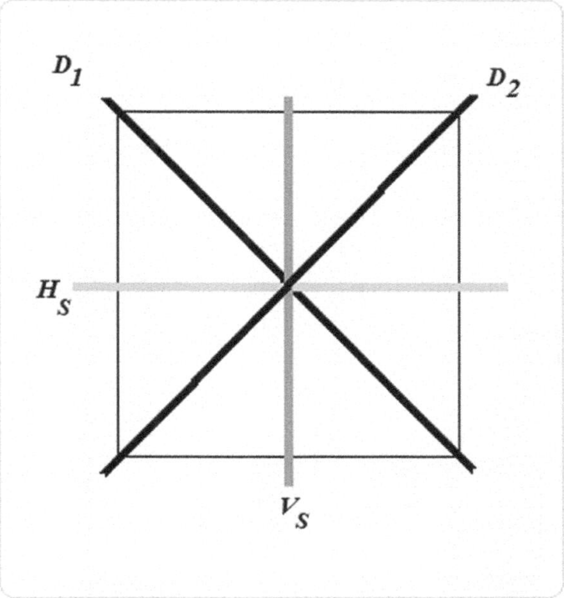

Fig. 10.7 Symmetry axes of the square

V_S = Reflection through the vertical
H_S = Reflection through the horizontal
D_S = Reflection through the diagonal D_1
D'_S = Reflection through the diagonal D_2
I_S = identity (no rotation)
R_{90} = rotation 90° clockwise about the center
R_{180} = rotation 180° clockwise about the center
R_{270} = rotation 270° clockwise about the center.

In this way, any new positions of cells are calculated from the multiplication table (Table 10.3):

The number of map configurations corresponding to each symmetry operation is given by the "*Burnside lemma*" (alternatively referred to as the "*Burnside–Polya theorem*"), which can be used to endow square partitions with topologically inequivalent positions, along with their associated symmetries. It yields configurations of symmetry-dependent maps, but the full set of map configurations is received only after the application of symmetry operations is applied to them.

Let G be a group of elements that permute vertices of objects. Two colorings are considered indistinguishable with respect to G if there is some element g belonging to G, such that g sends one coloring to another. Letting $\psi(g)$ be the number of colorings which are unchanged when affected by g and N_g the number of generic maps, then Burnside's lemma computes the number of generic maps N_g as:

Table 10.3 Multiplication table showing the results of multiplication of possible symmetries of the square

	H_S	V_S	D_S	D'_S	I_S	R_{90}	R_{180}	R_{270}
H_S	I_S	R_{180}	R_{90}	R_{270}	H_S	D_S	V_S	D'_S
V_S	R_{180}	I_S	R_{270}	R_{90}	V_S	D'_S	H_S	D_S
D_S	R_{270}	R_{270}	I_S	R_{180}	D_S	V_S	D'_S	H_S
D'_S	R_{90}	R_{270}	R_{180}	I_S	D'_S	H_S	D_S	V_S
I_S	H_S	V_S	D_S	D'_S	I_S	R_{90}	R_{180}	R_{270}
R_{90}	D'_S	D_S	H_S	V_S	R_{90}	R_{180}	R_{270}	I_S
R_{180}	V_S	H_S	D'_S	D_S	R_{180}	R_{270}	I_S	R_{90}
R_{270}	D_S	D'_S	V_S	H_S	R_{270}	I_S	R_{90}	R_{180}

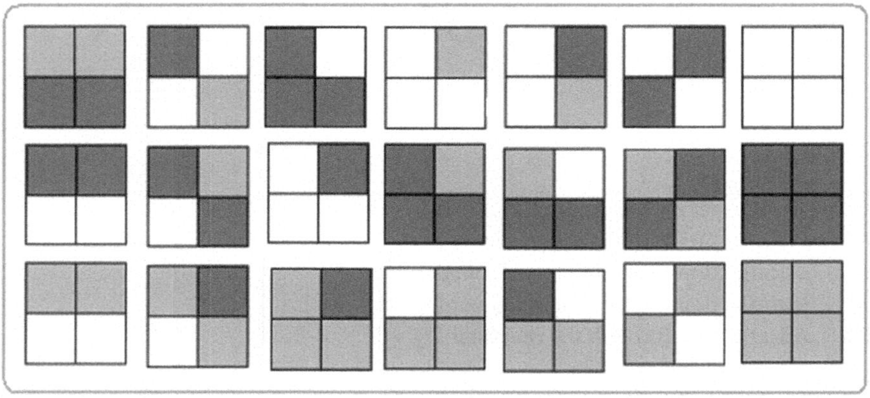

Fig. 10.8 The 21 symmetric 2 × 2 squares with 3 colors

$$N_g = \frac{1}{|G|} \sum_{g \in G} \psi(g) \tag{10.39}$$

For instance, for a 2 × 2 map with three colors the number of generic 2 × 2 square maps with three colors (Fig. 10.8) is (Table 10.4):

$$\begin{aligned} N_g &= \frac{1}{|G|} \sum_{g \in G} \psi(g) = \frac{1}{8}\left(3^4 + 3^1 + 3^2 + 3^1 + 3^2 + 3^2 + 3^3 + 3^3\right) \\ &= \frac{168}{8} = 21 \end{aligned} \tag{10.40}$$

Increasing the number of colors by one, the formula yields more possible map configurations for 2 × 2 maps with 4-colors, the total number of which is (Fig. 10.9):

Table 10.4 Calculation of symmetries of 2 × 2 squares with 3 colors

Symmetry	Cycle form	Number of configurations ψ(g)
I_S	(1), (2), (3), (4)	3^4
R_{90}	(1 2 3 4)	3^1
R_{180}	(1 3), (2 4)	3^2
R_{270}	(1 4 3 2)	3^1
V_S	(1 2), (3 4)	3^2
H_S	(1 4), (2 3)	3^2
D_S	(1), (2 4), (3)	3^3
D'_S	(1 3), (2), (4)	3^3

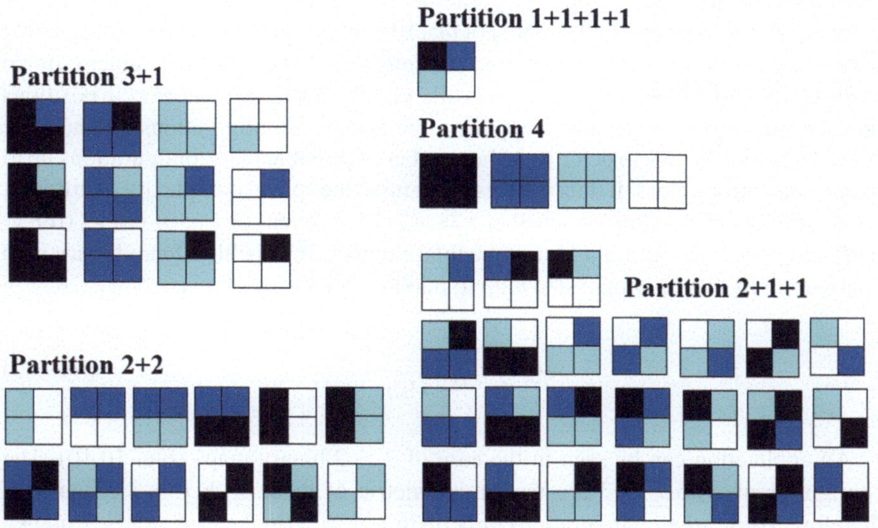

Fig. 10.9 The 53 possible 2 × 2 square maps with all possible combinations of 4 colors

$$
\begin{aligned}
N_g &= \frac{1}{|G|}\sum_{g\in G}\psi(g) \\
&= \frac{1}{8}\left(4^4+4^1+4^1+4^2+4^2+4^3+4^3+4^3\right) \\
&= \frac{424}{8} = 53
\end{aligned} \tag{10.41}
$$

10.4 Calculating Binary Map Configurations

"Nature is indeed a sum, but not a whole"
(Gilles Deleuze 2012, p. 304)

Given the partitions of space acting on square maps, entropy classes can be defined next. In fact, no more black cells need to be allocated on a map after attaining the maximum entropy class, since after exceeding the maximum entropy threshold, all binary map configurations repeat themselves as black-and-white mirror reflections of the configurations which were derived prior to attaining maximum entropy class. This is because for entropy classes higher than $r = n/2$ (if $n =$ even) or $r = (n - 1)/2$ (if $n =$ odd), the resulting binary map configurations are mirror-like repetitions of their n-r counterparts. So a simple replacement of black by white cells (or white by black cells) *at the same positions* of the map produces identical spatial complexity values (simple replacements of black cells by white cells yields exactly the same cell positions on each map and this symmetry applies to all possible configurations). Hence, the central question is how to determine the number of possible map configurations up to maximum entropy class. It thus suffices to examine the spatial complexity of different configurations, depending on whether $r = n/2$ (if $n =$ even) or $r = (n - 1)/2$ (if $n =$ odd) and hence, the formula giving the total number of possible square binary map configurations $N(n)$ per map size n up to maximum entropy class is:

$$N(n) = \sum_{r=1}^{r} \frac{n!}{r!(n-r)!} \tag{10.42}$$

An application can be seen in the case of 2 × 2 binary maps (Fig. 10.10). The configurations with $r = 3$ are mirror-symmetric of those with $r = 1$. It suffices therefore to consider configurations only up to $r_{\max} = 2$ (in the case of 2 × 2 maps). As $n =$ even, so $r = 2$ and hence the number of possible configurations $N(n)$ up to maximum entropy class ($r = 2$) is:

$$N(4)_{r=2} = \sum_{r=1}^{2} \frac{4!}{r!(4-r)!} = \frac{4!}{1!3!} + \frac{4!}{2!2!} = 10 \tag{10.43}$$

Similarly, the number of all possible 3 × 3 binary maps configurations from $r =$ 1 up to the maximum entropy class (which is $r = 4$) is 255:

$$N = \sum_{r=1}^{r=\frac{n-1}{2}} \binom{n}{r} = \sum_{r=1}^{r=\frac{9-1}{2}=4} \binom{9}{4} = \binom{9}{1} + \binom{9}{2} + \binom{9}{3} + \binom{9}{4} = 255 \tag{10.44}$$

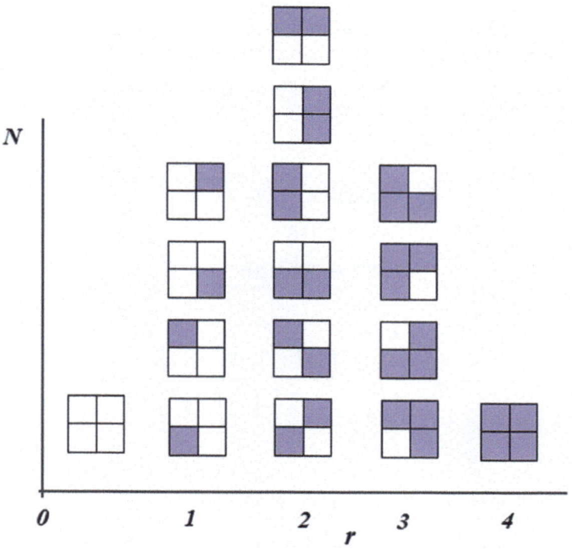

Fig. 10.10 All the possible configurations of 2 x 2 binary maps. When more than half of the cells are black, then the map configurations repeat themselves as exactly reversed, therefore without contributing any more to complexity beyond the state of maximum entropy, which is attained at the entropy class $r = 2$ for this map size

The number of possible configurations $N(n)$ up to maximum entropy class, depends on whether the total number of cells (n) is an even or an odd number:

$$N(n) = \sum_{r=1}^{r} \frac{n!}{r!(n-r)!} = \left\{ \begin{array}{ll} \sum\limits_{r=1}^{r=(n-1)/2} \frac{n!}{k!(n-r)!} & n = odd \\ \sum\limits_{r=1}^{r=n/2} \frac{n!}{r!(n-r)!} & n = even \end{array} \right\} \tag{10.45}$$

For $n =$ odd, we simply have:

$$N(n) = \sum_{r=1}^{r=\frac{(n-1)}{2}} \frac{n!}{r!(n-r)!} = 2^{n-1} - 1 \tag{10.46}$$

For $r = n/2$ (case where $n =$ even), the calculation of $N(n)$ is carried out by employing the Gaussian hypergeometric function ${}_2F_1$, so the formula giving the total $N(n)$ of binary maps configurations is:

$$N(n) = r \sum_{k=1}^{r=n/2} \frac{n!}{r!(n-r)!}$$

Table 10.5 Even in small binary maps (from 2 × 2 to 6 × 6 shown here), as the map size (n) increases, the sum of possible square binary map configurations $N(n)$ "explodes"

Binary map size (n)	Number of possible binary maps up to $r_{\max}$
4	10
9	255
16	36,493
25	16,777,216
36	38,897,306,020

$$= 2^n - 1 - \frac{n!_2F_1\left(1, 1 - \frac{n}{2}, 2 + \frac{n}{2}, -1\right)}{\left(\frac{n-2}{2}!\right)\left(\frac{n+2}{2}!\right)} \tag{10.47}$$

and therefore,

$$N(n) = \begin{Bmatrix} 2^{n-1} - 1 & n = odd \\ 2^n - 1 - \frac{n!_2F_1\left(1,1-\frac{n}{2},2+\frac{n}{2},-1\right)}{\left(\frac{n-2}{2}!\right)\left(\frac{n+2}{2}!\right)} & n = even \end{Bmatrix} \tag{10.48}$$

To get a glimpse of the "combinatorial explosion" of the number of possible binary map configurations $N(n)$ with increasing map size n, it suffices to consider the values of $N(n)$ with respect to n even only for some low values of n (Table 10.5). Hence, when embarking to carry out spatial analyses of any kind by using square binary maps with increasing map size, it always has to be considered that the number of *possible* configurations will increase very fast and so the computational complexity for examining the spatial complexity of *all* these configurations rapidly spirals out of computational control.

References

Abramowitz, M., & Stegun, I. A. (Eds.). (1972). *Handbook of Mathematical functions with formulas, graphs, and Mathematical tables*. New York: Dover.

Apostol, T. M. (1976). *Chapter 4 in Introduction to analytic number theory*. New York: Springer-Verlag.

Atkin, A. O. L., & Morain, F. (1993). Elliptic curves and primality proving. *Mathematics and Computation, 61*, 29–68.

Berndt, B. C. (1994). *Ramanujan's notebooks, Part IV*. New York: Springer-Verlag.

Borwein, J. M., & Borwein, P. B. (1987). *Pi & the AGM: A study in analytic number theory and computational complexity*. New York: Wiley.

Bruinier, J. H., & Ono, K. (2011, April 6). Algebraic formulas for the coefficients of half-integral weight harmonic weak mass forms. https://arxiv.org/abs/1104.1182/.

Deleuze, G. (2012). *The logic of sense*. London: Continuum.
Hardy, G. H. (1999). "Ramanujan's work on partitions" and "Asymptotic theory of partitions". Chapters 6 and 8 In *Ramanujan: Twelve lectures on subjects suggested by his life and work* (3rd ed, pp. 83–100 and pp. 113–131). New York: Chelsea.
Hardy, G. H., & Ramanujan, S. (1918). Asymptotic formulae in combinatory analysis. *Proceedings of the London Mathematical Society, 17,* 75–115.
Hardy, G. H., & Wright, E. M. (1979). *An introduction to the theory of numbers*. Oxford: Clarendon Press.
Katz, N. M. (1987). *Gauss sums, Kloosterman sums and Monodromy groups*. Princeton, NJ: Princeton University Press.
Kloosterman, H. D. (1926). On the representation of numbers in the form $ax^2 + by^2 + cz^2 + dt^2$. *Acta Mathematica, 49,* 407–464.
Kloosterman, H. D. (1946). The behavior of general theta functions under the modular group and the characters of binary modular congruence groups I. *Annals of Mathematics, 47*(317–375), 1946.
Rademacher, H. (1932). Zur Theorie der Modulfunktionen. *Journal für die reine und angewandte Mathematik, 167,* 312–336.
Rademacher, H. (1937). On the partition function. *Proceedings of the London Mathematical Society, 43,* 241–254.
Rademacher, H. (1943). On the expansion of the partition function in a series. *Annals of Mathematics, 44,* 416–422.
Uspensky, J. V. (1920). Asymptotic formulae for numerical functions which occur in the theory of partitions. *Bulletin of the Academy of Sciences of the USSR, 14,* 199–218.
Wilson, B. M. (1923). Proofs of some formulae enunciated by Ramanujan. *Proceedings of the London Mathematical Society, 21,* 235–255.

Chapter 11
The Spatial Complexity of 3 × 3 Binary Maps

The triad first constituted the beginning, the middle and the end
‘Η τριὰς πρώτη συνέστησεν ἀρχήν, μεσότητα καὶ τελευτή
(Okellos, first century b.C.)

Abstract A complete complexity characterization of 3 × 3 binary maps is possible, from which it is revealed that for all such maps: (i) higher entropy class means higher spatial complexity in most cases (but not always); (ii) spatial complexity is "generated" fast but "slows down" as the entropy increases; (iii) high spatial complexity coincides with high patchiness (confirming the widely held belief about this); (iv) the higher the clumpiness, the higher the spatial complexity; (v) the number of generic maps per entropy class decreases as the entropy class grows (implying that identifying generic forms is equivalent to focusing on a number of possible configurations at a scale of reduction which is particularly important for spatial complexity: a reduction from exponential to polynomial growth); (vi) even a slight introduction of "otherness" (that is dark cells) in a binary map induces a high increase in spatial complexity. So "invading species" in an ecosystem (or, equivalently, "invading" black cells in a binary map), regardless of their particular location in the map or how few they may be, immediately create a substantial difference in the map's spatial complexity; (vii) some values of spatial complexity concentrate disproportionately larger numbers of map configurations than other ones.

Keywords Spatial Complexity · 3 × 3 maps · Spatial Computation · Binary maps · Map Complexity · Entropy and Complexity · Geocomputation

11.1 Parameters of Spatial Complexity of Binary Maps

"Difference is of two kinds… the first is called a difference of number the other of kind"
(David Hume, "A treatise of human nature", 1740)

Besides entropy class, additional parameters (*not* measures) of a map's spatial complexity (aside of entropy/diversity) are its "patchiness" (P), "adjacency" (A)

F. Papadimitriou, *Spatial Complexity*, https://doi.org/10.1007/978-3-030-59671-2_11

and "clumpiness" (B). The role of these parameters in spatial complexity has been identified earlier (Papadimitriou 2002; Papadimitriou 2009; Papadimitriou 2012).

Patchiness (P) is the number of non-adjacent (disjoint) patches of the same color dispersed over the dominant cover of the binary map (which, in the language of landscape ecology is called "the matrix").

Adjacency (A_d) is the number of edges shared by cells of the same patch type (it refers to patches only; not to the dominant color). It is defined as the number of edges between cells of the same cover type V_i in the map:

$$A_d = \sum_{i=1}^{n} \partial_i V_i \tag{11.1}$$

and refers only to the cells that are defined as "colored" (not belonging to the dominant population of cells).

A typical assumption is that the more patchy a map, the more complex it is expected to be. Spatial complexity is expected to decrease if patches of the same color are clumped together and at this point enters the parameter "clumpiness".

Clumpiness (B) is defined here to be the maximum size of the largest edge-continuous block (aggregate) of cells of the same patch type: $B = \max\{\text{patch size of the same color}\}$ and it is examined because clustering plays a central role in self-organization and complexity (Bormashenko et al. 2020).

As an example, consider the calculation of r, P, B, C_{P1} and C_{P2} for the three flag-like example binary maps of the same size (5 × 5), resembling the flags of three European countries, shown in Fig. 11.1. The results of the calculations are given in Table 11.1.

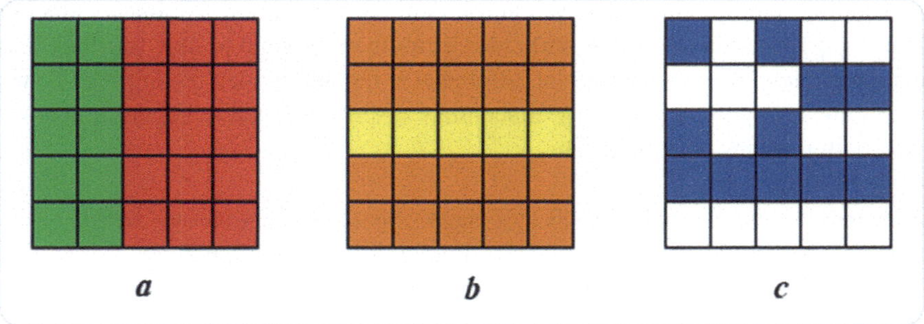

Fig. 11.1 Calculating the spatial complexity on three simplified flag-like example binary 5 × 5 maps resembling the flags of three southern European countries

Table 11.1 Results of the calculations of the metrics and parameters of spatial complexity of three simplified flag-like example binary 5 × 5 maps shown in Fig. 11.1

Map	r	P	B	A	C_{PI}	C_{P2}
a	10	1	10	13	6	5
b	5	1	5	4	8	10
c	11	4	7	7	15	23

11.2 The Spatial Complexity of 3 × 3 Binary Maps

"…the origin of numbers from 0 and 1, which I have observed is the most beautiful symbol of the continuous creation of things from nothing, and of their dependence on God"

(Gottfried Leibniz, 1646–1716, written on 29-3-1698.

Source:Leibniz,G. 1863, "Mathematische Schriften"

ed.C.I.Gerhardt, volume V, Berlin: A.Ascher. page 239)

Let us now return to the 3 × 3 binary maps. Of the 255 total binary 3 × 3 map configurations up to maximum entropy class, some are topologically equivalent therefore yielding the same spatial complexity (because symmetric configurations produce topologically equivalent positions of cells). Calculations of C_{PI} and C_{P2} on isomorphic binary maps yield the same results. Thus, non-isomorphic binary maps can be identified that will be called here "*generic*" binary maps. These are specific for each entropy class. For $r = 1$, 2 and 3 these are shown in Fig. 11.2 and for $r = 4$ are shown in Fig. 11.3.

With $r = 6$, the 3 × 3 binary maps yield the same results in calculations as the binary maps with $r = 3$. Equivalently, for $r = 7$ the results are the same as in the case of $r = 2$. For this reason, the calculations are carried out only up to $r = (n\text{-}1)/2$, that is up to $r = 4$. Consequently, C_{PI} and C_{P2} are calculated for all 3 × 3 generic maps, up to maximum entropy class, as shown in Table 11.2.

The average C_{PI} per entropy class r (for all entropy classes) can be described by the following cubic formula (Fig. 11.4):

$$C_{P1} = 2.03129 + 2.31226r - 0.392857r^2 + 0.0341667r^3 \quad \text{(corr.coeff.} = 0.996). \tag{11.2}$$

Notice that the rate of increase of spatial complexity *decreases with increasing entropy class.*

The number of map configurations per C_{PI} complexity value (Fig. 11.5) can be modeled by a polynomial:

$$N_{C_{P1}} = 512.143 - 434.091C_{P1} + 104.857C_{P1}^2 - 6.19444C_{P1}^3 \quad \text{(corr.coeff.} = 0.969) \tag{11.3}$$

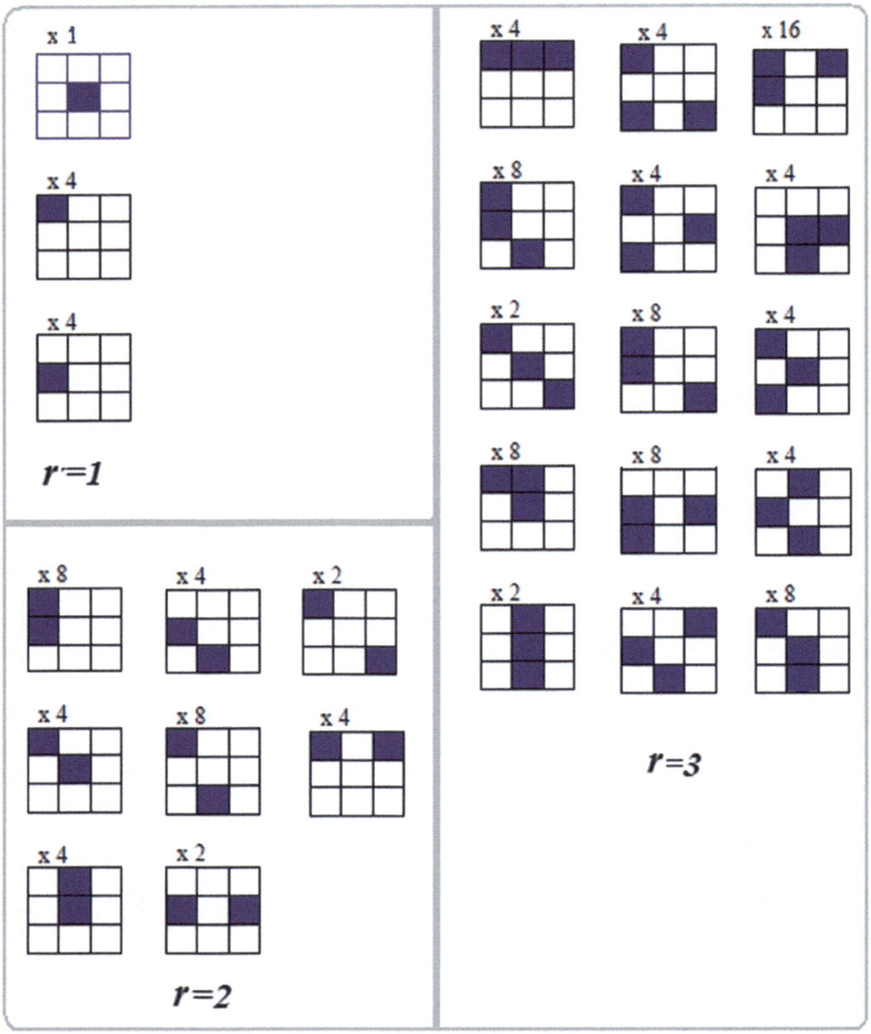

Fig. 11.2 Generic 3 × 3 binary maps for entropy classes $r = 1,2$ and 3 black cells per binary map. Numbers atop of each binary map show the multiplicity of each generic map

This function shows that it is more likely to encounter average and high C_{PI} values in 3 × 3 binary maps than lower complexity values.

The average C_{P2} per entropy class r is given by the following model (Fig. 11.6):

$$C_{P2} = 0.0192857 + 2.5556r + 0.122143r^2 - 0.104167r^3$$
$$\text{(with correlation coefficient} = 0.9989), \quad (11.4)$$

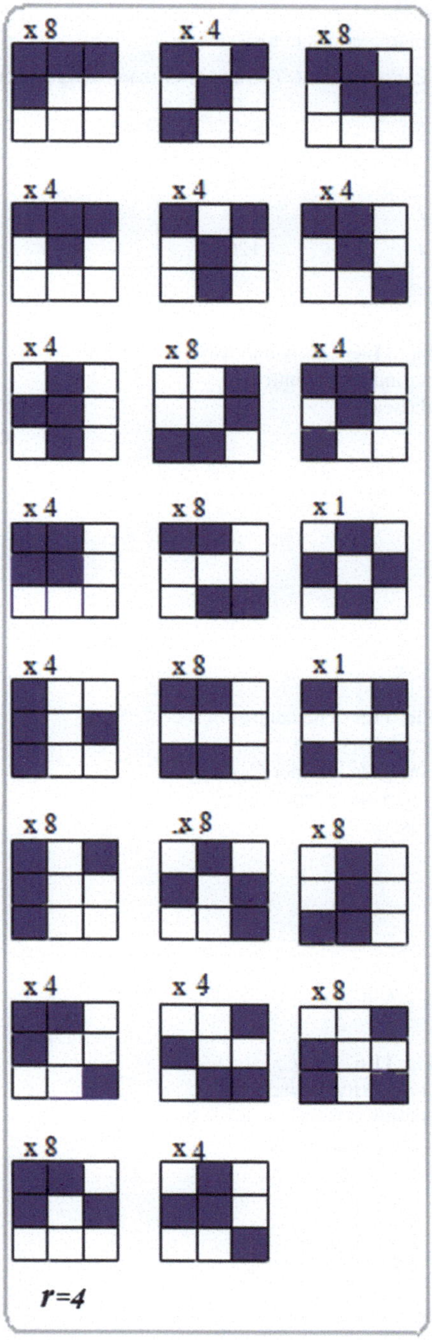

Fig. 11.3 Generic 3 × 3 binary maps of all the map configurations for entropy classes with $r = 4$ black cells per binary map withmultiplicity of each generic configuration

Table 11.2 There are 49 generic binary maps of size 3 × 3 coresponding to 255 possible configurations in total

Entropy class (r)	Number of map configurations (N) per entropy class r	Of which generic maps
1	9	3
2	36	8
3	84	15
4	126	23
total	255	49

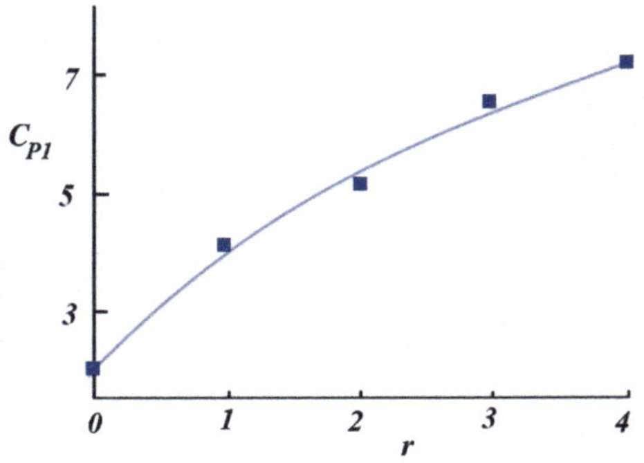

Fig. 11.4 C_{P1} complexity per individual entropy classes r_i

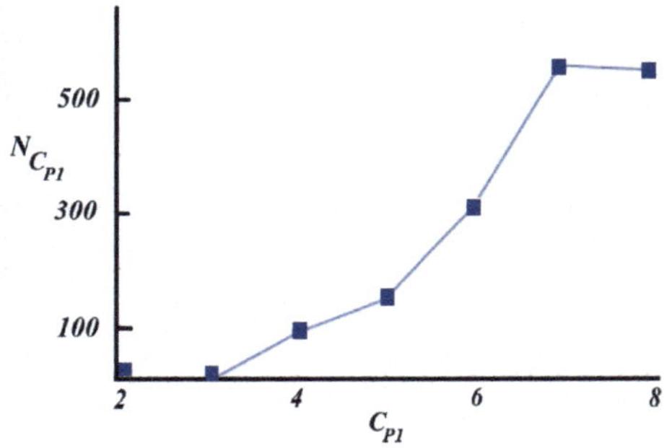

Fig. 11.5 The number of 3 × 3 binary map configurations per C_{P1} value (up to maximum entropy class)

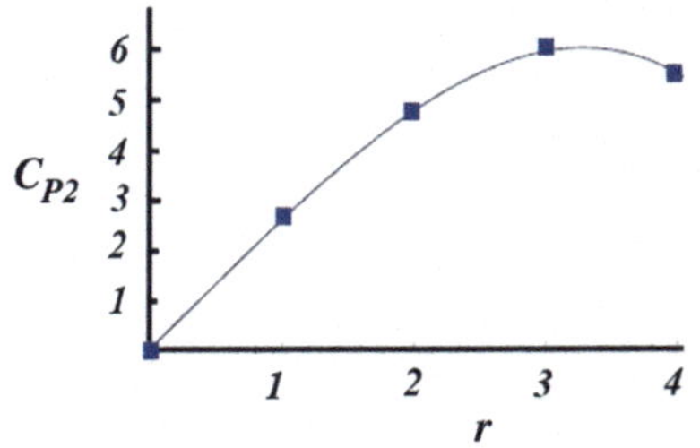

Fig. 11.6 The C_{P2} complexity per individual entropy classes r_i is given by a cubic relationship

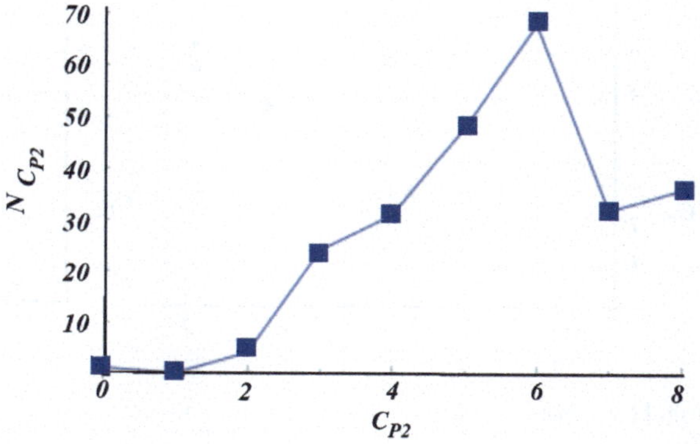

Fig. 11.7 The number of 3 × 3 binary map configurations per C_{P2} value (up to maximum entropy class)

while an approximation to the number of 3 × 3 binary maps $N_{CP2}=f(C_{P2})$ per C_{P2} value (Fig. 11.7) can be described by the model (corr.coeff. 0.86):

$$N_{C_{P2}} = 0.585859 - 7.31866C_{P2} + 6.64358C_{P2}^2 - 0.658249C_{P2}^3 \quad (11.5)$$

11.3 Patchiness, Adjacency and Clumpiness

> "Nature is Harlequin's cloak made entirely of solid patches and empty spaces; she is made of plenitude of void, beings and nonbeings, with one of the two posing itself as unlimited while limiting the other"
>
> (Gilles Deleuze, 2012, p. 304)

The relationships of C_{P1} with patchiness P and clumpiness B are shown in Fig. 11.8 and the selected formulas best fitting the values of C_{P1} for both P are (corr.coeff. 0.996):

$$C_{P1} = 2.16082e^{0.558367P} \quad (11.6)$$

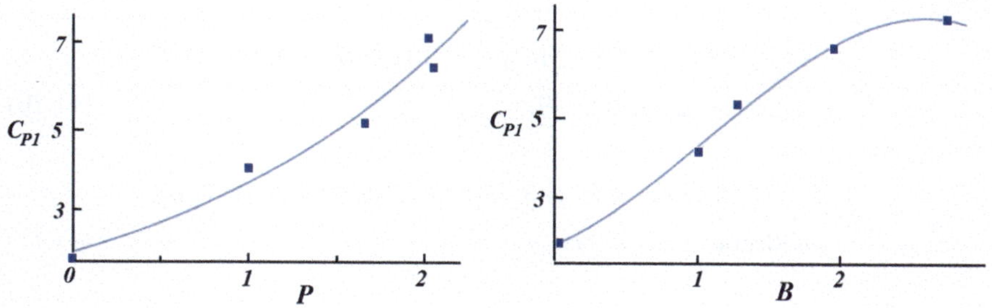

Fig. 11.8 C_{P1}—complexity with respect to patchiness P and clumpiness B in 3 × 3 binary maps

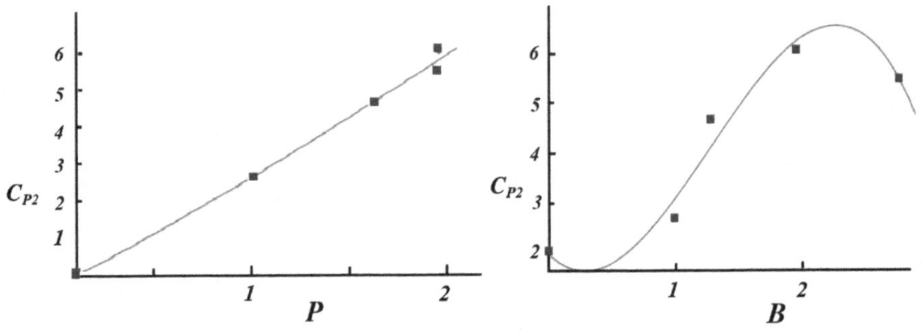

Fig. 11.9 Spatial complexity C_{P2} with respect to patchiness (*P*) and clumpiness (*B*) in 3 × 3 binary maps

and for *B*:

$$C_{P1} = 1.99101 + 1.4249B + 1.20611B^2 - 0.37946B^3$$
$$\text{(with corr.coeff.0.997)} \quad (11.7)$$

Thus high clumpiness and high patchiness are both associated with high C_{PI} complexity.

The relationships best fitting the values of C_{P2} with *P* and *B* classes respectively are both polynomial (Fig. 11.9):

$$C_{P2} = -0.00053822 + 2.50646P + 0.175495P^2 - 0.00625255P^3$$
$$\text{(corr.coeff. 0.995)} \quad (11.8)$$

and

$$C_{P2} = 1.97166 - 2.55623P + 4.97499P^2 - 1.30259P^3$$
$$\text{(corr.coeff. 0.959)} \quad (11.9)$$

Further, the values of can be best described by polynomial models also (Fig. 11.10):

$$P = 0.00247143 + 1.12461r - 0.131214r^2 - 0.00575r^3$$
$$\text{with corr.coeff.0.9999,} \quad (11.10)$$

$$B = 0.0197429 + 1.25588r - 0.425429r^2 + 0.0708333r^3$$
$$\text{with corr.coeff.0.994} \quad (11.11)$$

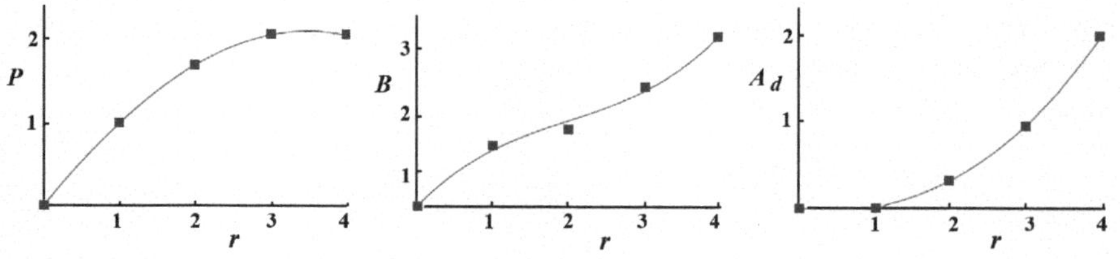

Fig. 11.10 Patchiness (P), clumpiness (B) and adjacency (A_d) with respect to entropy class (r) in 3×3 binary maps

and

$$A_d = -0.00282857 - 0.114476r + 0.120286r^2 - 0.00833333r^3$$
$$\text{with corr.coeff. } 0.9998 \qquad (11.12)$$

It consequently becomes apparent that patchiness, as well as clumpiness and adjacency, all increase with increasing entropy class, although patchiness shows a somewhat different behavior for the higher values of r: the maximum P ($P = 2.08474$) is at $r = 3.486$, that is at an entropy class lower than the maximum.

11.4 Generic 3 x 3 Binary Maps and Their Multiplicities

"You feel the hidden calculation... an elemental maze, unfathomable forest"

(Osip Mandelstam, 1891–1938, "Notre Dame", 1912)

The percentage of generic maps with respect to the total number of possible map configurations *decreases* with increasing map size. For instance, in 2×2 binary maps, there are three generic configurations out of the ten possible configurations (hence 33% are generic maps) and in 3×3 binary maps there are 49 generic maps out of the 255 configurations (that is 19.2%). Furthermore, the number of generic maps N_g by entropy class *decreases* as the entropy class grows. A simulation model can be obtained by calculating the total number of generic map configurations per multiplicity ξ for all classes r (ξ_r), which is shown in Table 11.3 and defined as the sum:

$$N_g = \sum_{r=1}^{r=4} \xi_r \qquad (11.13)$$

The relationship between N_g and ξ can be simulated by a polynomial function, of the form (Fig. 11.11):

Table 11.3 Numbers (N_g) of the 49 generic 3 × 3 binary maps per entropy class (r), with their respective multiplicity factors (ξ_r)

Multiplicity (ξ)	$r = 1$	$r = 2$	$r = 3$	$r = 4$	Ng
1	1	0	0	2	3
2	0	2	2	2	6
4	2	4	7	8	21
8	0	2	5	11	18
16	0	0	1	0	1

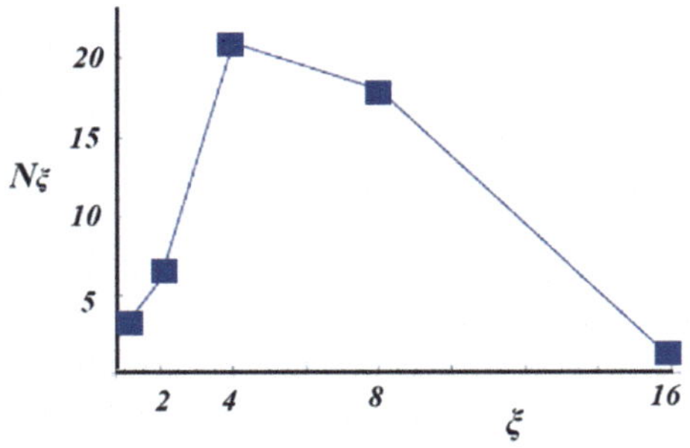

Fig. 11.11 The number of generic 3 × 3 binary maps (N_g) per multiplicity class (ξ) of generic maps, for all entropy classes *r*, can be described by a polynomial function

$$N_g = -9.7381 + 12.3443\xi - 1.47621\xi^2 - 0.0466582\xi^3$$
$$\text{(correlation coefficient 0.942)}. \quad (11.14)$$

Thus, the number of multiplicity of binary map configurations of 3 × 3 binary maps increases *polymomially* with the number of generic maps.

The generic maps are endowed with different topologies and so they can produce different symmetries. Spatial complexity can therefore be "tamed" by identifying topologically inequivalent configurations. This is evident, even at the simplest level of 3 × 3 maps with $r = 1$, where three different maps, all with the same entropy, but with different topologies, yield different spatial complexities (Fig. 11.12). If the exact generic forms are known, then a polynomial relationship (instead of an exponential model of allocations which would be expected from the formula $2^{n-1} - 1 = 255$) suffices to produce all the 255 − 49 = 206 non-generic map configurations. This means that identifying generic forms is equivalent to focusing on a number of possible configurations at a scale of reduction which is important: a reduction from exponential to polynomial growth. And this reduction is only possible thanks to the identification of symmetric map configurations that can be produced by the generic maps.

Yet, spatial complexity depends on border effects and this dependence is more significant for smaller maps. The boundaries between cells can range from 2 to 3 or 4, depending on the location of the cell in the map (cells located at the border areas

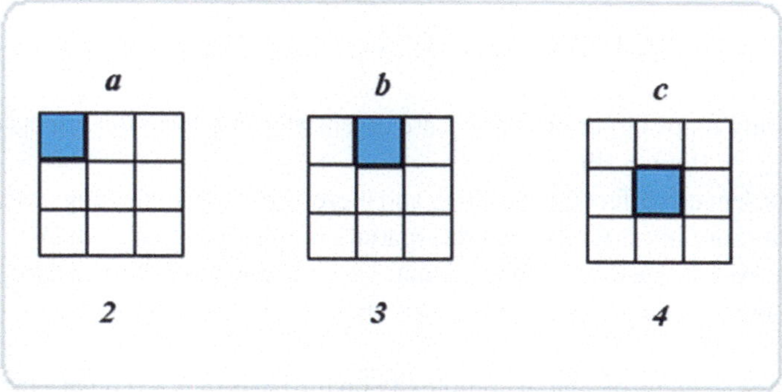

Fig. 11.12 Generic configurations of three simple 3 × 3 binary maps. In the case of map *a*, the colored cell has only 2 neighboring cells (one east, one south), in the case of *b* it has 3 (east, west, south) and in *c* it has 4 (all four directions)

of a square map have 2 or 3 boundaries, while those located at the map's central areas have four boundaries).

With the number of cells of a square map with two, three or four boundaries (Fig. 11.13) symbolized respectively as ∂_2, ∂_3, ∂_4, the border cells ∂_2, ∂_3 and the central cells assume the following values (where n is the total number of cells of the square map):

$$\begin{aligned} \partial_2 &= 4 \\ \partial_3 &= 4\sqrt{n} - 8 \\ \partial_4 &= (\sqrt{n} - 2)^2 \end{aligned} \tag{11.15}$$

Consequently, it is easy to verify that the larger the map size, the more insignificant the number of outer border cells becomes in comparison to the inner cells:

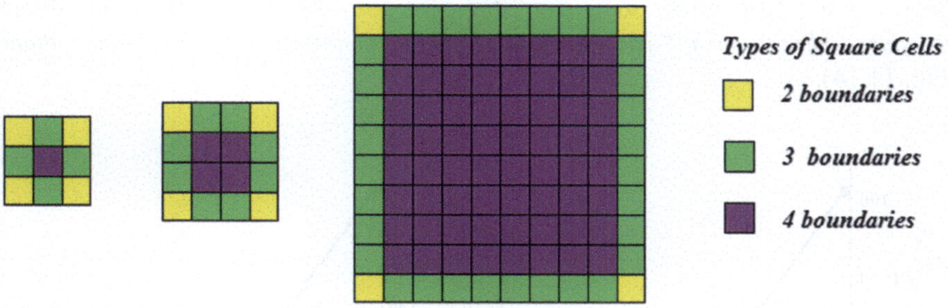

Fig. 11.13 There are three topological types of square cells according to the number of boundaries they share with their neighbouring cells. As the map size n increases, the number of 4-boundary cells increases at the expense of the other types of cells

$$\lim_{n\to\infty}\left(\frac{\partial_2+\partial_3}{\partial_4}\right)=\lim_{n\to\infty}\left(\frac{4\sqrt{n}-4}{n+4-4\sqrt{n}}\right)=0 \tag{11.16}$$

In fact, only maps larger than 7 × 7 are essentially free from the effects of border cells: $\partial_4 - \partial_3 > 0\ \forall n > 49$.

Thus, border cells affect the topology and therefore the spatial complexity of small maps, much more importantly than the spatial complexity of large maps.

It is however interesting to notice that 3 × 3 are the smallest square maps for which any other cells are more numerous than the four corner cells: $\partial_3 - \partial_2 > 0$ $\forall n > 9$

Hence, assessments of spatial complexity on the basis of 3 × 3 binary maps can be made, although, for safer estimates, the analysis should be made on square domains at least 7 × 7 large.

Various ecological observations have confirmed correlations between increased "diversity" and "habitat complexity"; a notion which entails spatial complexity and functional complexity (Dean and Connell 1987; Poggio et al. 2010; Keith et al. 2006). Besides ecology, complexity correlates with "diversity" in various fields: in materials science (Gleitzer 1980), neurosciences (Blaustein and Golovina 2001), biochemistry (Okazaki et al. 1998) etc. Interestingly, from an ecological study in Switzerland (Lischke 2005), empirical observations showed that spatial complexity was high when the *first* species colonized the region, when, at the fronts of the *Picea abies* immigrations, "spots of increased diversity appeared" (Lischke 2005, p. 159). Thus, "invading species" in an ecosystem (or, equivalently, "invading" black cells in a binary map), regardless of their particular location in the map or how few they may be, immediately create a substantial difference in the map's spatial complexity. As shown from the present study, *even a slight introduction of "alterity" or "otherness" (that is dark cells) in a binary map induces a high increase in spatial complexity.* This might be anticipated, although it could not have been verified without a complete complexity characterization of 3 × 3 binary maps. In fact, in both C_{P1} and C_{P2}, *the highest rate of increase in spatial complexity takes place in between the entropy classes 0 and 1*, therefore suggesting that *some entropy classes (the lower ones) are endowed with more "dynamism" to "produce" spatial complexity than others* (Fig. 11.14).

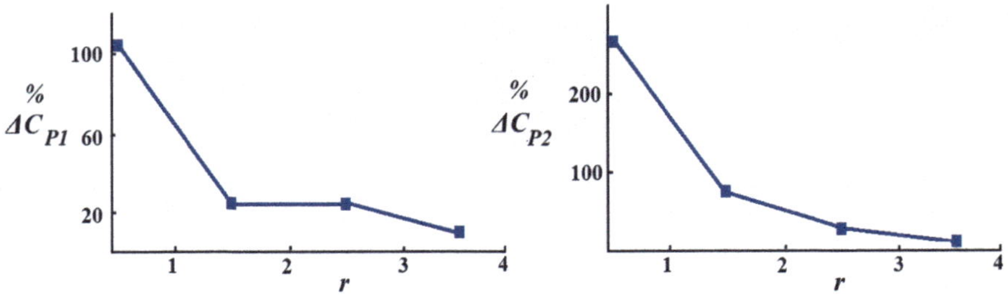

Fig. 11.14 The rate of increase in spatial complexity increases rapidly for both C_{P1} and C_{P2} as the entropy class rises from 0 to 1 and then decreases with higher entropy class

Yet, although the maximum average C_{P1} is at $r = 4$, the maximum of the average C_{P2} is attained (approximately) at the entropy class $r = 3$ (not at $r = 4$). Thus, the change of C_{P2} with entropy indicates that *maximum spatial complexity may occur in between minimum and average entropy.* The range of values that the metric C_{P2} yields for 3 × 3 binary maps conforms with the anticipated variations of complexity in "between order and chaos", or "between order and randomness", as complexity (not *spatial* complexity) is thought to be by many researchers, although with a somewhat different expression here: *spatial complexity maximizes in between minimum and maximum entropy*. There is a strong cross-disciplinary need for methods to calculate and compare the spatial complexities of two-dimensional images, maps, pictures, landscapes and other 2d representations and surfaces. This aim can not be satisfied without having a measure to compare spatial complexity with. For this reason it was necessary to calculate all the expected possible spatial complexity values for all possible binary spatial configurations. In the present case, this requirement was met at the level of 3 × 3 maps. Evidently, if larger than 3 × 3 binary maps were used, the results might differ. Although this may seem obvious, it has to be considered that we are still short of a formula that would enable us to estimate the spatial complexity of all the i.e. 36,493 configurations that would be required to calculate for the 4 × 4 binary maps (let alone for larger maps).

11.5 Spatial Analysis at "King's Neighborhood"

The Pythagoreans decorated with meanings the numbers and the shapes of Gods
"Οἱ δὲ Πυθαγόρειοι καὶ ἀριθμοὺς
καὶ σχήματα θεῶν ἐκόσμησαν προσηγορίαις"
(Plutarch, 46-119 a.D., "De Iside et Osiride", 381f)

Square arrangements keep puzzling scientists and artists for their geometric simplicity, which, in the digital era, carry some of the most complex information structures, such as big data of satellite imagery, GIS maps, and all forms of large geospatial databases, which are some of the most complex kind of observational data ever collected by humans.

More than twenty centuries ago, Pythagoras discovered the immense power of squares, by revealing that three squares can always be imagined to correspond to each side of a triangle. Seen from this perspective, he essentially proved that the simplest of all spatial forms, the triangle, is intimately related to squares. The Pythagoreans claimed that the number 3 is the most important of all numbers and attributed to the triad a central role in the cosmos, because it summarizes the triplet beginning-middle-end. In praise of the number nine, the neoplatonic philosopher Plotin wrote his famous "Enneads" and, centuries earlier, according to Hesiod's "Theogony" there were 9 days and nights separating the sky from the earth. The Pythagoreans were particularly inclined to study figurative numbers.Take, for instance, the *tetraktys* (τετρακτύς). It would be difficult not to observe that the sacred symbol of the

Pythagoreans conceals a small (3 × 3 may be?) map (Fig. 11.15). Tetraktys formed the basis of the Pythagorean oath, but quaternities were not uncommon in philosophy and theology throughout the ages, with most widely known among them the "quadrivium" in Neoplatonic epistemology: a sign converting point data to area is an important (inherent?) process of human perception.

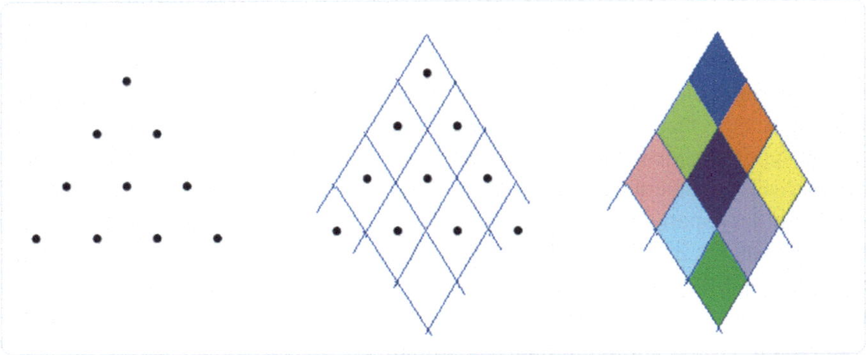

Fig. 11.15 Tetraktys (left) formed the basis of the Pythagorean oath. The ten points might as well be perceived as defining a 3 × 3 map (right)

Simple 3 × 3 maps have been used by humans for planning the geographical space since the antiquity, while archaeological evidence suggests that the 3 × 3 game tic-tac-toe was played in many places all over the world. Probably invented in ancient Egypt, we now know that, as deceivingly "simple" as it looks, it can nevertheless be played in 9! = 362,880 different ways.

Two millennia ago, well-fields in ancient China applied a management scheme called "jing-tian-zhi" (well-land-system), consisting in 3 × 3 square fields with a communal area in the middle, surrounded by eight ownerships. Delving into some classic old chinese texts can be rewarding in ideas about the central role of the square number 9 in spatial analysis (3 × 3 squares). A legendary planner and geometer of ancient China, "Yu the Great", measured nine mountains, nine rivers, nine marshes and arranged the lands to be cultivated within such areas. The "Yugong" (the book of the tribute to Yu) describes how Yu marked out the nine provinces of China. Characteristically, the classic Chinese text "The Tribute of Yu" (禹贡) reads as follows in the beginning: "Yu marked the nine provinces. Then, the hills, increased the rivers' depth; defined the borders of the land". The master plan for spatial divisions, the "Hong Fan" (洪 範) allegedly described the division in nine regions ("ch'eou"), and was brought to Yu by a turtle. The 3 × 3 arrangement had the central cell at the center and eight surrounding cells defining the 8 trigrams of the "Book of Changes", the "I-Ching", nine defined the shape of the royal residence (the "Ming T'ang"), and, as a matter of fact, was also the defining number for royal ceremonies in China. The

Taoist ceremonies also have a base of nine, as those of the Mayas and the Aztecs. And, for Christians, it is the simplest square map that can host the shape of a cross.

As it turns out, 3×3 neighbourhoods are fundamental for the arts also, given the old and well known method of aesthetic appreciation that is called "*rule of thirds*". This practical rule was first named so by John Thomas Smith in his book "Remarks on Rural Scenery", quoting the work of Sir Joshua Reynolds (1783), discussing the balance of dark and light in a painting. Although not a mathematically proven rule, it is nevertheless a "rule of thumb" for evaluating and appreciating the emotional and artistic power emitted to the viewer by paintings, images, photographs, sceneries. The rule suggests the subdivision of a picture on the basis of a 3×3 grid, in order to determine the location in the grid so as to highlight what is more important in the entire scenery. The rule of thirds suggests that placing the picture's most important element along the thirds lines (or close to their intersections) produces always a higher aesthetic impact than if it were placed anywhere else in the picture. Empirical observations also show that a much more powerful impression impact is conveyed to a spectator if the prominent figure of the image (the form of a human, a tree, a house etc.) is *not* located at the central square of the 3×3 grid, but in the area of the "thirds" instead (Fig. 11.16).

Given these, it is expected that 3×3 maps would have not escaped the attention of visual artists. Kazimir Malevich painted (1915) his "Black Cross": a simple black cross formed by joining the five central cells of a 3×3 black-and-white map. Similarly, Ad Reinhardt (1913–1967) painted uniformly black pictures, i.e. his "Abstract Painting" (1963) is essentially a grid of 3×3 black squares. Sol Le Witt painted (1967) his "Serial Plan", a work of conceptual art created by using 3×3 squares. Clearly, these are only few representative examples among countless other ones in art, but what is more important is that 3×3 neighborhoods are a key to unlock the secrets of spatial complexity of small square binary maps.

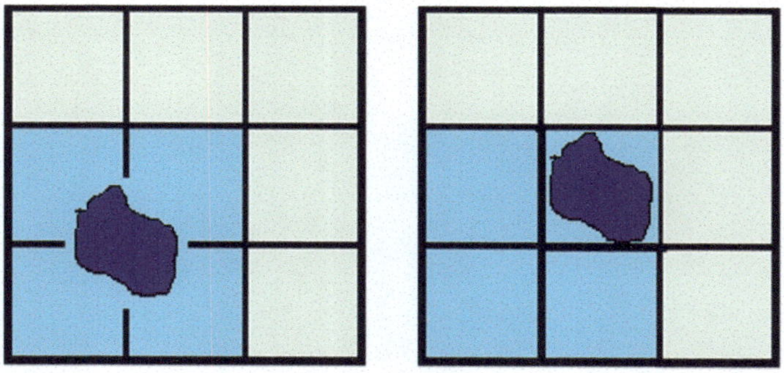

Fig. 11.16 The "rule of thirds". Locating the key feature of a painting on the intersection of the lines of thirds lines on the lower left part of the image creates a higher visual impact (that is makes the image more "interesting") than if it were located at the centre of the image

References

Blaustein, M. P., & Golovina, V. A. (2001). Structural complexity and functional diversity of endoplasmic reticulum Ca^{2+} stores. *Trends in Neurosciences, 24*(10), 602–608.

Bormashenko, E., Fedorets, A. A., Frenkel, M., Dombrovsky, L. A., & Nosonovsky, M. (2020). Clustering and self-organization in small-scale natural and artificial systems. *Philosophical Transactions of the Royal Society A, 378,* 20190443.

Dean, R. L., Connell, J. H. (1987). Marine invertebrates in an algal succession. III. Mechanisms linking habitat complexity with diversity. *Journal of Experimental Marine Biology and Ecology, 109* (3–18), 249–273.

Deleuze, G. (2012). *The logic of sense*. London: Continuum.

Gleitzer, C. (1980). Diversity and complexity of the wustite solid solutions I-tentative rationalization of the miscibility data and classification of the wustite ternary fields and of the postsaturation reactions. *Materials Research Bulletin, 15*(4), 507–519.

Keith, A. M., van der Wal, R., Brooker, R. W., Osler, G. H. R., Chapman, S. J., & Burslem, D. F. R. P. (2006). Birch invasion of heather moorland increases nematode diversity and trophic complexity. *Soil Biology & Biochemistry, 38*(12), 3421–3430.

Lischke, H. (2005), Modeling tree species migration in the Alps during the Holocene: What creates complexity?. *Ecological Complexity, 2*(2), 159–174.

Okazaki, T., Kondo, T., Kitano, T., & Tashima, M. (1998). Diversity and complexity of ceramide signalling in apoptosis. *Cellular Signalling, 10*(10), 685–692.

Papadimitriou, F. (2002). Modelling indicators and indices of landscape complexity: An approach using GIS. *Ecological Indicators, 2,* 17–25.

Papadimitriou, F. (2009). Modelling spatial landscape complexity using the levenshtein algorithm. *Ecological Informatics, 4*(1), 51–58.

Papadimitriou, F. (2012). The algorithmic complexity of landscapes. *Landscape Research, 37*(5), 599–611.

Poggio, S. L., Chaneton, E. J., & Ghersa, C. M. (2010). Landscape complexity differentially affects alpha, beta, and gamma diversities of plants occurring in fencerows and crop fields. *Biological Conservation, 143*(11), 2477–2486.

Chapter 12
Complexity of Binary Maps of Primes and Transcendentals

Number became the first principle and this is indefinite and incomprehensible.
The number has in itself all the infinite possible numbers that may come.
And of the numbers the first entity was the unity, which fatherly generated all the other numbers
"Ἀριθμός γέγονε πρῶτος ἀρχή, ὅπερ ἐστίν ἀόριστον ἀκατάληπτον,
ἔχων ἑαυτῶ πάντας τούς ἐπ' ἄπειρον δυναμένους ἐλθεῖν ἀριθμούς κατά τὸ πλῆθος.
Τῶν δὲ ἀριθμῶν ἀρχή γέγονε καθ'υπόστασιν η πρώτη μονάς,
ήτις ἐστί μονάς ἄρσην γεννῶσα πατρικῶς πάντας τούς ἄλλους ἀριθμούς"
(Pythagoras, 580-496 bC)

Abstract Different map sizes and different approximations of π produce interestingly different binary map representations with substantial differences in their spatial complexity. Sometimes, knowing the numbers behind the spatial structures not only explains a map's structure, but can also be used to predict how a map might look like if it extended in space (such is the case of transcendental numbers). Assigning black cells to prime numbers produces square binary maps of primes-and-composites from which it can be seen that: (a) even-numbered binary such maps "produce" clumps, while odd-numbered ones do not; (b) C_{P2} decreases with increasing map size n in binary maps of primes-composites; (c) both the author's C_{P1} and C_{P2} complexity metrics are always higher in odd-numbered binary maps than in even-numbered maps of primes-and-composites.

Keywords Spatial complexity · Prime numbers · Binary maps · Map complexity · Transcendental numbers and Complexity · pi · Number theory and Complexity

F. Papadimitriou, *Spatial Complexity*, https://doi.org/10.1007/978-3-030-59671-2_12

12.1 Numbers Defining Spaces

"So Nature deals with us, and takes away our playthings one by one, and by the hand leads us to rest so gently, that we go scarce knowing if we wish to go or stay, being too full of sleep to understand how far the unknown transcends the what we know"

(Henry Wadswarth Longfellow, 1807–1882, "Nature")

The title of this section may at first appear surprising, particularly considering that only very limited research has been hitherto carried out in this particular field. The truth is that besides geometry and topology, another branch of mathematics, number theory, most often passes unnoticed in the study of spatial arrangements. The repercussions of this for spatial scientists may be too early to anticipate, but it is worth noticing them.

Some number-related aspects of spatial complexity emerge from simulations and experiments with binary maps. They can be interesting not only in terms of mathematics, but also for future research, as some of them may prove valuable for gaining some deeper insights in spatial complexity. Such number-theoretic considerations may relate to various kinds of numbers (integers, reals, transcendentals, complex etc.). The emergence of π in "Buffon's needle problem" for instance (a spatial probability problem which is examined in another chapter) leads us to question whether this is a unique case of number-theoretical interest emerging from a problem of spatial probability, or it might as well imply that number theory is essential to understand spatial complexity. Before opting for the first case, we should rather consider a few more facts, since, surprisingly, despite the fact that some problems of spatial combinatorics have revealed the presence of π in their solutions, this seems to have passed more or less unnoticed.

Counting the number of ways a square can be covered by domino tilings can be illustrating: according to Matousek (2010, p. 85) there are 12,988,816 possible tilings of the 8 × 8 chessboard by 2 × 1 rectangle dominoes (Fig. 12.1), while the formula giving the number of domino tilings of an *mxn* chessboard involves trigonometric functions and yet, yields integers as results (Matusek 2010, p.93):

$$\sqrt{\prod_{k=1}^{m}\prod_{l=1}^{n}\left(2\cos\frac{k\pi}{m+1}+2i\cos\frac{\pi l}{n+1}\right)} \tag{12.1}$$

But this is not the only case that π shows up unexpectedly in spatial analysis. Consider, for instance this problem: Choose one cell of a square map, with an arithmetic regularity, with ever decreasing size. Beginning with an 1 × 1 square map (that is the entire area) and proceeding to a 2 × 2 square map (thus ¼ of the map's area has been chosen), choosing one cell again from a 3 × 3 square map (meaning that 1/9 of this map's area has been selected) and carrying on the same way down to $(1/n) \times (1/n)$ (Fig. 12.2), the total area accumulated from all these choices is tantamount to calculating the following series:

Fig. 12.1 One of the 12,988,816 possible tilings of the 8 × 8 chessboard that can be made by 2 × 1 rectangle dominoes

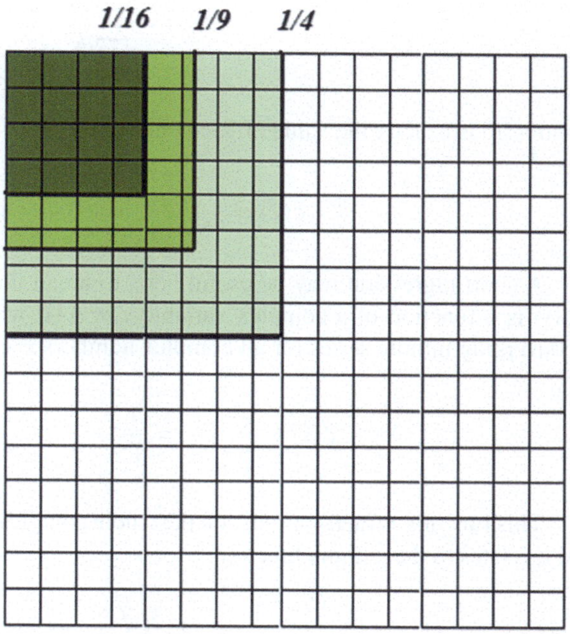

Fig. 12.2 A hint about the possible role of number theory in spatial complexity. Calculating the total area of the sum $\frac{1}{2^2} + \frac{1}{3^2} + \frac{1}{4^2} + \ldots + \frac{1}{n^2}$ leads to the rather unexpected result $\pi^2/6$, or, otherwise stated, deriving a solution to a spatial problem involving squares leads to a transcendental number which is not usually associated to square shapes

$$\frac{1}{2^2} + \frac{1}{3^2} + \frac{1}{4^2} + \ldots + \frac{1}{n^2} \tag{12.2}$$

The result of this calculation may only be obtained through a "magic" trick due to Euler, and is:

$$\frac{1}{1^2} + \frac{1}{2^2} + \frac{1}{3^2} + \frac{1}{4^2} + \ldots + \frac{1}{n^2} = \frac{\pi^2}{6} \cong 1.6449340668\ldots \tag{12.3}$$

What is far more interesting to notice here, is that $\pi^2/6$ is a transcendental number. It does seem noteworthy that the solution of a spatial problem *involving squares only* (not circles or other shapes), can *only* be expressed in terms of a transcendental number that is usually associated to a different shape.

But behind the solution to this simple spatial problem might possibly lie some other interesting relations (in terms of number theory) as well, as, for instance, $\pi^2/6$ relates to Riemann zeta function:

$$\zeta(s) = \sum_{n=1}^{\infty} \frac{1}{n^s} = \frac{1}{1^s} + \frac{1}{2^s} + \frac{1}{3^s} + \ldots \tag{12.4}$$

that is one of the most intensely studied functions.

As well known, the value of the Riemann ζ function of a positive integer n is defined as:

$$\zeta(n) = \sum_{k=1}^{\infty} \frac{1}{k^n} \tag{12.5}$$

and $\pi^2/6$ is exactly the value of Riemann zeta for the integer 2:

$$\zeta(2) = \frac{\pi^2}{6} \cong 1.6449340668\ldots \tag{12.6}$$

A short digression may be useful here, to recall that the Riemann zeta function $\zeta(s)$ is a function of a complex variable $s = \sigma + it$ which can also be written as a converging infinite series for all complex numbers s with real part greater than 1:

$$\zeta(s) = \sum_{n=1}^{\infty} \frac{1}{n^s} = \frac{1}{1^s} + \frac{1}{2^s} + \frac{1}{3^s} + \ldots \tag{12.7}$$

This function is meromorphic on the whole complex s-plane, but for real numbers x, it relates to the gamma function:

$$\zeta(x) = \left(\frac{1}{\Gamma(x)}\right)\left(\int_0^{\infty} \frac{u^{x-1}}{e^u - 1} du\right). \tag{12.8}$$

Two characteristic values are $\zeta(0) = -1/2$ and $\zeta(1) = \infty$, with some of its values relating to other known numbers and constants: the harmonic numbers, the Euler-Mascheroni constant γ, and, most surprisingly, to physical quantities, such as the critical temperature of the Einstein-Bose condensates $\zeta(3/2) = 2.612$, the integration of Planck's law to derive the Stefan-Boltzmann law in Physics: $\zeta(4) = \pi^4/90 = 1.0823\ldots$ among several other relations. Given these, it makes sense to question whether any more relations between different types of numbers (transcendental, primes, reals, complex etc.) and spatial complexity might exist. With partial series we have that up to a given n, the total area of squares is:

$$1^2 + 2^2 + 3^2 + 4^2 + \ldots + n^2 = \frac{n(n+1)(2n+1)}{6} \tag{12.9}$$

but the first term of this equation is equal to the $\zeta(-2)$ of the Riemann zeta function, and hence, the sum of the series

$$\zeta(-2) = 1^2 + 2^2 + 3^2 + 4^2 + \ldots + n^2 \tag{12.10}$$

also results from calculations that entail trigonometric functions again:

$$\zeta(-2) = -\frac{1}{56}\int_0^\infty \frac{(1+t^2)\cos\left(-2\tan^{-1}(t)\right)}{\cosh(\pi t/2)}dt \tag{12.11}$$

or even e:

$$\zeta(-2) = \frac{1}{6} + 2\int_0^\infty \frac{(1+t^2)\sin\left(-2\tan^{-1}(t)\right)}{e^{2\pi t} - 1}dt \tag{12.12}$$

(where $\tan^{-1}$ is the inverse tangent function and cosh is the hyperbolic cosine function). Equivalently, the sum of increasingly larger volumes up to infinity is $\zeta(-3) = 0.0083333\ldots$

12.2 Complexity of Binary Square Maps of Primes and Composites

> Good definitely scattered among the figures of evil. Anamorphosis of good. Evil definitely scattered among the figures of good. Anamorphosis of evil
>
> (Jean Baudrillard 2005, p.142)

The key question tackled here is: "What is the spatial complexity of binary maps if prime numbers are represented as colored cells on square maps?"

Binary maps resulting from the allocation of primes on $n =$ odd-numbered square maps display diagonal alignments, so cells sharing a common side are uncommon (Fig. 12.3).

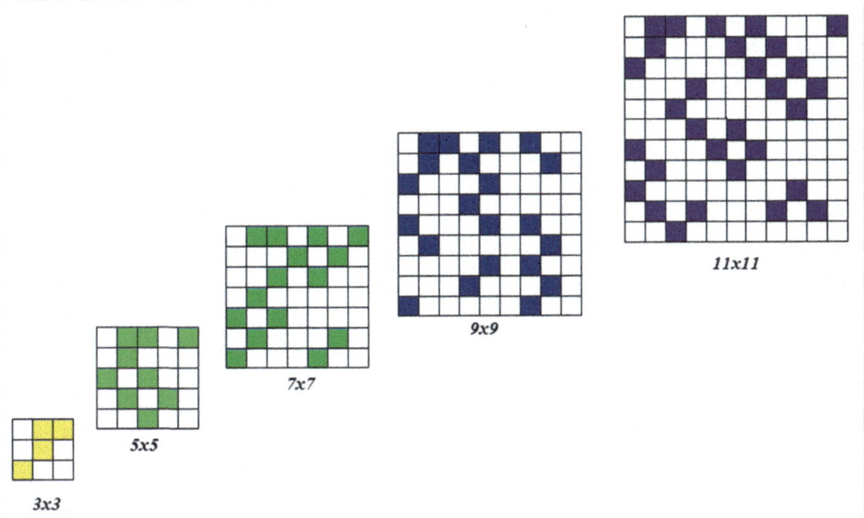

Fig. 12.3 Binary maps resulting from n=odd-numbered maps, with prime numbers allocated as darker cells on progressively larger maps. Notice the diagonal patterns emerging from these maps and the relative absence of clumps

Notice that the maps with n = odd have only one clump, even as their size increases from $n = 9$ to $n = 121$. On the contrary, when the primes are allocated on even-numbered maps, they produce more clumps with size of at least two cells (Fig. 12.4).

As a result of the fact that even-numbered maps produce vertical arrangements of black cells, while odd-numbered maps display alignments of black cells along diagonal directions, it is expected that odd-shaped maps would create higher spatial complexity, since they generate more dissimilar contacts between cells.

This can be verified from calculations of C_{P2} for maps 3×3 up to 10×10, in that the formula relating C_{P2} and map size n grows with a trigonometric pattern, in which *odd-numbered maps have higher complexity than even-numbered ones* (Fig. 12.5):

$$C_{P2} = 0.785n + \tan(0.04828n) - \tan(\tan(0.04459n)) \tag{12.13}$$

with correlation coefficient = 0.99973885

As may easily be verified, *even-numbered maps "produce" clumps while odd-numbered ones do not*. Indeed, the clump number B remains constant (equal to 1) in 3×3, 5×5, 7×7 etc. maps, but increases with map size in even-numbered maps. It can be verified that odd-numbered maps "keep" the number of clumps B to a minimum (counting as a clump a conglomerate of two cells or more, either horizontally or vertically), so a question raises as to the relationship of B with the number of primes p (corresponding to the entropy of the map) in n = even numbered maps. For square binary maps 4×4 up to 16×16 with n = even, a trigonometric

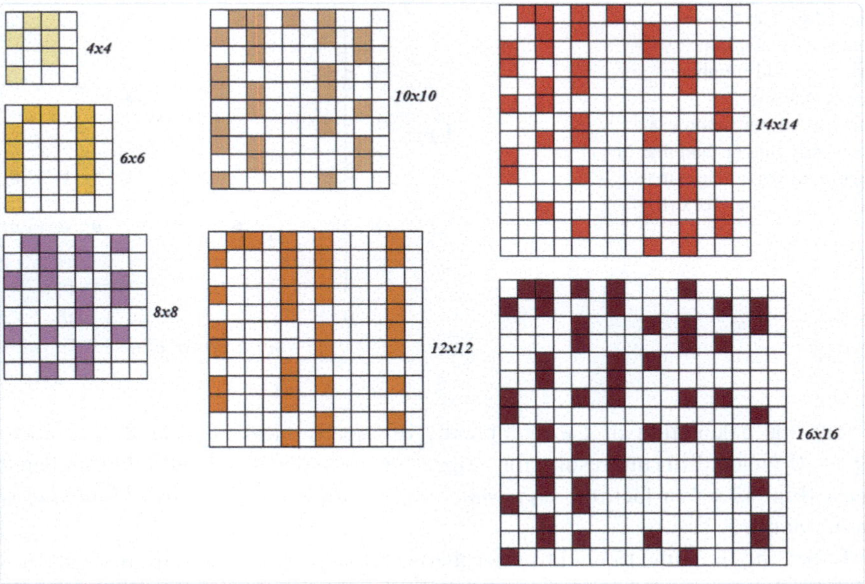

Fig. 12.4 Binary maps resulting from prime numbers allocated on progressively larger maps, but always on map sizes with n = even. Notice the presence of clumps (of at least two colored cells) and the presence of entire columns without any colored cell at all

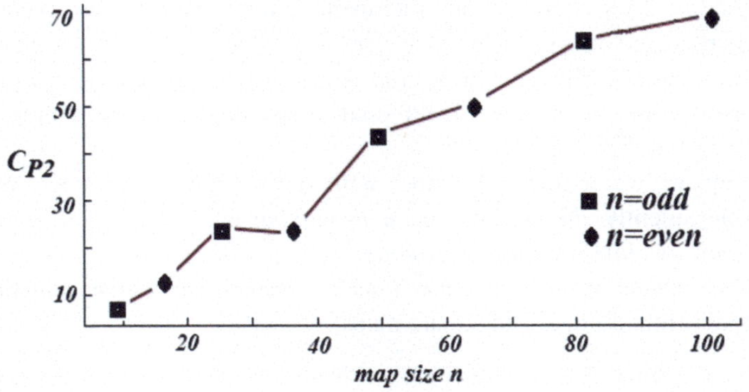

Fig. 12.5 Odd-numbered maps have higher C_{LP} complexity than even-numbered ones (measurements only on maps from $n = 3 \times 3$ to $n = 10 \times 10$)

relationship is a best fit for clumpiness B values with respect to p:

$$B = 0.292p - 0.4872\tan(0.5739p) - 0.6177\cos(0.5234 + p) \qquad (12.14)$$

with goodness of fit = 0.99978491 and correlation coefficient = 0.99989552.

Fig. 12.6 The C_{PI} complexity of compressed strings of odd-numbered binary maps of primes-and-composites is constantly higher than that of even-numbered maps (map sizes 2 × 2 up to 10 × 10)

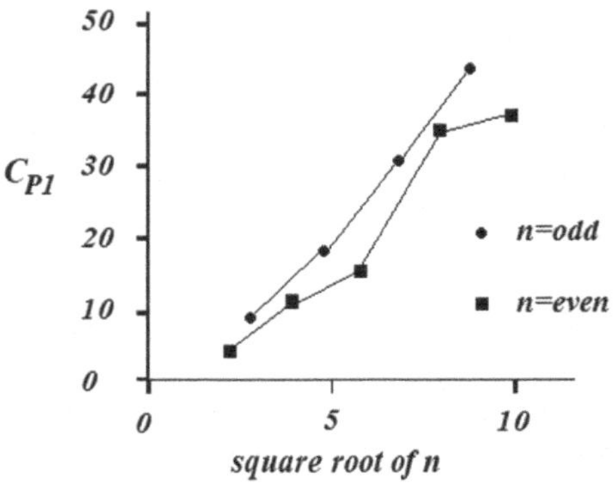

Yet, the calculation of C_{PI}-complexity of square maps of sizes 2 × 2 up to 10 × 10 yields differing results for strings for even-numbered and odd-numbered maps (Fig. 12.6). In fact, the *C_{PI}-complexity is always higher for odd-numbered binary maps.*

Expecting the same to hold true for map sizes larger than 10 × 10, it might be a strong indication (not a proof) that *sequences of odd-numbered square binary maps of primes-and-composites are "able" to "generate" higher spatial complexity than even-numbered ones.*

Some spatial aggregates that appear in even-numbered maps of primes-and-composites may be due to the presence of twin primes. Allocating primes as black cells along a line does not reveal any interesting pattern. But it does reveal striking patterns when carried out over a 2d space. The "*Ulam spiral*" is a known such pattern, but it can not reveal differences in spatial complexity as allocations of primes on square maps do (as, i.e. evidence from small maps suggests here, clumps tend to appear in even-numbered grids).

But all square binary maps of primes with size n higher than 3 × 3 have $r_{max} > p$ and, consequently, *the larger maps with number of primes equal to the map's maximum entropy class are the 3 × 3 maps.*

With these considerations, some questions for future research arise from the spatial allocations of prime numbers on square maps:

- Is there an optimal square size for n (with n = even), for which the number of clumps is maximized in binary maps of primes?
- What are the biggest clumps that can be created in binary maps of primes-and-composites, with size up to a specified value of n?
- Is there an upper barrier to clump size, whatever the value of n?

We currently do not possess answers to these questions. But answering them might give us hints about the ways that number theory underlies spatial complexity and the reverse: the ways by which spatial complexity may reveal (and lead to) number-theoretical problems. Understanding the relationships between prime numbers and spatial complexity can have repercussions in other fields related to spatial complexity.

For instance. knots can be "*primes*" or "composites" (nontrivial knots), just as integers are and the number of prime knots with n crossings increases fast: 0, 0, 1, 1, 2, 3, 7, 21, 49, 165, 552, 2176, 9988, 46,972, 253,293, 1,388,705 etc.

12.3 Square Maps from Transcendental Numbers

> God makes the world by calculating, but his calculations never work out exactly and this inexactitude or injustice in the result, this irreducible inequality, forms the condition of the world. The world "happens", while God calculates; if the calculation were exact, there would be no world
>
> (Gilles Deleuze 2010, p. 280)

In 1874, Cantor proved that the algebraic numbers are countable, but the real numbers are uncountable. The transcendental numbers are uncountably infinite. But *how* can they be identified? And, transposing this line of thought to spatial analysis of square maps, how might these uncountably infinite many transcendentals correspond to spatial forms? It would make sense to assume that some spatial patterns may correspond to numbers lurking almost imperceptible and this raises questions about possible limits to perceiving and understanding spatial complexity. Let us consider a simple example, by considering a string of binary square maps, of increasing size, corresponding to the following strings (from left to right):

011111011...
0111110110011101...
0111110110011101110011001...
011111011001110111001100101100111111...

At first sight, none of these strings (or, ultimately, the last string, which comprises all the previous ones) seems to reveal any known pattern. But they conceal one: if the digits 1,2,3,4,5 of another string are mapped to 0 s and the digits 6,7,8,9,0 are mapped to 1 s, the following correspondence between the two strings is derived:

0.011111011001110111001100101100111111
|| |...
0.207879576350761908546955619834978770

The latter string corresponds to the decimal digits of the number i^i, since

$$i^i = e^{i \ln i} = e^{-\left(\frac{\pi}{2}+2k\pi\right)} = 0.207\ldots \tag{12.15}$$

Expectedly, arbitrarily many digits of the binary string can be isolated and converted to a spatial representation (Fig. 12.7).

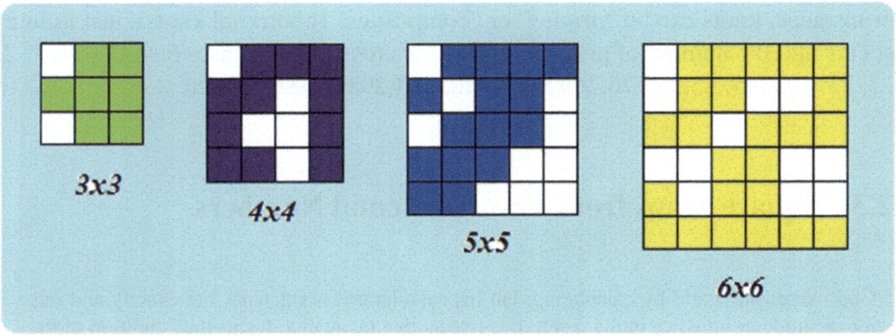

Fig. 12.7 Changing patterns and clusters of colored and white cells can mislead human perception: in fact, all these square binary maps are generated by the same number: they represent the first decimal digits of the number $i^i = 0.207\ldots$ with the decimals 1,2,3,4,5 represented by white cells and the decimals 6,7,8,9,0 by colored cells

Interestingly, following *Bellard's formula*, a binary description of π can be calculated:

$$\pi = \frac{1}{2^6} \sum_{n=0}^{\infty} \frac{(-1)^n}{2^{10n}} \left(-\frac{2^5}{4n+1} - \frac{1}{4n+3} + \frac{2^8}{10n+1} - \frac{2^6}{10n+3} - \frac{2^2}{10n+5} - \frac{2^2}{10n+7} + \frac{1}{10n+9} \right) \tag{12.16}$$

which yields a binary string for $\pi = 3.141$:

11.001001000011100101011000000100000110001001001101110100101111 00...

but if the precision of π were allowed to increase to two more decimals, that is 3.14159, then the binary string is altered after the 13th decimal:

11.00100100001111110011111000000011011100001100110111001000011101...

Thus, depending on the *precision* of the description of π, Bellard's formula produces different binary strings. As in other cases, precision-dependence brings about significant changes in spatial representations (which, in turn, results in differences in spatial complexity).

If we take, for instance, the first 49 digits of 3.14... and arrange them in a 7×7 map (Fig. 12.8), we would produce a completely different map than if we took the first 49 digits with the higher precision $\pi = 3.14159$. In turn, this means there are significant differences in the spatial complexities of the two maps.

The largest black block on the left map is 13 while on the right one it is 8 (61% lower). Yet, the white blocks are comparable: the largest white block on the left is

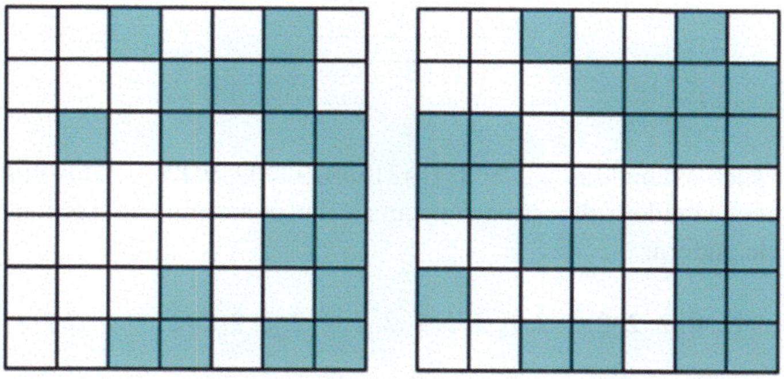

Fig. 12.8 The spatial representation of the first 49 digits of $\pi = 3.141$ arranged on a 7×7 binary map is different than the representation of the same number of digits of $\pi = 3.14\mathbf{159}$ (right)

27 (it covers more than half of the 7×7 map's area) while the one on the right is 22 (that is 81% as large). Also, the number of patches on the left map is 4, while on the map on the right it is 7 (75% higher). The C_{PI} complexity of the left map is 37 (with the compression $\lambda = W^2BW^2B$), while the C_{PI} complexity of the right map is 32 (with the compression $\lambda = WB^2$). The C_{P2} complexity of the left map is 34 and of the right is 38 and hence the differences in spatial complexity range from +17% to −13.5% for only two decimals of π.

Hence, *different configurations producing different spatial complexities, may all be derived from the same transcendental number.* And this can not be perceived initially without knowing the underlying mathematical process or property that has generated the spatial configuration. Moreover, knowing the transcendental number behind a map's spatial structure, not only explains the spatial allocation, but can also be used to predict a map's configuration, if enlarged to extend in space.

Recalling that any real decimal number x belonging to the interval [0,1] can be represented by a continuous fraction and using continued fractions to encode spatial properties, it is easy to verify that different encodings of the same spatial elements may produce varying descriptions of spatial complexity and this without changing the scale of observation. But, identifying an optimal encoding is a computationally hard undertaking in both binary and multicolored maps. Continued fractions however, are "base-invariant", meaning that some numbers which may appear random eventually present unexpected "beautiful" patterns. One such is the "golden section" (the first 50 digits are given here):

$$1.6180339887498948482045868343656381177203091798057 6\ldots,$$

This apparently "messy" series of decimals surprisingly corresponds to the simplest continued fraction possible:

$$\cfrac{1}{1+\cfrac{1}{1+\cfrac{1}{1+\cfrac{1}{1+\cfrac{1}{1+\dots}}}}}$$

Another such example is $\sqrt{2} = 1.4142135623730950488\ldots$ This number also appears to have a random allocation of decimals, but as a continued fraction, it reveals a very simple pattern:

$$[1, 2, 2, 2, 2, 2, 2, 2, 2, 2, 2, 2, 2, 2, 2, 2, 2, 2, 2, \ldots\ldots].$$

Similarly, the transcendental number e appears random in its decimals

$$2.71828182845904523536028747135266249775724709369995\ldots$$

but in continued fraction form it presents a logical procession of even numbers separated after every two 1 s:

$$[2, 1, 2, 1, 1, 4, 1, 1, 6, 1, 1, 8, 1, 1, 10,$$
$$1, 1, 12, 1, 1, 14, 1, 1, 16, 1, 1, 18, 1, 1,$$
$$20, 1, 1, 22, 1, 1, 24, 1, 1, 26, 1, 1, \ldots.]$$

Simple patterns with continued fractions have other numbers also with infinite decimals; in example π which is equal to to:

$$\cfrac{4}{1+\cfrac{1^2}{2+\cfrac{3^2}{2+\cfrac{5^2}{2+\cfrac{7^2}{2+\cfrac{9^2}{2+\dots}}}}}}$$

or

$$3+\cfrac{1^2}{6+\cfrac{3^2}{6+\cfrac{5^2}{6+\cfrac{7^2}{6+\cfrac{9^2}{6+\dots}}}}}$$

These examples show that there may exist a hidden structure in a spatial form and this structure may correspond to some number, but we have no clue as to how to identify the number behind the structure.

References

Baudrillard, J. (2005). *The intelligence of evil or the lucidity pact*. London: Bloomsbury.
Deleuze, G. (2010). *Difference and Repetition*. London: Continuum.
Matousek, J. (2010). *33 Miniatures*. Providence: American Mathematical Society.

Part IV
Understanding Spatial Complexity

Chapter 13
Enigmas of Spatial Complexity

It is a riddle, wrapped in a mystery,
inside an enigma; but perhaps there is a key
(Sir Winston Churchill, 1874–1965, "The Russian Enigma", 01-20-1939)

Abstract From Thurston's geometrization theorem to the Banach-Tarski paradox, spatial complexity can be enigmatic, due to singularities, immersions, infinities (e.g. Alexander's wild knots, Antoine's necklace, Ford circles, Hilbert curves). Measuring spatial complexity in already "complex" spatial settings constitutes a major challenge for future research, as it should combine algorithmics, computation and topology within a single whole. Further, completely homogeneous maps can still be complex if their complexity is measured by C_{PI}, because topological differentiation (i.e. division in square cells) creates spatial complexity: a completely undifferentiated square map without cells defined on it has a different spatial complexity than a square map with cells identified in it. And yet, the same allocations of numbers to cells of binary maps can produce different configurations with different spatial complexities.

Keywords Spatial complexity · Map Complexity · Homogeneity and Complexity · Algorithmic Complexity · Complexity and Geometry · Complexity and Topology · Geocomputation

13.1 Geometrization, Singularities and Immersions

Your deep narrow spaces developed into geography;
The altitude was "the mountain of the earth" and to climb down it was a horror
"Tus contadas varas de fondo se nos volvieron geografía;
un alto era "la montaña de tierra" y una temeridad su declive"
(Jorge Luis Borges, 1899–1986,"Curso de los recuerdos"/The flow of memories)

Extending the analysis of spatial complexity to higher than two-dimensional spaces entails not only difficulties, but surprises also. Measuring spatial complexity

F. Papadimitriou, *Spatial Complexity*, https://doi.org/10.1007/978-3-030-59671-2_13

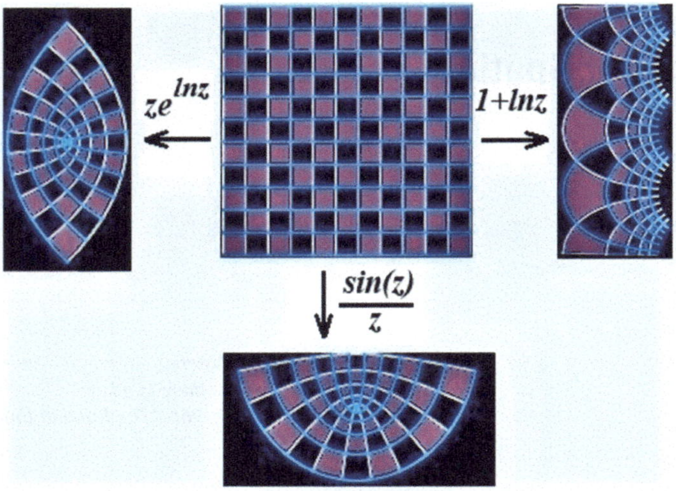

Fig. 13.1 Some mappings of a binary map by complex functions produce binary maps on which spatial complexity can be calculated by algorithmic means

on mappings of binary maps which are defined by using complex functions is a challenge. In some simple cases, it is possible to calculate spatial complexity (Fig. 13.1), but some mappings of complex functions may produce binary maps on surfaces on which spatial complexity can be very difficult to calculate algorithmically. Since the nineteenth century, and after the works of Riemann, Möbius and Klein, we know that every closed 2d surface is topologically equivalent to orientable spheres (with a finite number of handles) or to non-orientable surfaces (such as the projective plain and the Klein bottle) and that three geometries (euclidean, hyperbolic, spherical) suffice for describing such surfaces. Yet, one century later, it was discovered that the possible geometries on 3d manifolds are determined by Thurston's "Geometrization Conjecture", which, after its proof by Perelmann (2002, 2003a, b) became the "Geometrization Theorem", asserting that every smooth, compact, orientable 3d manifold can be cut along a set of 2d spheres and 1-holed tori, in a way that each one of the resulting parts has one of precisely eight possible geometries (Thurston 1982; Thurston 1997).

Before proceeding to Thurston's geometrization, it is essential to recall that as the distance $d(x,y)$ between two points x and y varies, the path from x to y will differ depending on the distance metrics of curve lengths over the manifold. Following the Geometrization Theorem, aside from the euclidean, spherical and hyperbolic geometries of the 3d space, there are yet another five less widely known geometries of 3d manifolds; specifically these are the geometry S^2xR (where S^2 is the 2-sphere), H^2xR (where H^2 is the hyperbolic plane), the geometry of the universal cover of the Lie group $SL^2(R)$, the "sol geometry" (that is the geometry of the Lie group R in semidirect product with R^2) and the "nil geometry" (which is the geometry of the Lie group of real matrices of the form:

$$\begin{pmatrix} 1 & x & y \\ 0 & 1 & z \\ 0 & 0 & 1 \end{pmatrix}$$

If x_1, x_2,x_3 are the three spatial coordinates, the metric for the sol geometry is (Lopez and Munteanu 2011):

$$ds^2 = e^{2x_3}dx_1^2 + e^{-2x_3}dx_2^2 + dx_3^2 \tag{13.1}$$

and for the nil:

$$ds^2 = dx_1^2 + dx_2^2 + (dx_3 - x_1 dx_2)^2 \tag{13.2}$$

These geometries are not easy to visualize. Perceiving shapes in sol geometry for instance, can be facilitated by solving a set of trigonometric equations in three dimensions, so as to calculate geodesics within this geometry (Bölcskei and Szilágyi 2006):

$$\begin{aligned} x(\theta, \phi) &= -\cot\theta\cos\phi(e^{-r\sin\theta} - 1) \\ y(\theta, \phi) &= \cot\theta\sin\phi(e^{r\sin\theta} - 1) \\ z(\theta, \phi) &= r\sin\theta \end{aligned} \tag{13.3}$$

Back in 1833, Bolyai proved that every polygon can be divided in triangles, in such a way that when re-arranged, these triangles will compose a square. Later, Hilbert's third problem questioned whether the same applies to 3d solids with tetrahedra instead of triangles. After Dehn disproved it, it follows that shapes in the 3d space are significantly more difficult to analyse by means of "tetrahedrizations" than shapes in the 2d space by means of triangulations. In fact, the 3d space is significantly trickier than we could possibly expect. With respect to points scattered in a 3-space, Borsuk's conjecture for 3-space is that every n-dimensional point set can be decomposed in n + 1 pieces of smaller diameter (Borsuk 1933). It would have been convenient to hold true, but it was disproved for large n (Kahn and Kalai 1993). And a counter-intuitive result proven by Hausdorff in 1914 asserts that it is possible to divide a sphere into a finite set of non-measurable parts, which, when re-united, will make two spheres, each one with area equal to that of the first one. A decade later, Banach and Tarski proved the same for volumes, with their famous "*Banach-Tarski paradox*" (Fig. 13.2). Meanwhile, *eversions* may also present surprises. A sphere eversion (turning the sphere inside out) is possible by a homotopy of immersions from S^2 to R^3 (Smale 1958), following a peculiar procedure (Phillips 1966; Francis and Morin 1980).

The calculation of C_{PI} or C_{P2} on some surfaces is impossible, i.e. on the surface described by

$$f(x, y) = \tan\big((y - x) - \tan(\log x)\sin(y\pi)\sqrt{xy - \pi} - \tan x + 2y\big) \tag{13.4}$$

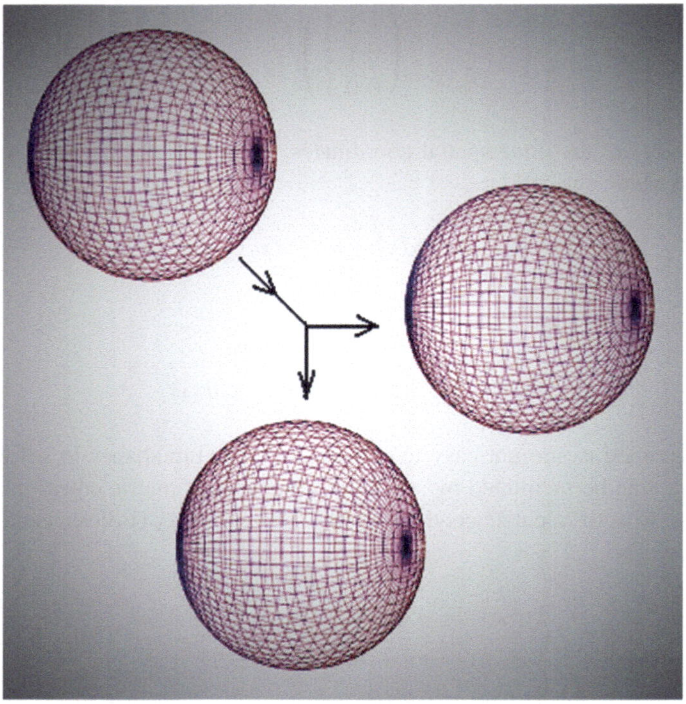

Fig. 13.2 A sketch for the visualization of the Banach-Tarski paradox in 3d space

In cartographic, ecological and geographical spatial analyses using 2d maps however, one *mostly* deals with surfaces and volumes describable by integrable functions, that is with the provision that a surface or a volume can be described by an integrable function f. In two dimensions (i.e. in cases of simple planar square maps) C_{P1} and C_{P2} are applicable on bounded regions R_{xy} of the xy-plane for which the integral can be calculated because the limit of the sum of infinitesimal areas A_i exists:

$$\iint_{R_{xy}} f(x, y)dA = \lim_{\|\Delta\| \to 0} \sum_{i=1}^{n} f(x_i, y_i)\Delta A_i \tag{13.5}$$

Equivalently, for 3d regions (volumes), the triple integral of a function f of three variables that is continuous over a bounded solid region R_{xyz} is defined (in terms of elementary volumes Vi) as

$$\iiint_{R_{xyz}} f(x, y, z)dV = \lim_{\|\Delta\| \to 0} \sum_{i=1}^{n} f(x_i, y_i, z_i)\Delta V_i \tag{13.6}$$

These calculations can yield results only if the respective limits exist (when they do not, it should be one additional cause of concern for spatial complexity). If the function describing a surface has singularities, then some assessments of spatial complexity may not be feasible at singularities. Possibly, it would make sense to recall that one of the most complex spatial objects is the surface described by the classic *Weierstrass function*, known since 1872, which is everywhere continuous, but nowhere differentiable:

$$f(x) = \sum_{k=0}^{\infty} a^k \cos(b^k \pi x) \tag{13.7}$$

with a being a real number $0 < a < 1$, b a positive integer and $ab > (3\pi + 2)/2$.

At this point, a divergence can be detected between the strictly mathematical and the "applied" (or "engineering") approaches to spatial complexity assessments: while mathematicians may find it difficult to assess spatial complexity of a surface with singularities, in which case, they might proceed to consider what type a singularity is and whether it might be removable. The surface created by the potential of the "*swallowtail catastrophe*"

$$V = x^5 + ax^3 + bx^2 + cx \tag{13.8}$$

is one of the basic unfolding types of singularities in R^3 (Zeeman 1977; Thom 1989; Arnold 1992), combining two minima and two maxima that meet at a single value of x (Fig. 13.3). Practically however, either C_{P1} or C_{P2} can be applied to the entire surface that is visible from above (even as it is folded), disregarding the singularity lurking beneath.

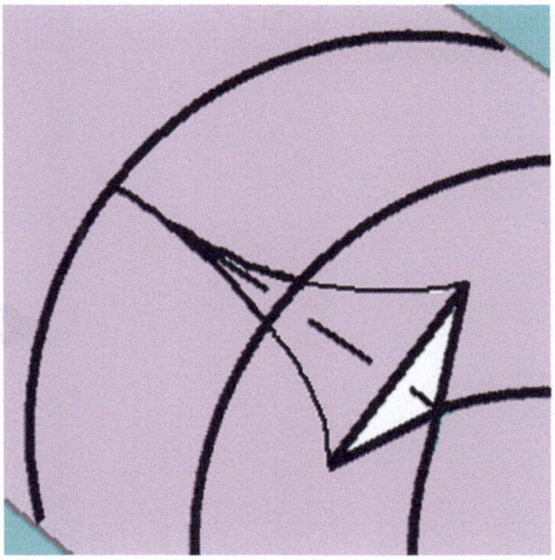

Fig. 13.3 A sketch of the "swallowtail catastrophe"

In fact, even small differences in the equations describing two surfaces may result in extremely different plots, different surfaces and hence there is no guarantee that either C_{P1} or C_{P2} would apply to the resulting surface. Compare for instance the simple smooth surface resulting from the equation

$$f(x, y) = \cos((x - \sin y)\cos y) \tag{13.9}$$

with the extremely densely folded surface resulting from an equation that differs only by its last component:

$$f(x, y) = \cos((x - \sin y)\cos y - \tan(x + y)) \tag{13.10}$$

Aside of singular points, *immersions* constitute another enigma of spatial complexity. If a surface is immersed in a higher dimensional space, it is increasingly more complex than a surface that is simply embedded in it (Fig. 13.4). In principle, it would not be possible to apply C_{P1} or C_{P2} on such surfaces without discontinuous leaps from one part of such a surface to another, due to the self-intersections. *Enneper's surface* for instance, is such an injective immersed surface $f(u,v)$ following a mapping $f(u, v) : R^2 \rightarrow R^3$ with the general parametric formula

$$f(u, v) = \left(u - \frac{u^3}{3} + uv^2, v - \frac{v^3}{3} + vu^2, u^2 - v^2\right) \tag{13.11}$$

Another immersed surface is "*Boy's surface*" that is an immersion of the real projective plane in R^3, with a parametrization by Kusner (1987), by which the coordinates (u,v,w) are expressed in terms of the complex number z with magnitude $\|z\| \leq 1$:

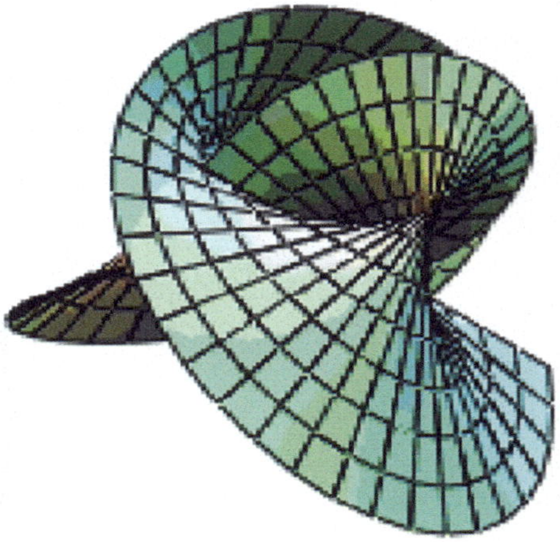

Fig. 13.4 Example of a 2d surface immersed in R^3

$$\begin{pmatrix} u \\ v \\ w \end{pmatrix} = \frac{1}{g_1^2 + g_2^2 + g_3^2} \begin{pmatrix} g_1 \\ g_2 \\ g_3 \end{pmatrix} \tag{13.12}$$

where

$$\begin{aligned} g_1 &= -\tfrac{3}{2}\mathrm{Im}\left(\frac{z(1-z^4)}{z^6+\sqrt{5}z^3-1}\right) \\ g_2 &= -\tfrac{3}{2}\mathrm{Re}\left(\frac{z(1+z^4)}{w^6+\sqrt{5}z^3-1}\right) \\ g_3 &= -\tfrac{1}{2} + \mathrm{Im}\left(\frac{z+z^6}{z^6+\sqrt{5}z^3-1}\right) \end{aligned} \tag{13.13}$$

13.2 Spatial Complexity and Infinity

One should not attribute the idea of the infinite before confirming the entire set of its numbers that are in between one and infinite.

"τὴν δὲ τοῦ ἀπείρου ἰδέον πρὸς τὸ πλῆθος μὴ προσ-

φέρειν πρὶν ἄν τις τὸν ἀριθμόν αὐτοῦ πάντα κατίδῃ τὸν

μεταξὺ τοῦ ἀπείρου τε καὶ τοῦ ἑνός"

(Plato, 428–348 b.C.,"Philebus", 16c)

Besides singularities, shapes produced by infinite repetitions of elementary geometric or topological shapes point towards a fundamental problem in spatial analysis. Following *Cantor's theorem* (1878), the number of points in any part of the line of reals equals the number of points in the entire line of R and, passing on to the second dimension where things become "really" spatial (at least as conventionally defined in advance, in the context of this book), it is also equal to the number of points of any plane in two dimensions, and, by consequence, also equal to the number of points of any finite-dimensional space. By breaking the connectedness (i.e. by puncturing) and therefore changing the cardinalities of the corresponding sets, we are able to identify classes of infinity. But we are still short of procedures for estimating spatial complexity of objects in infinite-dimensional and related spaces such as the Eilenberg–MacLane spaces, the infinite-dimensional complex projective spaces etc. In the previous sentence, the word "in" leads to further riddles (i.e. assessing spatial complexity on immersed surfaces in infinite dimensional spaces?). Perhaps, a first approach in this respect is to *measure* the spatial complexity of fractal objects such as the Koch curve or the *Hilbert space-filling curve* (Fig. 13.5), since these objects are self-similar across scales, ad infinitum. Typically, Hilbert's curve is 2d, but it can be drawn in 3d cubes also (Fig. 13.6) and, in 3d many surfaces can get a lot more complex, as, indicatively, three characteristic examples from topology reveal:

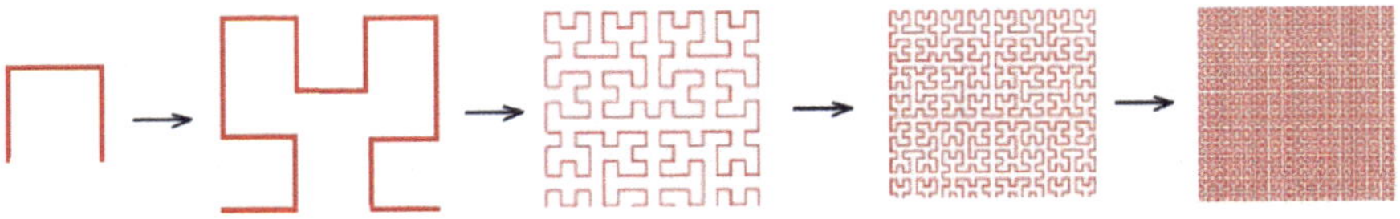

Fig. 13.5 The creation of Hilbert's space-filling curve: from a simple three-segments shape and with iterative repetitions, the plane is eventually covered

Fig. 13.6 The first step for the creation of Hilbert's space-filling curve in 3d (in a cube)

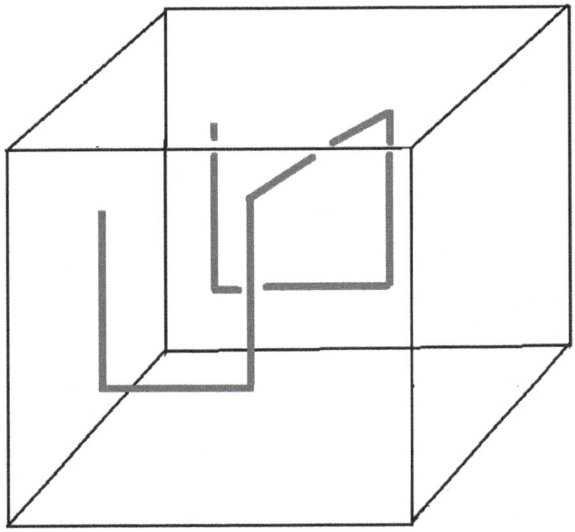

(i) *Alexander's horned spheres* (intertwined 3d surfaces extending to infinity): while C_{P1} or C_{P2} can be applied to spheres with calculations on strips on grids meridian-per-meridian, would they also apply on horned spheres?

(ii) *Antoine's necklace* is an infinite chain of chains of tori, with diameters decreasing to zero; it is a "*wild embedding*" of a set in R^3. Wild embeddings involve shapes defined up to infinity. What might the spatial complexity of such shapes be? And, even more intriguingly, *might there exist shapes with infinite spatial complexity*?

(iii) *Ford circles* on the plane (Fig. 13.7) are created by drawing initially two circles with radii 1/2 k^2 each (where *m,k* are integers), centered at (*m/k*, 1/2 k^2) and continuing with circles of progressively minor sizes for different values of *m,k*. The conditions are that *m,k* are coprime integers and *m/k* is irreducible fraction. Essentially, Ford circles are a modern form of the old greek *Apollonian gasket* and, expectedly, there can be 3d Ford spheres also (Northshield 2015).

The possibility that Ford circles essentially yield a visualization of Cantor's transfinite numbers should not be precluded (Pickover 2009). Given that the smallest such number (the $\aleph_0$) represents the cardinality of the integers, it is an undecidable problem (under the Zermelo-Fraenkel theory and the axiom of choice) whether the

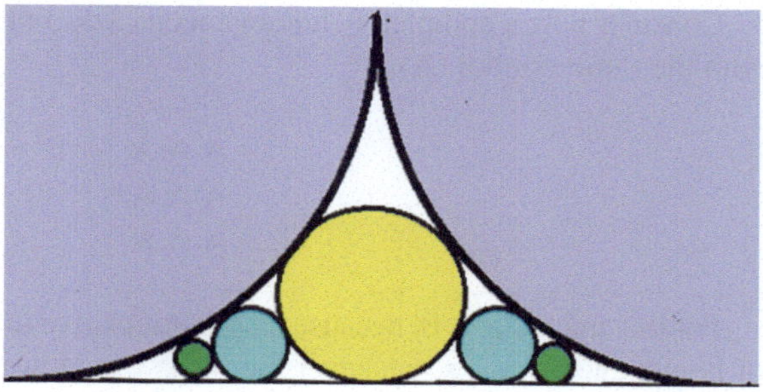

Fig. 13.7 The first steps for making Ford circles on the plane

next transfinite $\aleph_1$ represents the cardinality of real numbers: $\aleph_1 = 2^{\aleph_0}$. Besides, the cardinality of the set of (geometrically) "constructible numbers" is also $\aleph_0$. In this way, Ford circles lead us to a *"numbers behind spatial complexity"* hypothesis, now with an additional question: *Might spatial complexity be transfinite (or infinite)*?

To tackle this question, one perhaps needs to observe that two maps of different size n, with exactly the same spatial pattern in them, may have different C_{PI} complexities. In Fig. 13.8 the small map a is characterized by the irreducible string

$$W^5B^2W^2BW^2BWBW$$

and hence has an C_{PI} complexity equal to 13, while the map b has the string

$$W^{19}B^2W^6BW^6BWBW^{27}$$

and C_{PI}= 16.

Yet, both maps have exactly the same spatial allocation of their five dark cells, with exactly the same spatial extent and size (confined in a 3 × 3 area).

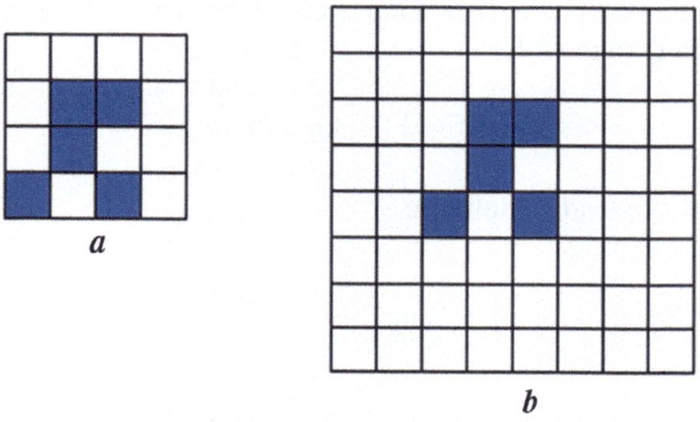

Fig. 13.8 The C_{PI} complexity of the map a is 13, while that of the map b is 16. Yet, both maps have exactly the same spatial differentiation pattern. So, even if a map is undifferentiated, its *size* affects C_{PI} complexity. The C_{P2} is the same for both maps: $C_{P2} = 12$

Consider now a completely homogeneous 3 × 3 map with each one and all cells with the same symbol (A):

A A A
A A A
A A A

For this map, C_{P2}= 0, because C_{P2} is insensitive to the global spatial complexity of homogeneous maps, whatever the map size. However, the same does not apply to C_{PI} and this prompts to make some further observations. Evidently, the previous simple map has C_{PI} equal to $C_{PI}\{A^9\} = 2$. Consider now a larger map, i.e. 10 × 10, whose cells are, again, all the same (A):

A A A A A A A A A A
A A A A A A A A A A
A A A A A A A A A A
A A A A A A A A A A
A A A A A A A A A A
A A A A A A A A A A
A A A A A A A A A A
A A A A A A A A A A
A A A A A A A A A A
A A A A A A A A A A

The C_{P2} of this map is also zero, but $C_{PI}\{A^{100}\} = 4$. Equivalently, the uniform 100 × 100 map has $C_{PI}\{A^{1000}\} = 5$, and eventually, as the map size n grows, the C_{PI} of homogeneous maps is given by C_{PI}(h), that is the C_{PI} complexity which is necessarily endowed with a homogeneous map without any heterogeneity at all, *so long as a topological differentiation is defined* on the map (definition of square cells, that is definition of boundaries among cells). With growing map size n, C_{PI}(h) increases according to the rule:

$$C_{P1}(\mathrm{h}) = 1 + \lceil \log_{10}(n+1) \rceil \tag{13.14}$$

As the map size n tends to infinity:

$$\lim_{n\to\infty}\left(\frac{C_{P1}(\mathrm{h})}{n}\right) = \lim_{n\to\infty}\left(\frac{1 + \lceil \log_{10}(n+1) \rceil}{n}\right) = 0 \tag{13.15}$$

Hence, in terms of C_{PI} complexity, there is a (very small) price to be paid (in terms of complexity, or "difficulty of understanding" a map), if spatial uniformity expands to large sizes. And this makes sense, by considering that a 3 × 3 square map with only 9 (identical) cells is a different entity than a 100 × 100 square map with

10,000 identical cells: if C_{PI} complexity is adopted, then *topological differentiation creates spatial complexity*. This may be negligibly small (depending on the map size), but certainly *non-zero*. The increase in spatial complexity of uniform maps is an altogether different issue, beyond the (obvious) scale-dependence of complex spatial forms, which is well-known from the literature on fractal objects. In fact, it touches upon the very delicate *relationship between topology and information*, in the sense that an *undifferentiated map* (without divisions into cells) has a different C_{PI} complexity than a map with cells defined on it.

13.3 Same Numbers—Different Spatial Complexities

> Anyone who considers arithmetic methods of producing random digits is, of course, in a state of sin
>
> (attributed to John von Neuman, 1903–1957)

Being able to measure spatial complexity however (even under many restrictions and initial assumptions), does not guarantee that we are able to understand it or even perceive the mechanisms that have created it. There are reasons to suppose that behind spatial complexity processes may lurk imperceptible "works of numbers". For this reason, besides geometry, topology and algorithmic theory, number theory might be essential in understanding how spatial complexity is generated. Consider, for instance, a binary square map, as shown in Fig. 13.9.

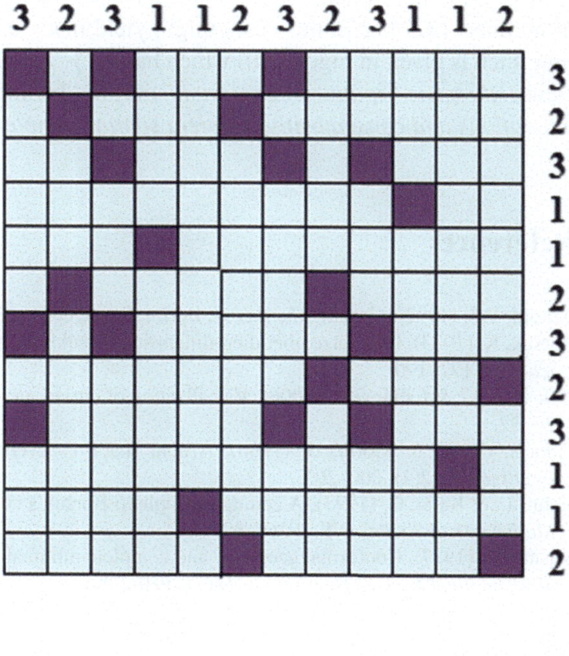

Fig. 13.9 A binary square map. Numbers of black cells per column and per row are shown atop and on the right side of the map respectively. Notice the regularity in number sequences repeated both horizontally and vertically: 323112

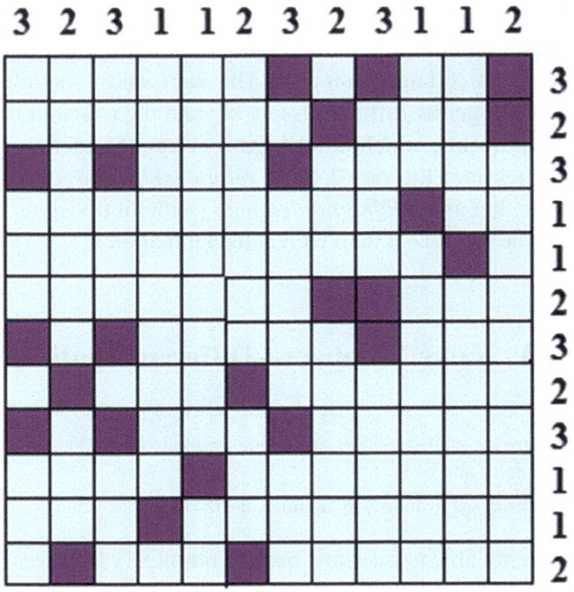

Fig. 13.10 An alternative and different-looking binary map, produced by the same sequence of numbers: 323112

Without conducting any measurement of spatial complexity, and based only on a snap visual judgement, this map appears like a random one. But this map conceals a precise regularity that is not geometric; it is a number-related regularity emerging from the number of cells that contain black cells per column and per row. The C_{P2} of this map is 68.

It is even more interesting to consider that exactly the same example sequence of numbers (323112 in this case) might yield many different spatial configurations (one such is given in Fig. 13.10) which has a C_{P2} equal to 78 (that is 14.7% higher). Hence, *the same "spatium numerorum"* may be "behind" *different alternative spatial allocations and consequently, different spatial complexities.*

References

Arnold, V. I. (1992). *Catastrophe theory*. Berlin: Springer-Verlag.

Borsuk, K. (1933). Drei satze iiber dien-dimensionale euklidische sphare. *Fundamenta Mathematicae, 20,* 177–190.

Bölcskei, A., & Szilágyi,B. (2006). Visualization of curves and spheres in sol geometry. *KoG-10,* 26–32.

Francis, G. K., & Morin, B. (1980). Arnold shapiro's eversion of the sphere. *Mathematical Intelligencer, 2*(4), 200–203.

Kahn, J., & Kalai, G. (1993). A counterexample to Borsuk's conjecture. *Bulletin of the American Mathematical Society, NS, 29,* 60–62.

Kusner, R. (1987). Conformal geometry and complete minimal surfaces. *Bulletin of the American Mathematical Society, New Series, 17*(2), 291–295.

Lopez, R., & Munteanu, M. I. (2011). Surfaces with constant mean curvature in sol geometry. *Differential Geometry and its Applications, 29,* S238–S245.

Northshield, S. (2015). Ford circles and spheres. arxiv:1503.00813v1math.NT, 3 March.

Perelmann, G. (2002). The entropy formula for the Ricci flow and its geometric applications. *arXiv:math.*DG/0211159 v1 November 11, 2002.

Perelmann, G. (2003a). Ricci flow with surgery on three manifolds. *arXiv:math.*DG/0303109 v1 March 10,2003, preprint.

Perelmann, G. (2003b). Finite extinction time to the solutions to the Ricci flow on certain three manifolds. *arXiv:math.*DG/0307245 July 17, 2003, preprint.

Phillips, A. (1966). Turning a surface inside out. *Scientific American* May, 112–120.

Pickover, C. (2009). *The maths book.* London: Sterling.

Smale, S. (1958). A classification of immersions of the two-sphere. *Transactions of the American Mathematical Society, 90*(2), 281–290.

Thom, R. (1989). *Structural stability and morphogenesis: An outline of a general theory of models.* Reading, MA: Addison-Wesley.

Thurston, W. P. (1982). Three dimensional manifolds, Kleinian groups and hyperbolic geometry. *Bulletin of the American Mathematical Society, 6,* 357–381.

Thurston, W. P. (1997). *Three-dimensional geometry and topology* (Vol. 1). Princeton NJ: Princeton University Press.

Zeeman, E. C. (1977). *Catastrophe theory-selected papers –1977.* Reading, MA: Addison-Wesley.

Chapter 14
Taming Spatial Complexity

And so always driven towards new shores,
brought there without return in the eternal night,
will we ever be able to set an anchor
in the ocean of ages, even for one day ?
"Ainsi toujours poussés vers de nouvaux rivages
dans la nuit éternelle emportés sans retour,
ne pourrons-nous jamais, sur l' océan des âges
jeter l'ancre un seul jour?"
(Alphonse de Lamartine, 1790–1869, "Le Lac")

Abstract Buffon's needle, Varignon's theorem, Jung's theorem, Sylvester's problem, are all hints that random allocations of spatial objects often obey to geometric restrictions. Further, it can be speculated that, so long as population distributions on surfaces might be represented by strings of symbols, there is no apparent reason why the arcsine law and Khinchin's bound should not apply to such strings (i.e. strings representing binary maps). From an algebraic point of view, spatial symmetries may also serve as an extremely powerful tool by which spatial complexity can be deciphered. From D_4 to Redfield polynomials, symmetries help us understand and create spatial complexity. Their central role in "taming" spatial complexity brings associations with important results from graph theory and network analysis (i.e. from Ramsey theory, the Szemerédi regularity theorem etc). Taming spatial complexity may be expected by applying a "sudoku" analysis (9 x 9) exploiting the results on the spatial complexity of 3 x 3 maps,and from Grothendieck's inequality.

Keywords Spatial complexity · Symmetry and Complexity · Map Complexity · Cartography and Complexity · Geocomputation · Sudoku method · Grothendieck

14.1 Taming Spatial Randomness?

"Even in the valley of the shadow of death,
two and two do not make six"
(Leo Tolstoy, 1847–1910)

F. Papadimitriou, *Spatial Complexity*, https://doi.org/10.1007/978-3-030-59671-2_14

Fig.14.1 "Buffon's needle problem" is about tossing a needle of length L over a space that is divided by parallel lines drawn at equal lengths, a. If the needle's length L is equal to a, then the probability of intersection is $P = 2/\pi$

"Buffon's needle" is a well known case in the mathematical literature, which also reveals how seemingly unrelated disciplines of mathematics can unexpectedly be related. In his "Essai d' Arithmétique Morale" (1777), the Comte de Buffon, posed a problem of geometric probability: "What is the probability that a needle falling randomly over a floor divided by parallel lines crosses anyone of them?" (Fig. 14.1).

Indeed, tossing a needle of length L over a space divided by parallel lines of distance a between them (with $L < a$), the probability that the needle intersects anyone of the parallel lines is equal to:

$$P_r = \frac{2L}{\alpha} \tag{14.1}$$

and if the length of the needle is $L = a$, then the solution leads to a ratio involving, surprisingly, the transcendental number π:

$$P_r = \frac{2}{\pi}. \tag{14.2}$$

Further, if $L > a$, then the probability of intersection is:

$$P_r = 1 - \frac{2}{\pi\alpha}\left[a \sin^{-1}\left(\frac{a}{L}\right) + L\left(\sqrt{1 - \left(\frac{a}{L}\right)^2} - 1\right)\right] \tag{14.3}$$

so as the length L tends to infinity, the probability of intersection tends to certainty.

Laplace extended Buffon's work in 1812 to a rectangular grid, demonstrating that if the distances are a and b apart respectively, then the probability of a needle length 1 intersecting the grid is:

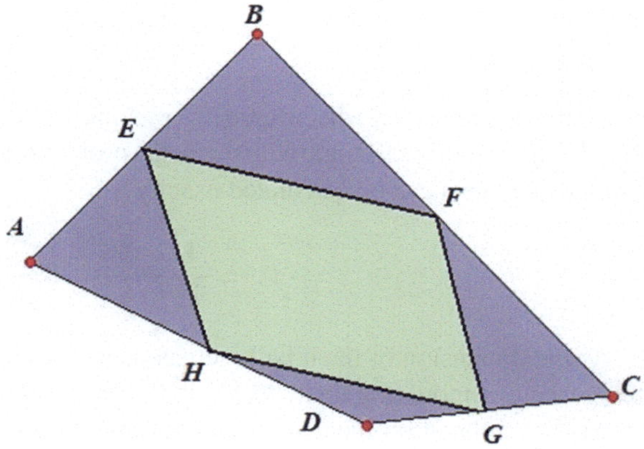

Fig. 14.2 "Varignon's theorem" as an example of "hidden" spatial laws. If four randomly selected points *A*,*B*,*C*,*D* define a convex quadrangle on the plane, then the midpoints of any such convex quadrangle will always define a parallelogram (*E*,*F*,*G*,*H*)

$$P_r = 1 - \frac{2L(a+b) - L^2}{\pi \alpha b}. \tag{14.4}$$

The main interest in Buffon's needle problem lies in the fact that the transcendental number π pops up from spatial probability problems to the extent that π might even be evaluated to several decimals, simply by conducting repeated experiments with different values of L and a. So an experiment in spatial randomness eventually renders an unanticipated solution involving the transcendental number π in it and hinting that spatial randomness may obey to geometric laws.

But Buffon's needle problem is just one example. Surprisingly, locations of geometric objects in 2d very often obey to geometric "rules" governing the relationships among them. These rules may nevertheless pass unnoticed, even during everyday geometric and cartographic measurements and applications.

One characteristic example is "*Varignon's theorem*": let there be four randomly selected points allocated on the plane, defining an arbitrary (convex) quadrangle; then, the theorem ascertains that the midpoints of the sides of this quadrangle form a parallelogram (Fig. 14.2). The unexpected result here is that the shape resulting from joining the midpoints of the quadrangle's sides will *always* be a parallelogram.

Along the same lines, "*Jung's problem*" asks whether it is always possible to enclose a finite set of scattered points in a circle (Jung 1901, 1910). The theorem states that from a finite set of scattered points, with largest possible distance h between them, no matter how scattered the points may be in space, it is always possible to enclose them in a circle with maximum length equal to:

$$a = \frac{h}{\sqrt{3}}. \tag{14.5}$$

This length is the 2-dimensional case of the general Jung length for n-dimensions (Dekster 1995, 1997):

$$a \leq h\sqrt{\frac{n}{2(n+1)}} \tag{14.6}$$

Similarly, Hall (1982) discovered that picking any three random points from within a circle and defining a triangle with them, the probability of obtaining an acute triangle from these points can be calculated exactly as:

$$P_r = \frac{4}{\pi^2} - \frac{1}{8} \approx 28\% \tag{14.7}$$

As the dimension of the n-ball increases, so does the probability of obtaining an acute triangle, mounting up to 90.5% in the 9th spatial dimension.

Another related problem is to find the number of incidences between m points and L lines on the plane. The solution is (Bollobas 2010):

$$2(2mL)^{2/3} + L \tag{14.8}$$

while these m points determine at most.

$$(4m)^{4/3} + m \tag{14.9}$$

unit distances among them (Szekely 1997).

Similarly, the average distance between two points on the surface of a circle is (Weisstein 2011a):

$$D_c = \frac{\int_0^\pi \sqrt{2 - 2\cos\theta}d\theta}{\int_0^\pi d\theta} = \frac{4}{\pi} \tag{14.10}$$

and the probability of circle and line picking is also a constant (Weisstein 2011b):

$$2 - \frac{16}{\pi^2} \tag{14.11}$$

For further information, the reader may refer to the relevant texts relating to geometric probability i.e. (Kendall and Moran 1963; Kendall et al. 1998).

Thus, laws and constants are involved in random geometric allocations. Now let us consider two further examples of spatial probability, beginning with "*Sylvester's problem*" (posed in 1865): Let E be a bounded, everywhere convex, planar domain. The problem then is to compute the probability of four points being located in E, such that one of them lies in the triangle formed by the other points. The formula giving the probability is:

$$P_r = \frac{4}{A_E^4} \iiint\limits_E A(x_1, x_2, x_3)dx_1dx_2dx_3. \tag{14.12}$$

If the spatial domain E is a triangle, then $P = 1/3$ and if it is an ellipse, then

$$P_r = \frac{35}{12\pi^2}. \tag{14.13}$$

More generally, as shown by Blaschke in 1917, for any arbitrary shape of the domain R, the following inequality holds:

$$\frac{35}{12\pi^2} \leq P_r \leq \frac{1}{3} \tag{14.14}$$

A similar argument applies to the 3-dimensional space: the probability that 5 points randomly allocated in a 3-dimensional bounded domain E can form a polyhedron with 5 vertices is (Berger 2010):

$$P_r = \frac{5}{A_E^4} \iint \iint_E A(x_1, x_2, x_3, x_4) dx_1 dx_2 dx_3 dx_4 \tag{14.15}$$

Hence, *even random allocations of spatial objects often obey to geometric restrictions and formulas.* Unfortunately however, the existence of such restrictions may not be as helpful as we would like in order to explain spatial order and disorder.

Aside of these, refreshing the very basics of probability theory may be rewarding for gaining an even deeper understanding of spatial complexity and for this we need to revert back to the basic stochastic behavior of one-dimensional series of symbols. In fact, blocks of cells of either 0 s or 1 s are expected to repeatedly appear in a long binary string of stochastic events, if we allow a "sufficiently" long observation time. Specifically, in 1d, for every $0 < a < 1$, the probability that after n-trials the sequence of the number of coin tosses that one or zero occurs repeatedly $S_n < na$ is given by the "arcsine":

$$\lim_{n\to\infty} (P_r(T_n < a)) \to \frac{1}{\pi} \int_0^a \frac{1}{\sqrt{x(1-x)}} dx = \frac{2}{\pi} \arcsin\left(\sqrt{a}\right) \tag{14.16}$$

This is the well-known *arcsine law* (discovered by Paul Levy in 1939). There is no apparent reason why this law should no apply to a string of symbols representing the consecutive cells of one 2d strip only (of fixed width), suggesting that *blocks of same type of cells are expected to appear in any sufficiently large map that consists in one strip only.* Applying the *Borel-Cantelli lemma* (1917) to a such a one-strip-only map i.e. with black and white cells, it is almost certain that both tossings (black and white) can be synchronized only finitely many times ("synchronized" means that there will be as many 0 s as 1 s after n-tosses). However long or disorganized the tossing experiment may be, its fluctuations of the sequence S_n-np' will be bound (where p' is the apriori probability given by the law of large numbers as stated by Borel in 1909). This is guaranteed by the "*Khinchin theorem bound*" (1924), assuring that for every $\varepsilon > 0$ and for large enough n, almost surely, the sequence S_n minus nq will be bound from above by the bound:

$$(1+\varepsilon)\sqrt{2p'(1-p')n \ln \ln n} \tag{14.17}$$

or, as often alternatively expressed,

$$\lim_{n\to\infty} \sup \frac{S_n - np'}{\sqrt{2p'(1-p')n \ln \ln n}} = 1 \tag{14.18}$$

14.2 Taming Spatial Complexity with Symmetries?

I reach to my center, to my algebra and to my key,
my mirror. Soon I will know who I am.
"Llego a mi centro, a mi álgebra y mi clave,
a mi espejo. Pronto sabré quién soy"
(Jorge Luis Borges, 1899–1986,"Elogio de la sombra" /In praise of darkness)

It appears normal for humans to try to understand spatially complex forms by rapidly spotting symmetries on them (Fig. 14.3), but, despite our fairly advanced

Fig. 14.3 One normally first looks for symmetries in order to understand what a spatially complex shape might be (or represent). Failing to figure out what this shape is supposed to be, it may be satisfactory nevertheless to notice that it is symmetric, so there is a (minimum) self-assurance that, *at least*, there is some logic in it

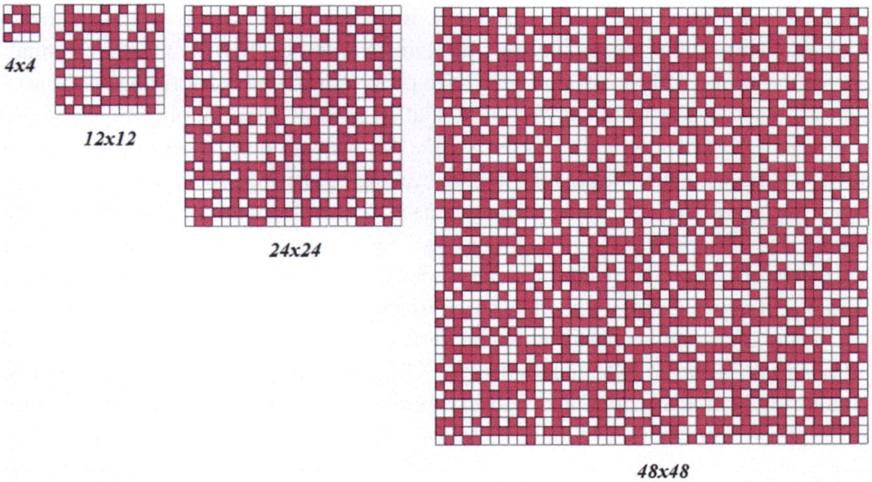

Fig.14.4 Beginning with a 4 × 4 binary map, multiplying it so as to create a 12 × 12 map, then proceeding to create a 24 × 24 and, eventually, a 48 × 48 binary map. The underlying simplicity of the 48 × 48 map (the fact that it is composed by symmetry operations on a basic 4 × 4 map) is hardly perceptible at first glance

knowledge of mathematical symmetries, psychological studies of symmetry detection by humans (Huang and Pasher 2002) revealed that the visual detection of spatial symmetry is often imprecise. This however does not preclude the possibility that, in all likelihood, humans may fail to perceive the presence of small symmetric patterns if they are arranged so as to form a larger *non-symmetric* figure. Consider, for instance, a figure composed of various symmetric repetitions (inversions, rotations left, right, etc.) of the same basic 4 × 4 map (Fig. 14.4).

It appears that the answer to the question whether symmetry and structure are essential to understand spatial complexity by using algebraic methods (Papadimitriou 2012) could only be sought at a fundamental level. "*Ramsey theory*" is probably relevant in this respect. This theory stems from combinatorics and graph theory and explores the question of how big should the original system be, so that at least one of its subsystems has a certain property. Further to Ramsey's theorem, the reader is also referred to the *Van der Waerden, Erdös- Szekeres and Hales-Jewett theorems* (Van der Waerden 1927; Erdös and Szekeres 1935) and to the related literature (Harary 1983; Landman and Robertson 2004). To those unacquaintened with Ramsey Theory however, it suffices to say that it essentially ascertains that *structures inevitably appear* in subsystems and there is always a subsystem with a higher degree of organization than the original system, to be found in every partition of any large structured system or object (although it does not specify which subclass is the one that contains the structured sub-system). By virtue of Ramsey's theorem (Ramsey 1930), any large enough finitely colored complete graph is necessarily endowed with some large monochromatic substructure. Such graphs have k vertices and each vertex is connected to every other by an edge. The smallest complete graph which, if colored with two colors only (i.e. black B and green W), must contain either a black or a white subgraph and defines the Ramsey numbers R(B, W). Even for small graphs, the values of Ramsey numbers have proven difficult to determine and few of them are known. For instance, we know that R(3, 3) = 6 that applies to triangular graphs, R(4, 4) = 18, R(3, 4) = 9, R(5, 3) = 14. We still do not have an evaluation for R(5, 5) because that would require the drawing of an awesomely large number of graphs (approximately 2^{900}). Making only a slight shift from graphs to networks, the "*Szemerédi regularity theorem*" applies to all networks and ascertains that, however big a network may be, it can still be understood by adopting finite approaches. Formally, if G is a graph with at least V vertices, then there is an integer k with $m \leq k \leq M$ and an ε-regular partition of the vertex set of G into k sets, for every $\varepsilon > 0$ and positive integer m (Szemerédi 1975,1978). The profound relationship between symmetries in mathematics and conservation laws in physics proven by Noether (1918) showed that conservation laws follow symmetry properties of nature: energy corresponds to symmetry with respect to time shift, momentum reflects geometric (spatial) shifts and angular momentum corresponds to spatial rotation. Both Ramsey Theory and Noether's theorem imply that symmetry and structure lie at the heart of natural phenomena and given that maximum entropy implies equal probabilities, it has been claimed (Rosen 2008, p. 280) that "the entropy death of the universe can also be called its symmetry death".

Symmetries play a key role in calculating all possible spatial configurations from generic ones. By exploring symmetries, we are not only able to understand spatial complexity of square maps, but also to create spatially complex forms from them. But how does a spatial pattern emerge from a repetition of spatial symmetries? Take, for instance, D_4 (the symmetry group of the square). This can be described by a set of 8 matrices, each one representing a symmetry type of the square (vertical, horizontal):

$$I_S = \begin{pmatrix} 1 & 0 \\ 0 & 1 \end{pmatrix}, H_S = \begin{pmatrix} 1 & 0 \\ 0 & -1 \end{pmatrix}, V_S = \begin{pmatrix} -1 & 0 \\ 0 & 1 \end{pmatrix}, \tag{14.19}$$

the rotational symmetries:

$$R_{90} = \begin{pmatrix} 0 & -1 \\ 1 & 0 \end{pmatrix}, R_{180} = \begin{pmatrix} -1 & 0 \\ 0 & -1 \end{pmatrix}, R_{270} = \begin{pmatrix} 0 & 1 \\ -1 & 0 \end{pmatrix} \tag{14.20}$$

and the symmetries with respect to the left (D_S) and right (D_S') diagonals of the square:

$$D_S = \begin{pmatrix} 0 & -1 \\ -1 & 0 \end{pmatrix}, D_S' = \begin{pmatrix} 0 & 1 \\ 1 & 0 \end{pmatrix}. \tag{14.21}$$

It is easy to verify that multiplying anyone of these eight matrices with any other one yields the equivalent result of the composition of symmetries; in example

$$R_{90}D_S = \begin{pmatrix} 1 & 0 \\ 0 & -1 \end{pmatrix} = H_S \tag{14.22}$$

In 3d, the enumeration of all possible spatial configurations by means of symmetries is possible by extending the Burnside lemma to three dimensional objects by making use of the Redfield polynomials (Redfield 1927) which give the group symmetry G of the cube:

$$\frac{1}{|G|}\sum_{g\in G} s_1^{a_1(g)} s_2^{a_2(g)} s_3^{a_3(g)} \dots \tag{14.23}$$

So the number of ways of placing i.e. 4 squares and 4 circles at the 8 corners of a cube (with only one ball at each corner) is calculated from the composition of groups G_I and G_2: $G_1 \otimes G_2$.

where

$$G_1 = \frac{1}{24^2}\left(s_1^4 + 3s_2^2 + 8s_1s_3 + 6s_1^2s_2 + 6s_4\right)^2 \tag{14.24}$$

and

$$G_2 = \frac{1}{24^2}\left(s_1^8 + 9s_2^4 + 8s_1^2s_3^2 + 6s_4^2\right)^2 \tag{14.25}$$

Hence, the product has a sum of coefficients equal to:

$$\frac{1}{24^3}\left(1^8 8! + (512 \times 1^2 2! 3^2 2!) + (81 \times 2^4 4!) + (216 \times 4^2 2!)\right) = 7 \tag{14.26}$$

and thus result seven distinct configurations of the cube (Fig. 14.5).

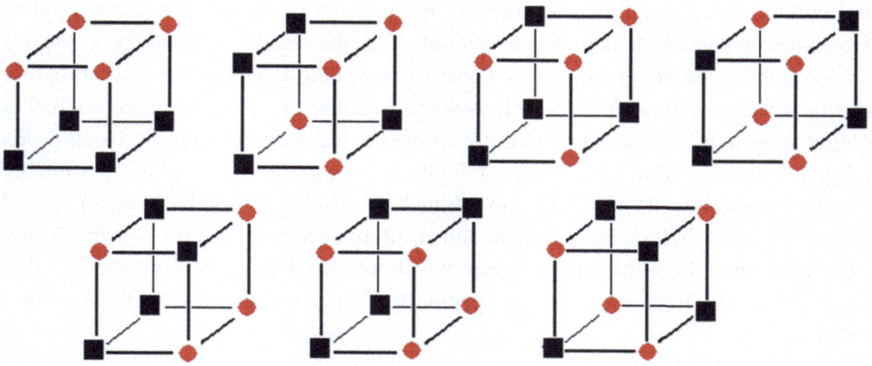

Fig. 14.5 The seven possible cube configurations, with four squares and four circles allocated to each one of its eight corners

Considering the wealth in artworks with friezes and symmetric patterns all over the world, it seems plausible to assume that the exploration of the total number of all possible symmetries of certain geometric shapes shuld have been of interest to humans since very early. But it was as late as 1891, when Fedorov proved that there are only 7 types of linear symmetries in friezes, 17 types of symmetries for planar shapes, 230 types for 3d symmetry groups. We also now know there are 4783 symmetry groups in the 4d space. Hilbert questioned whether might exist a finite number of symmetries for each n-dimensional space (this was his 18th problem), but Bieberbach proved that there is indeed a finite number of symmetries for each spatial dimension, but we still do not know which formula gives them.

14.3 Taming Spatial Complexity with Grothendieck's Inequality?

"To every problem there are two extreme versions:
the Russian version that nobody can simplify
without making it trivial, and the French version
that nobody can generalize any further"
(Vladimir Arnold, 1937–2010,"Real Algebraic Geometry", 2013, p.63).

With some degree of generalization admittedly, it can be said that what lies behind any distribution on a square map is, eventually, numbers. These numbers are linked to spatial patterns and the best way to verify this is by exploring the resulting spatial complexity that is generated as the "substratum" or numbers varies. As unexpected connections of number theory with spatial complexity may pass unnoticed, the connections between spatial objects and algebraic methods can be quite revealing. Following the *Tarski-Seidenberg theorem* for instance, there exists a finite algorithm for classifying the topological properties of curves defined by polynomial equations (although this requires a high number of calculations), and applies to even low values of the degree polynomial. And, the relation of spatial complexity with *geometric complexity theory* (a theory of computational complexity that mainly aims to prove that $P \neq NP$ as well as some hypotheses of algebraic geometry) is another issue to be explored in the future, as are the links between knots, 3-manifolds and prime numbers that have been established from within the field of "*arithmetic topology*" (Reznikov 1997, 2000; Morishita 2009). In our efforts to decipher spatial complexity nevertheless, both algebra and combinatorics can definitely help. Trying to decipher spatial complexity however, we should not lose sight from the fact that surprising results may show up unexpectedly. Consider for instance, the Lambert function:

$$x^a - x^b = (a - b)x^{a+b}n \tag{14.27}$$

weaving a real variable x and its powers (a and b) with a positive integer n. The solution of this equation is given by the Lambert series $W(x)$:

$$\begin{aligned} W(x) &= \sum_{n=1}^{\infty} \frac{(-1)^{n-1} n^{n-2}}{(n-1)!} x^n \\ &= x - x^2 + \frac{3}{2}x^3 - \frac{8}{3}x^4 + \frac{125}{24}x^5 \\ &- \frac{54}{5}x^6 + \frac{16.807}{720}x^7 + \ldots \end{aligned} \tag{14.28}$$

which yields as a result a real number for every $x \geq -1/e$.

This seemingly uninteresting function has the unexpected property (Eisenstein 1844) that the closed form:

$$\frac{-W(-\ln z)}{\ln z} \tag{14.29}$$

is the exact result of the calculation of the tower of infinite powers of the complex number z:

$$z^{z^{z^{\cdots}}} \tag{14.30}$$

and so this "intimidating" tower has an unexpected closed form solution:

$$z^{z^{z^{\cdots}}} = \frac{-W(-\ln z)}{\ln z} \tag{14.31}$$

Indeed, as the history of mathematics and physics has demonstrated time and again, breakthroughs await us at every turn of the road. They should not be underestimated, nor their possible impacts on gaining a better understanding of nature or advancing technology to fields that might have been considered prohibitive before. One characteristic example is Ramanujan's truly amazing formula linking e, π, and the golden section (φ), all together within a single expression:

$$\frac{1}{1 + \frac{e^{-2\pi}}{1 + \frac{e^{-4\pi}}{1 + \ldots}}} = (\sqrt{2+\varphi} - \varphi)e^{\frac{2\pi}{5}} \tag{14.32}$$

Another is the multiplication of two nxn matrices. If performed in the usual way, it is of the order of $2n^3$ additions and multiplications. But using *Strassen's method*, the number of operations is analogous to $n^{\log 7}$. Observing that $\log_2 7 < 3$, it is evident that Strassen's algorithm reduces the complexity of calculations. Spatial analysis has not and could not be left unscathed by such computational shortcuts. Consider for instance, that a matrix may represent the values of a square map. Then, the *Grothendieck inequality* states that if $a_{i,j}$ is an nxn (real or complex) matrix with

$$\left| \sum_{i,j} a_{i,j} s_i t_j \right| \leq 1 \tag{14.33}$$

for all (real or complex) numbers s_i, t_j of absolute value at most 1, then there exists a constant k_G, the so called "Grothendieck's constant", with the property that

$$\left| \sum_{i,j} a_{i,j} \langle S_i T_j \rangle \right| \leq k_G \tag{14.34}$$

for all vectors S_i, T_j in the unit ball $B(H)$ of a real or complex Hilbert space.

In fact, the Grothendieck constant is the smallest constant k_G that satisfies this inequality for all nxn matrices. Grothendieck himself tried to determine the value of this constant and found that:

$$1.57 \approx \frac{\pi}{2} \leq k_G \leq \sinh\left(\frac{\pi}{2}\right) \approx 2.3 \tag{14.35}$$

Eversince this discovery, the exact value of k remains a mystery. For instance, it was conjectured (Krivine 1979) that:

$$1.67696... \leq k_G \leq \left(\frac{\pi}{2 \ln(1+\sqrt{2})} \right) = 1.7822139781... \tag{14.36}$$

but this conjecture was disproved (Braverman et al. 2011).

The implications of Grothendieck's inequality for physics, computer science and human knowledge are profound and can be far reaching. In quantum mechanics for instance, k_G may be interpreted as an upper bound for the deviation in the Bell inequalities, marking the shift from a classical to a quantum state. In network analysis, it is possible to detect communities from within random graphs (i.e. of the Erdös-Rényi type) by using Grothendieck's inequality.

Examined from a topological perspective, one of the major issues emerging from Grothendieck's work is the fact that it is directly related to Gelfand's concept of "nuclear spaces". These are locally convex vector spaces of infinite dimension which are endowed with properties of finite-dimensional spaces (Pajot 2015). Thus, the inequality is able to relate a mathematical expression that is non-calculable (or, at least *NP*-hard) by an equivalent expression that is calculable by means of "semidefinite programming" and therefore solvable in polynomial time (Pisier 2012). Consequently, it won't be too far from the truth to claim that Grothendieck's work paved the way to render the infinite and non-calculable into finite and calculable. And this may very well have numerous favorable implications for spatial analysis and, by consequence, for the analysis of spatial complexity.

14.4 Taming Spatial Complexity with a "Sudoku Method"

"Take no thought of the harvest,
but only of proper sowing"
(Thomas Stearns Eliot, 1888–1965, "The Rock")

When a map's spatial complexity is measured, it can be compared against the maximum values of C_{P1} and C_{P2} that are readily calculated from the map's size n. But this measurement of complexity is, in a way, self-referential as the resulting value of C_{P1} or C_{P2} refers to the map's size only; *not to the possible spatial distributions that different colors might assume on that specific map size*. For map size 3×3, we now know the exact formulas giving C_{P1} and C_{P2} per entropy class. Given this, the spatial complexity of any map may also be compared to what its complexity would be expected *according to its entropy class*. This method can be called a "sudoku method", because it requires the division of a map in 9×9 squares. Consider, for instance, a 9×9 binary map (Fig. 14.6).

Subdividing it into 9 quadrants (T_1 to T_9, from the upper left corner to the down right corner of the image), of size 3×3 each, the allocation of the dark cells in the image (the entropy class r) is given by the following matrix:

$$r(T_i) = \begin{pmatrix} T_1 & T_2 & T_3 \\ T_4 & T_5 & T_6 \\ T_7 & T_8 & T_9 \end{pmatrix} = \begin{pmatrix} 5 & 3 & 9 \\ 1 & 6 & 2 \\ 2 & 4 & 0 \end{pmatrix} \qquad (14.37)$$

which is equivalent to the following matrix of *different* cells:

Fig. 14.6 An example 9×9 binary map

$$r'(T_i) = \begin{pmatrix} 4 & 3 & 0 \\ 1 & 3 & 2 \\ 2 & 4 & 0 \end{pmatrix} \tag{14.38}$$

because all higher than 4 values of complexity are the same as those remaining to 9. The expected spatial complexity depends on the total number of different cells allocated on the landscape (as "different" are considered the cells of matrix **r**'), because the (3 × 3) quadrants with numbers of black cells (entropy class r) from 5 to 9 have the same complexity values as the quadrants with entropy classes from 1 to 4, since they are simply symmetric opposites.

The calculation of C_{P1} for each 3 × 3 quadrant (calculated with rotations allowed also), symbolized here as C_{P1} (Ti) yields the following matrix:

$$\begin{pmatrix} C_{P1}(T_1) & C_{P1}(T_2) & C_{P1}(T_3) \\ C_{P1}(T_4) & C_{P1}(T_5) & C_{P1}(T_6) \\ C_{P1}(T_7) & C_{P1}(T_8) & C_{P1}(T_9) \end{pmatrix} = \begin{pmatrix} 6 & 4 & 2 \\ 4 & 8 & 8 \\ 4 & 6 & 2 \end{pmatrix} \tag{14.39}$$

Summing up all C_{P1} complexity values for all quadrants of the map:

$$C_{P1total} = \sum_{T_i=1}^{T_i=9} C_{P1}(T_i) = 44 \tag{14.40}$$

it follows that the average complexity of the map per quadrant T_i is:

$$C_{P1\,\text{average}} = 44/9 = 4.89 \tag{14.41}$$

This *"average complexity"* of the map at the 3 × 3 quadrant level can be compared with the *"expected complexity"* of 3 × 3 binary map configurations that is derived from the already known formula:

$$C_{P1} = 2.03129 + 2.31226r - 0.392857r^2 + 0.0341667r^3 \tag{14.42}$$

Since the map has 19 different cells in total, anyone of the 3 × 3 quadrants is expected to have (on the average) $r_{\text{average}} = $ (19 different cells)/(9 quadrants) = 2.111 different cells per quadrant. From this average entropy class, and from the curve of $C_{P1}(r)$ of the expected complexity, the average expected complexity per quadrant can be calculated. The C_{P1} corresponding to $r_{\text{average}} = 2.111$ different cells is calculated from the formula discovered for all 3 × 3 binary maps: $C_{P1\,\text{expected}} = 5.4832$. This is the *"expected C_{P2} complexity"* of the map at the 3 × 3 quadrant level that is derived from the model of all 3 × 3 binary square maps. We define the "expected complexity" to be the "*a priori complexity*", *that would on the average be expected from a random allocation of the image's dark cells in it.* Clearly, the a priori spatial complexity depends on the average entropy class per quadrant. Therefore, the ratio of the expected to the average complexity is:

Fig. 14.7 A more complex 9 × 9 binary map

$$C_{P1\,\mathbf{relative}} = C_{P1\,\mathbf{average}}/C_{P1\,\mathbf{expected}} = 4.89/5.4832 = 89.18\% \tag{14.43}$$

This is the *relative complexity* of the map, that is *a measure of the comparison of its spatial complexity compared to the expected spatial complexity that corresponds to the map's entropy class*. In this case, it means that the observed complexity of this map is approximately 89% of the spatial complexity that would be expected from a random allocation of as many dark cells in this 9 × 9 binary map. Adopting exactly the same procedure for C_{P2} and on the basis of the formula

$$C_{P2} = 0.0192857 + 2.5556r + 0.122143r^2 - 0.104167r^3 \tag{14.44}$$

it follows that $C_{P2\mathbf{average}} = 30/9 = 3.33$, $C_{P2\,\mathbf{expected}} = 4.945$ so the result is $C_{P2\,\mathbf{relative}} = 67\%$.

Simularly, for the map of Fig. 14.7., $C_{P1\,\mathbf{relative}} = 117.5\%$ and $C_{P2\,\mathbf{relative}} = 103.6\%$ and this allows us to conclude that this map is more spatially complex.

There is a strong cross-disciplinary need for methods to calculate and compare the spatial complexities of two-dimensional images, maps, pictures, landscapes etc. in 2d representations. This aim can not be satisfied without having a measure to compare spatial complexity against. For this reason it was necessary to calculate all the expected possible complexity types for all possible binary spatial configurations. In the present book, this objective was met at the level of 3 × 3 maps. The C_{P1} and C_{P2} of "large" maps are clearly difficult to calculate as they were for all 3 × 3 binary maps. But, with the sudoku method shown here for small maps, it is possible to calculate the spatial complexity of any "small-sized" map. Furthermore, any map can be divided in a grid of cells that would be multiples of 3 × 3 elementary cells, as shown here, so that nested sudokus can be created to cover larger maps and each sudoku can be considered as the quadrant of its immediately larger sudoku, in a fractal-like self-similar spatial division. And then, it is possible to use the same formulas to derive apriori complexity measurements, by comparing the observed (the measured) map complexity with the theoretically expected from a random allocation of dark (different) cells on the map. This comparison provides a measure of the map's spatial complexity with respect to its a priori expected spatial complexity. Evidently, if larger than 3×3 binary maps were used, the results would differ. Although this is obviously true, it has to be considered that we are still short of a formula that would give us the C_{P1} complexities of all the i.e. 36,493 configurations that would be required to calculate for the 4×4 binary maps (or for larger maps, such as 5×5, 6×6 etc.).

References

Berger, M. (2010). *Geometry revealed*. Heidelberg, New Yorks: Springer.

Bollobas, B. (2010). *The art of mathematics*. Cambridge: Cambridge University Press.

Braverman, M., Makarychev, K., Makarychev, Y., & Naor, A. (2011). The Grothendieck constant is strictly smaller than Krivine's Bound. In *52nd annual IEEE symposium on foundations of computer science (FOCS)*. pp. 453–462.

Dekster, B. V. (1995). The Jung theorem for the spherical and hyperbolic spaces. *Acta Mathematica Hungarica, 67*(4), 315–331.

Dekster, B. V. (1997). The Jung theorem in metric spaces of curvature bounded above. *Proceedings of the American Mathematical Society, 125*(8), 2425–2433.

Eisenstein, G. (1844). Entwicklung von . *Journal für die reine und angewandte Mathematik, 28*, 49–52.

Erdös, P., & Szekeres, G. (1935). A combinatorial problem in geometry. *Compositio Mathematicae, 2*, 463–470.

Hall, G. R. (1982). Acute triangles in the *n*-ball. *Journal of Applied Probability, 19*, 712–715.

Harary, F. (1983). A tribute to Frank P Ramsey, 1903–1930. *Journal of Graph Theory, 7*, 1–7.

Huang, L., & Pashler, H. (2002). Symmetry detection and visual attention: A binary-map hypothesis. *Vision Research, 42*(11), 1421–1430.

Jung, H. (1901). Über die kleinste Kugel, die eine räumliche Figur einschließt. *Journal Für Die Reine Und Angewandte Mathematik, 123*, 241–257.

Jung, H. (1910). Über die kleinste Kreis, der eine ebene Figure einschließt. *Journal Für Die Reine Und Angewandte Mathematik, 137*, 310–313.

Kendall, M. G., & Moran, P. A. P. (1963). *Geometrical probability*. New York: Hafner.

Kendall, W. S., Barndorff-Nielson, O., & van Lieshout, M. C. (1998). *Current trends in stochastic geometry: Likelihood and computation*. Boca Raton, FL: CRC Press.

Krivine, J.-L. (1979). Constantes de Grothendieck et fonctions de type positif sur les sphères. *Advances in Mathematics, 31*(1), 16–30.

Landman, B. M., & Robertson, A. (2004). *Ramsey Theory on the Integers, Student Mathematical Library, 24*. Providence, RI: American Mathematical Society.

Morishita, M. (2009). *Knots and primes*. Heidelberg: Springer.

Noether, E. (1918). Invariante variationsprobleme. Nachr. d. Konig. Gesellsch. d. Wiss. zu Gottingen, Math-phys. Klasse (1918), 235–257; English translation by M. A. Travel, in *Transport Theory and Statistical Physics 1*(3)1971, 183–207.

Pajot, P. (2015). La revanche d'un theoreme oublie. *La Recherche, 501–502*, 96–100.

Papadimitriou, F. (2012). Modelling landscape complexity for land use management in Rio de Janeiro. *Brazil. Land Use Policy, 29*(4), 855–861.

Pisier, G. (2012). Grothendieck's theorem, past and present. *Bulletin of the American Mathematical Society, 49*(2), 237–323.

Ramsey, F. P. (1930). On a problem of formal logic. *Proceedings London Mathematical Society, 30*(1), 264–286.

Redfield, J. H. (1927). The theory of group-reduced distributions. *American Journal of Mathematics, 49,* 433–455.

Reznikov, A. (1997). Three-manifolds class field theory (Homology of coverings for a nonvirtually-positive manifold). *Selecta Mathematica, 3,* 361–399.

Reznikov, A. (2000). Embedded incompressible surfaces and homology of ramified coverings of three-manifolds. *Selecta Mathematica, 6,* 1–39.

Rosen, J. (2008). *Symmetry rules: How science and nature are founded on symmetry*. Berlin-Heidelberg: Springer.

Szekely, L. (1997). Crossing numbers and hard Erdös problems in Discrete Geometry. *Combinatorics, Probability and Computation, 6,* 353–358.

Szemerédi, E. (1975). On sets of integers containing no k elements in arithmetic progression. Polska Akademia Nauk. *Instytut Matematyczny. Acta Arithmetica, 27,* 199–245.

Szemerédi, E. (1978). Regular partitions of graphs. Problèmes combinatoires et théorie des graphes. Paris: Colloq. International CNRS, University Orsay, Orsay, 1976, 260, pp 399–401.

Van der Waerden, B. (1927). Beweis einer Baudetschen Vermutung. *Nieuw Archief Voor Wiskunde, 15,* 212–216.

Weisstein, E. W. (2011a). Circle line picking. From *MathWorld*--A Wolfram Web Resource. https://mathworld.wolfram.com/CircleLinePicking.html

Weisstein, E. W. (2011b). Geometric Probability. From *MathWorld*--A Wolfram Web Resource. https://mathworld.wolfram.com/GeometricProbability.html

Part V
Epistemological, Psychological, Geophilosophical and Aesthetic Perspectives on Spatial Complexity

Chapter 15
Spatial Complexity, Psychology and Qualitative Complexity

The streets are the corridors of the soul and obscure trajectories of memory.
"Les rues sont les couloirs de l' ame
et des obscures trajectoires de la mémoire"
(Paul Virilio, 2007, p.21)

Abstract How is spatial complexity perceived by humans? The study of spatial complexity is scale-dependent: a spatial entity that may look simple at some spatial level, may reveal many details if examined closer at finer spatial scales. To perceive complexity, we often resort to generalizations: thematic (grouping up similar themes, i.e. similar colors) and geometric (grouping up similar shapes). We are able to retain the basic (or prevalent) thematic and geometric properties of spatial objects so as to make sense of them without significant loss of information. The principles of gestalt psychology are particularly relevant, as is the concept of "visual complexity", which has attracted the attention of many researchers for almost a century. Examples of binary maps are presented, showing that spatial randomness can not always be recognized easily, while the reverse also holds: regularities in allocations of black cells in a binary map can easily be misperceived visually as random allocations. Yet, the extent to which spatial complexity is associated to qualitative complexity is hitherto unknown, but it relates to one of the uttermost problems of human existence: discovering and eliciting meanings from spatial entities. Visual material conveying strong sense of meaning does not always coincide with high spatial complexity.

Keywords Spatial complexity · Gestalt psychology · Qualitative complexity · Complexity and Psychology · Geography and Psychology · Map complexity

15.1 Gestalt Psychology and Spatial Complexity

"I dare not guess; but in this life of error, ignorance, and strife, where nothing is, but all things seem, and we the shadows of the dream"

(Percy Bysse Shelley, 1792–1822, "The Sensitive Plant")

F. Papadimitriou, *Spatial Complexity*, https://doi.org/10.1007/978-3-030-59671-2_15

Fig. 15.1 Hermann grid creating the illusion of small grey squares in between the black ones

The perception of complexity has attracted and keeps attracting the interest of experimental psychologists for more than fifty years (Attneave 1957; Taylor and Eisenman 1964; Day 1967; Chipman 1977; Nicki and Moss 1975; Hekkert and Wieringen 1990; Messinger 1998; Strother and Kubovy 2003). Experimental psychology in particular, is rich in research results relating to the perception of space, and many of these are intimately related to the spatial complexity of the object(s) viewed, from the way we perceive faces (Afraz et al. 2010) to the way we perceive landscapes and ordinary objects. The perception of spatial patterns is a complicated issue, replete with surprises, such as (among many) illusory contours, grey ghost images "seen" from within Hermann grids (Fig. 15.1), "Mach bands" of shadows, "*texture gradients*" (which are perceived when equally-spaced elements in a scene appear denser from longer distances—see Fig. 15.2) etc.

Experimental evidence has suggested (Popple and Levi 2005) that items placed close to one another distort the sense of order and increase spatial uncertainty to the viewer, although spatial thinking may actually reduce the complexity of several problems (Cain 2019). The "parahippocampal place area" of the temporal cortex has been identified by neurophysiology as the region of the brain that is activated upon looking at physical spaces (whether they are indoors or outdoors), in the same way as the "fusiform face area" under the temporal lobe is activated when viewing faces. By activating the parahippocampal gyrus, the brain automatically identifies particular objects it can use for navigation in space and use them as landmarks for guidance.

Perhaps no other psychological theory is more related to the perception of spatial complexity than gestalt psychology, the basic tenets of which are articulated in a number of "laws". With the exception of one that is basically time-related (the "law

Fig. 15.2 The "texture gradient": equally-spaced elements appear densely packed if viewed from longer distances

of common fate"), the remaining laws are space-related and, as such, they should clearly be brought in the context of spatial complexity. So let us briefly examine them one by one vis-a-vis with spatial complexity:

The "*law of simplicity*" (or "pragnanz") asserts that perception tends to eventually produce as simple structures as possible. In other words, this is equivalent to claiming that humans have an inherent mechanism for reducing complexity (including and most prominently) spatial complexity, so as to create a spatially simpler structure that the brain's aparatus can proccess as rapidly as possible. According to this law, similar spatial elements tend to be grouped together (Fig. 15.3). This is another way of expressing the thematic generalization of visual input and, quite probably, many animal species are endowed with this capacity. A further generaliation (and complexity reduction) is pertinent to the "*law of nearness*" asserting that the mind tends to group together elements that are near to one another (Fig. 15.4).

Essentially, this is a mental process of *spatial complexity reduction* and can be conceived along the same lines as the "*law of good continuation*", which ensures that our perception tends to connect spatial elements so as to get a sense of continuity (Fig. 15.5). Finally, the "*law of familiarity*" explains how spatial elements are grouped easier if the resulting groups are conducive to meaningful or familiar spatial shapes.

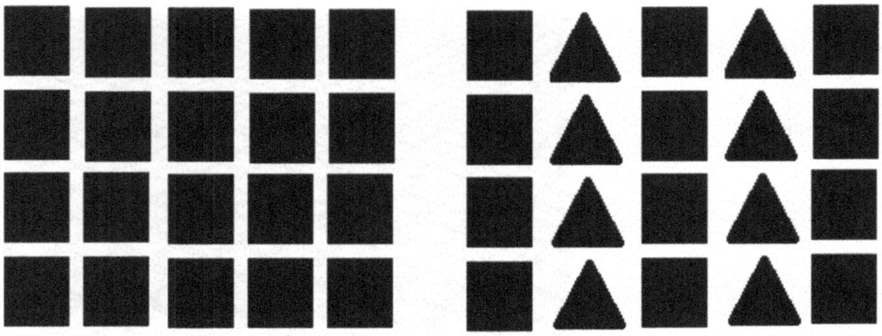

Fig. 15.3 The gestalt law of simplicity: an array of squares is perceived as either columns or rows of triangles. If some of them are changed to triangles, then the shape will be perceived as a spatial arrangement of vertical columns

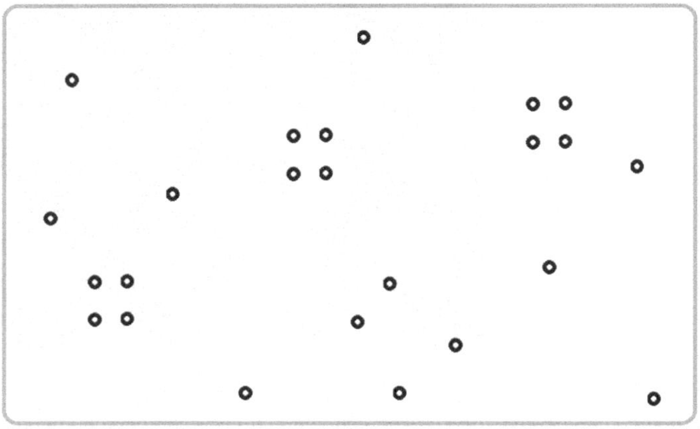

Fig. 15.4 The gestalt law of nearness: Three squares are easily discernible in this space of points, due to the spatial proximity (nearness) of four points in each one of these three sets of four points

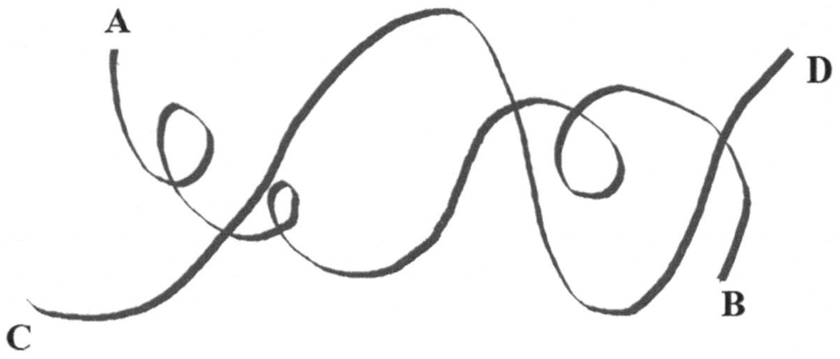

Fig. 15.5 The gestalt law of good continuation: Beginning a line from point A leads to B (not to D), as, according to this law, connected points of lines or curves follow the smoothest course from a start point to an end point. The same applies to the curve starting from C: it "normally" ends at D instead of B

Essentially, this law expands further the previous ones, by linking spatial complexity with qualities (meanings of spatial objects).

Aside of these, the "problem of *perceptual segregation*" consists in distinguishing a figure from its backdrop. It was accepted right from the beginnings of the theory of landscape ecology in which the main landscape cover (the backdrop in this case) is called "the matrix" that surrounds dispersed fragments and patches of other landscape covers. In the arts, the impressive impact of gestalt psychology on art is exemplified by the "Group de Recherche d' Art Visuel" that was founded in Paris by several artists with Victor Vasarely (1908–1997) among them, thus founding the "*optical art*".

Beyond these classic laws of gestalt psychology however, an alternative set of gestalt-type principles has been suggested (Palmer 1992; Palmer 1999; Palmer and Rock 1994). As in the gestalt laws, aside of one principle that is time-specific (the "principle of synchrony"), the remaining two are both space-specific. The "principle of uniform connectedness" asserts that connected regions according to some physical or geometric criterion (color, brightness, texture etc.) tend to be perceived as a single unit, while the "principle of common region" explains how spatial elements in the same region of space are grouped together. These principles seem to match exactly with the thematic generalization of maps and images, so as to reduce the perceived spatial complexity to levels affordable by the brain for processing. The latter principle also reveals the psychological importance of borders among spatial units.

15.2 Scale-Dependence of Spatial Complexity

> The image has become the final form of commodity reification
> (Guy Debord, 1931–1994,"The society of the spectacle", 1967)

The spatial complexities in the world around us are difficult for the human brain to handle, and this is precisely why spatial complexity is so much intricately linked to the psychology of perception. Indeed, unless we have to make a highly detailed analysis of an image for some particular purpose, it would be vain to consider its analysis at the pixel level, unless there is a good practical reason to do so. Analyzing spatial complexity for practical purposes however, will most often eventually involve some level of generalization. Truly, while an image may appear "complex" at some level of observation, it may as well appear "simple" at another (Fig. 15.6).

This observation points to the "generalization" of spatial information that is indispensable to the brain's internal mechanisms of information processing. Generalization helps the viewer (or the researcher) to derive meanings from visual objects of high spatial variability. This is a twofold process: thematic (unification of similar classes/colors/categories/themes etc.) and geometric (unification of similar shapes), so as to derive a simplified version of the initial spatial object, which nevertheless still makes sense to the viewer, although with less detail (Fig. 15.7).

Fig. 15.6 Spatial complexity is scale-dependent: a spatial entity that may appear simple at some spatial resolution may gradually reveal many chromatic details if examined closer, at finer spatial scales. Here, a photograph of the center of Sao Paulo, as examined at different levels of observation: the spatial complexities of the resulting maps are expectedly different

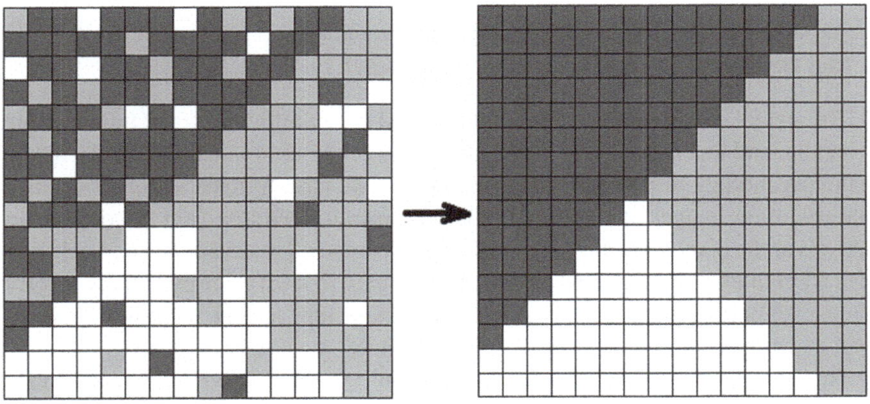

Fig. 15.7 The two types of "generalization" of an image: thematic (grouping up similar themes, i.e. similar colors) and geometric (grouping up similar shapes together). By instantly applying generalization procedures, the human mind is able to retain the basic (or prevailing) thematic and geometric properties of a spatial object so as to be able to make sense of it, by affording the least loss of important information

Not only *thematic*, but also *geometric* simplifications are necessary in order to process the information arising from highly complex spatial forms and surfaces. Using quadrants is one method for such complexity-sensitive spatial analyses (Papadimitriou 2013). The generalization process however, taken to the extremes, may lead to *over-simplification*, in which neither the original meaning of the spatial object is retained, nor its critical geometric details (Fig. 15.8). When it happens, the hallmark of over-simplification is the *impairment of meaning* or the loss of fundamental spatial and thematic characters of the original spatial object, to the extent that the simplified object loses completely the meaning or principal character or quality of the original object. These being said, it follows that there is a strong qualitative aspect of spatial complexity, which needs to be considered when embarking to carry out any analysis of spatial complexity. And these qualities are, quite often, inexorably linked to the spatial resolution of the object observed. Further, since the foveal vision is better than the peripheral vision, the further a symbol or pattern or spatial element lies from the viewer's focus of attention, the less it is discernible. The perception of

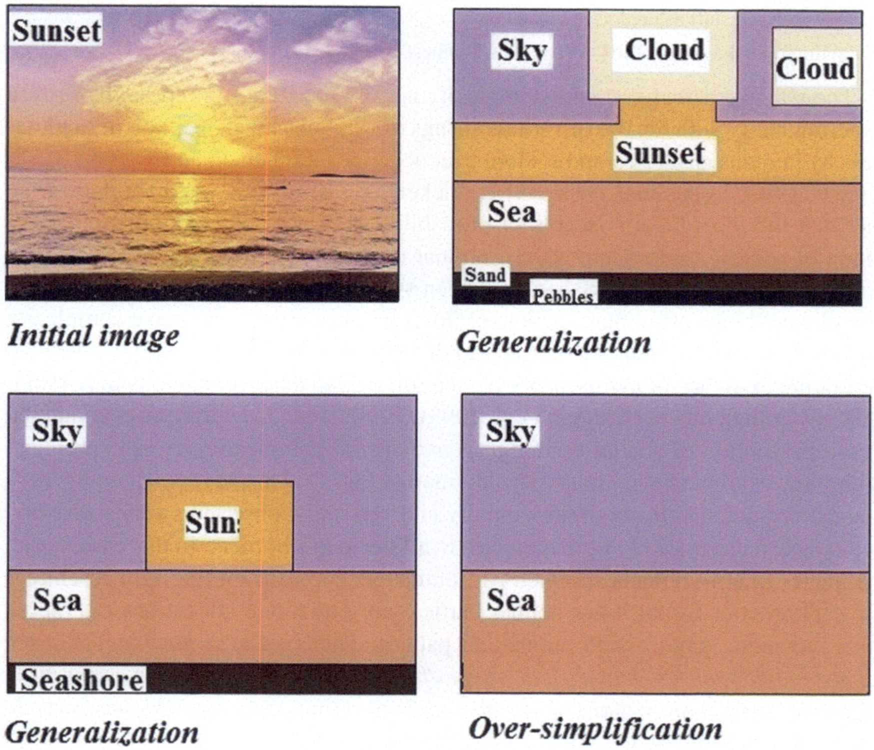

Fig. 15.8 A generalization of a spatial object should not lead to over-simplification, in which the original spatial complexity is lost by its largest part, along with the meaning of the original object. The overly simplified image at the bottom right bears almost none of the messages that the initial image emits to the viewer. Yet, there is no general rule defining when and how much should spatial complexity be simplified

spatial complexity is therefore *focus-dependent*: the more remote a spatial element is from the focus of the gaze, the less it is perceived by the viewer (although it may well be present).

It is for this reason, that "salience maps" have been created depicting fixations for the eye, as the gaze casts on some 2d or 3d object. Such maps reveal how the viewer focuses more on some particular regions of the scene or picture (Parkhurst et al. 2002) and they might be used as surrogate auxiliary indicators of the visual complexity of an object. Experimental evidence (Kidd et al. 2012) has suggested that 7–8 months old infants tend to focus on neither too simple nor too complex items, but whether this applies to adults also and under what conditions is still largely unknown.

15.3 Perception of Spatial Randomness and Spatial Complexity

"We are such stuff as dreams are made on"

(William Shakespeare, 1564–1616, "The Tempest", IV.i.148)

The way we perceive spatial complexity is truly puzzling. As repeatedly proven experimentally with the use of various strings of symbols, the perception of randomness by humans is skewed and seldom ever accurate (Kahneman and Tversky 1972; Kareev 1992; Lopes and Oden 1987; Nickerson 2002). This is partly due to the fact that the theoretical concepts of probability may not coincide with subjective views of what is random and what is regular (Beltrami 1999), to the extent that the term "*subjective complexity*" has even been suggested (Falk and Konold 1997). To complicate things even more, people of different ages seem to perceive complexity and randomness differently (Martial Van der Linden et al. 1998), while people who accept the existence of extrasensory perception tend to discover patterns in randomness easier than others (Brugger et al. 1993). To illustrate the surprises encountered in the perception of spatial randomness and spatial complexity, let us experiment with some example binary maps.The left map of Fig. 15.9 displays equal numbers of counts of colored cells, both horizontally and vertically, observing a repetitive and symmetric pattern: 4111114 that applies to all the map's borders. In this case, visual symmetry simply reflects symmetry in numbers. The map on the right resembles the capital letter B, but it has neither vertical nor horizontal allocations of colored cells indicating any possible numerical pattern. Thus, *patterns meaningful to the human mind such as geometrical shapes/figures are not expected to necessarily follow patterns in allocations of numbers also:* meanings from spatial allocations do not necessarily follow numerical patterns in allocations.

Different interesting problems appear when allocations of colored cells appear as if they were random. The maps of Fig. 15.10 are two such cases, although map (*a*) appears less random than map (*b*), deemed by its vertical set of allocations of cells. While both maps appear random, map (*a*) has a striking normality in the number of colored cells per row. Furthermore, map (*b*) has half of its colored cells concentrated

Fig. 15.9 Two 8 × 8 binary maps, with 14 colored cells in each one of them. The map on the left has a symmetric allocation of colored cells, either if considered horizontally or vertically and this symmetry is visible in the numbers of cells per row and per column. The map on the right shows a pattern reminiscent of the capital letter B but without any numerical pattern corresponding to it

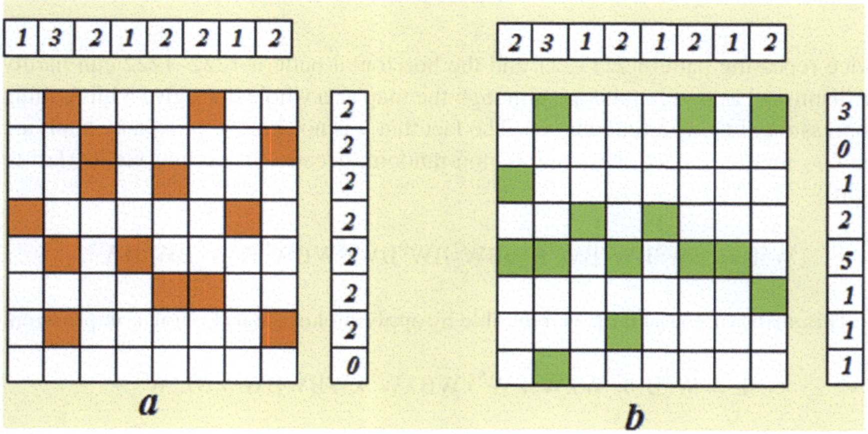

Fig. 15.10 Two 8 × 8 binary maps, with entropy class 14 each one, and with different partitions of the number 14, either if counted horizontally or vertically. Counts show numbers of colored cells per row and per column. While both maps appear random, map (*a*) has a striking normality in the allocation of colored cells per row

in only two rows (the 4th and 5th), and as such, they are more aggregated than in map (*a*) which displays a higher dispersion of its colored cells. Given this, map (*a*) would appear to the viewer as the random one rather than map (*b*), but, as the regularity in its counts per row suggests, it is not so.

It is possible to check such divergences between perception and reality, by using one of the spatial complexity metrics. So let us examine yet another case (Fig. 15.11). It is also difficult to decide whether this map is random or not. Both the vertical

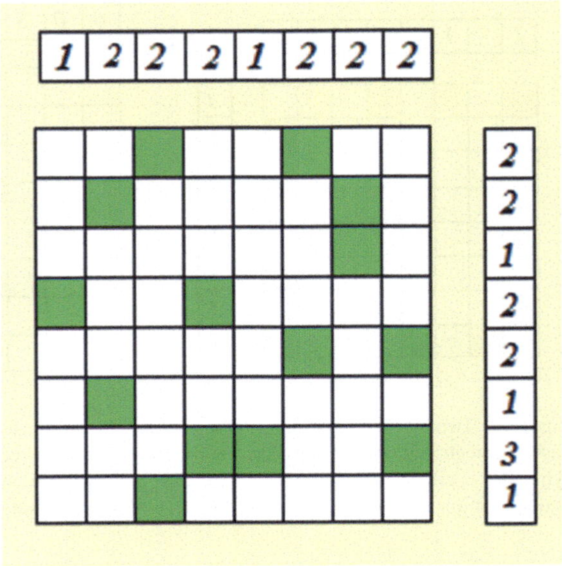

Fig. 15.11 An 8 × 8 binary map with horizontal and vertical partitioning that displays a repetitive pattern (counts show numbers of colored cells per row and per column), either if counted horizontally (1222–1222) or vertically (221-221-31). By applying a spatial complexity metric, it can be proven that it is non-random

twice repeating pattern 221–221 and the horizontal pattern 1222–1222 can hardly be dismissed as purely random, although the map as a whole does give a misleading impression of being a random one. The fact that it is not, can be proven by applying the C_{PI} metric. Indeed the string is non-random, because its coding yields (left to right):

$$W^2BW^2BW^3BW^4BW^7BWBW^2BW^9BWBWBW^9B^2W^2BW^2BW^5$$

This string of 39 symbols is reducible by applying i.e. the following compression:

$$\lambda = W^2B,\ \lambda^2W\lambda W^2\lambda W^5\lambda WB\lambda W^7\lambda WBWBW^7\lambda B\lambda^2W^5$$

which results to 34 symbols.

Similarly, in 3d it is not always visible or otherwise perceptible that a knot may be "isotopic" to another, simpler knot. For instance, the "Figure 8" knot (resembling the figure of the number 8) can be twisted so as to become unrecognizable, although it can still be simplified by typical "unknotting" moves. But reality is quite often far more complex than these simple cases, to the extent that the degrees of linkedness or knottedness can be intractable. Indeed, a 3d surface can be so much twisted, knotted and braided that becomes inexorably complex and impossible to visually perceive how much complex it is without modifying its structure either manually or mechanically (Fig. 15.12). It therefore follows that any limitations to visually perceive (decode, untangle, compress etc.) spatial complexity do *not* necessarily imply *cognitive* limitations also.

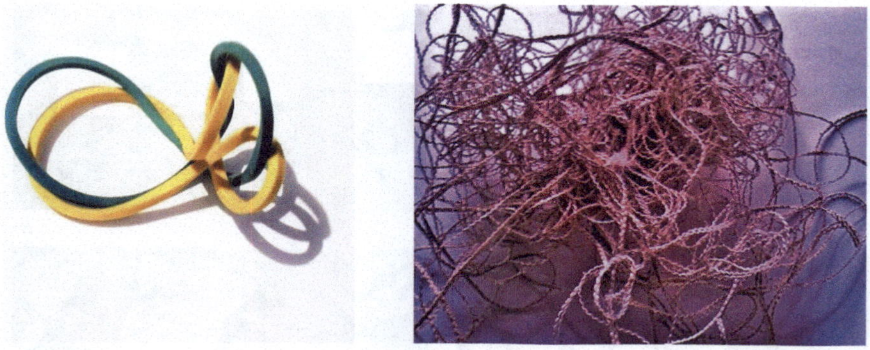

Fig. 15.12 It may not be easy to visually discriminate even the two disjoint rubbers (left). But strands, strings or surfaces can be so much knotted, inter-linked and inter-braided, that it is impossible to visually solve the resulting riddle (right) without a hands-on intervention to simplify it

15.4 Qualitative Complexity Versus Spatial Complexity

"Every sign by itself seems dear. What gives it life?"

(Ludwig Wittgenstein 1953, p. 128)

It is not certain however, that gestalt laws encapsulate the *qualitative complexity* that is sometimes associated to spatial complexity (or simplicity). This is a special kind of complexity that relates to meanings, concepts, ideas, beliefs, feelings, affects associated with a particular piece of space (Papadimitriou 2010). Besides, following Carl Gustav Jung's theory of the collective unconscious, an archetype is a rather simple shape that summarizes a very large set of meanings, which may also be culture-specific. In the context of spatial complexity, an archetype can be regarded as a simple geometric shape with *a huge qualitative complexity encapsulated in a strikingly small spatial complexity*.

Consider a geological map for instance: its complexity does not only depend on the areas of rock classes, but also on linear features, such as faults. Furthermore, despite occupying relatively insignificant portions of the map's area, some map symbols (letters, numbers, characters) may nevertheless convey meanings and connotations that can attract the viewer's attention far more than the backdrop, which may also be spatially very complex. In fact, the semantic complexity associated to the spatial complexity can sometimes be assessed quantitatively (Papadimitriou 2012).

At this point, let us elaborate further on the relationships between spatial complexity and meanings, by means of the set of example photographs of Fig. 15.13.

The original photograph is *D*. Its size is 324 kb. If the figures of cats are subtracted from it, the resulting image *A* is meaningless, but it nevertheless has a size of 260 kb. Allowing for the cats' bodyshapes to show up (but not their faces), then image *B* begins to make some sense (an area with bodies of some unrecognizable animals which could be cats, but also other animals) even with higher spatial complexity,

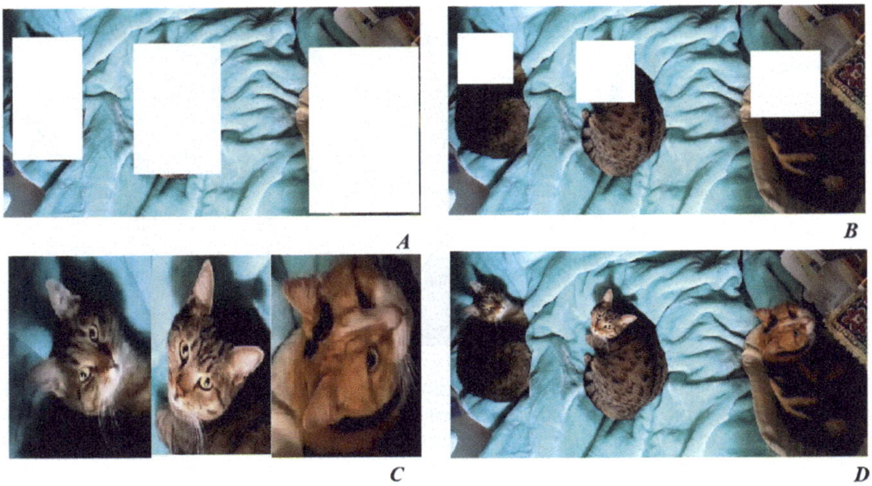

Fig. 15.13 A case of spatial complexity vs meaning: *D* is the original image. Photograph *C* is the most meaningful to the viewer, although it has the least spatial complexity

as it requires 304 kb memory. Yet, while the size of image *C* is only 142 kb, it conveys the most important part of the meaning of the original image (*D*), as it allows (the author only!) to identify faces that are unlike any other: Fafy (left), Mikraki (middle) and Mati-Mati (right). Hence, although subtracting details from the original image *D* results in images of lower spatial complexity as anticipated, photographs *A* and *B* are still meaningless as the cats' faces are not shown. But it is precisely the faces that convey the meaning which makes a difference for the entire (meaningless) images *A* and *B*, although with *less* spatial complexity than either *A*, *B* or *D*. Hence, *highly meaningful visual material does not necessarily coincide with high spatial complexity*. This is because meanings are purely subjective and overcome the algorithmic, computational, geometric and topological determinants of spatial complexity. This example shows why spatial complexity is different than qualitative complexity. At the landscape level, it has been shown that a spatially highly complex landscape is not necessarily endowed with high qualitative complexity and vice versa (Papadimitriou 2010). Yet, qualitative complexity is not only personal; it may also result as a social or cultural construct, recognizable or acceptable by many people. Consider, for instance, four photographs (Fig. 15.14), invoking different relationships between spatial complexity and meanings associated to them by different viewers.

A low complexity seascape in a sunset can be vested with *any meaning*. Oppositely, a widely known landscape, such as the rock of the Athens Acropolis, has been associated with more or less *the same set of meanings* that are understandable by billions of peoples, regardless of this landscape's spatial complexity (Parthenon, the center of ancient Athens, associated with the city that has been the cradle of democracy and philosophy). Yet, the spatially complex details appearing on the face

Any meaning ***Universal meaning*** ***Personal meaning***

Fig. 15.14 Searching for meaningful interpretations of spatially complex objects. Perceiving spatial complexity is most likely unrelated to the mathematical properties of the spatial objects observed (see text for explanation)

of Mati-Mati has a *personal meaning* (to the author only), that distinguishes this cat from all other similar ones.

These simple examples give glimpses of the problem of attributing and associating meanings and qualities to spatial complexity. As yet, philosophy does not seem to possess a universally acceptable general theory of meaning. Furthermore, how the human mind forms meanings from images has long been a field of intensive research and, therefore, the mechanisms by which the mind associates meanings to spatial forms with regard to their spatial complexity (or to their algorithmic length) is hitherto unknown. Thus, spatial complexity is intricately connected to one of the paramount problems of human existence: *discovering and eliciting meanings from anything spatial*.

References

Afraz, A., Pashkam, M. V., & Cavanagh, P. (2010). Spatial heterogeneity in the perception of face and form attributes. *Current Biology, 20,* 2112–2116.

Attneave, F. (1957). Physical determinants of the judged complexity of shapes. *Journal of Experimental Psychology, 53,* 221–227.

Beltrami, E. (1999). *What is random? Chance and order in mathematics and life*. New York: Springer-Verlag.

Brugger, P. (1997). Variables that influence the generation of random sequences: An update. *Perception and Motor Skills, 84,* 627–661.

Cain, A. J. (2019). Visual thinking and simplicity of proof. *Philosophical Transactions of the Royal Society A, 377,* 20180032.

Chipman, S. F. (1977). Complexity and structure in visual patterns. *Journal of Experimental Psychology: General, 106,* 269–301.

Day, H. (1967). Evaluations of subjective complexity, pleasingness and interestingness for a series of random polygons varying in complexity. *Perception and Psychophysics, 2,* 281–286.

Falk, R., & Konold, C. (1997). Making sense of randomness: Implicit encoding as a basis for judgment. *Psychological Review, 104*(2), 301–318.

Hekkert, P., & Wieringen, P. C. W. (1990). Complexity and prototipicality as determinants of the appraisal of cubist paintings. *British Journal of Psychology, 81,* 483–495.

Kahneman, D., & Tversky, A. (1972). Subjective probability: A judgment of representativeness. *Cognitive Psychology, 3*(3), 430–454.

Kareev, Y. (1992). Not that bad after all: Generation of random sequences. *Journal of Experimental Psychology: Human Perception and Performance, 18*(4), 1189–1194.

Kidd, C., Piantadosi, S. T., & Aslin, R. N. (2012). The Goldilocks effect: Human infants allocate attention to visual sequences that are neither too simple nor too complex. *PLoS ONE, 7*(5), e36399.

Lopes, L. L., & Oden, G. C. (1987). Distinguishing between random and nonrandom events. *Journal of Experimental Psychology. Learning, Memory, and Cognition, 13*(3), 392–400.

Martial Van der Linden, M., Beerten, A., & Pesenti, M. (1998). Age-related differences in random generation. *Brain and Cognition, 38,* 1–16.

Messinger, S. M. (1998). Pleasure and complexity: Berlyne revisited. *The Journal of Psychology, 132,* 558–560.

Nickerson, R. S. (2002). The production and perception of randomness. *Psychological Review, 109*(2), 330–357.

Nicki, R. M., & Moss, V. (1975). Preference for non-representational art as a function of various measures of complexity. *Canadian Journal of Psychology, 29,* 237–249.

Palmer, S. E. (1992). Common region: A new principle of perceptual grouping. *Cognitive Psychology, 24,* 436–447.

Palmer, S. E. (1999). *Vision Science.* Cambridge MA: MIT Press.

Palmer, S. E., & Rock, I. (1994). Rethinking perceptual organization: The role of uniform connectedness. *Psychonomic Bulletin & Review, 1,* 29–55.

Papadimitriou, F. (2010). Conceptual Modelling of Landscape Complexity. *Landscape Research, 35*(5), 563–570.

Papadimitriou, F. (2012). Artificial intelligence in modelling the complexity of mediterranean landscape transformations. *Computers and Electronics in Agriculture, 81,* 87–96.

Papadimitriou, F. (2013). Mathematical modelling of land use and landscape complexity with ultrametric topology. *Journal of Land Use Science, 8*(2), 234–254.

Parkhurst, D., Law, K., & Niebur, E. (2002). Modeling the role of salience in the allocation of overt visual attention. *Vision Research, 42,* 107–123.

Popple, A. V., & Levi, D. M. (2005). The perception of spatial order at a glance. *Vision Research, 45,* 1085–1090.

Strother, L., & Kubovy, M. (2003). Perceived complexity and the grouping effect in band patterns. *Acta Psychologica, 114,* 229–244.

Taylor, R. E., & Eisenman, R. (1964). Perception and production of complexity by creative art students. *The Journal of Psychology, 57,* 239–242.

Wittgenstein, L. (1953/1963). *Philosophical investigations.* (G.Anscombe transl.). Oxford: Blackwell.

Chapter 16
Spatial Complexity, Visual Complexity and Aesthetics

Woe betide him who relies solely on mathematics
(Wassily Kandinsky, 1931, p. 31)

Abstract The concept of "visual complexity" is related to spatial complexity. From landscapes to paintings and fractals, aesthetic appeal is directly related to spatial complexity (or, at least, to its determinants). In this chapter, it is examined how and why: (a) neither spatial complexity nor simplicity guarantee aesthetic appeal; (b) quite often, neither too complex nor too simple forms are relatively more preferable; (c) the aesthetic appeal of a spatial form can be scale-dependent on its spatial complexity; (d) spatially complex forms may be aesthetically pleasant only as parts themselves of a larger spatial arrangement; (e) the viewer's perspective of a spatial extent is critical: what seems unordered and complex from one perspective, may appear ordered and simple if viewed from another; (f) spatial orientation, dispersal and aggregation affect spatial complexity and, by consequence, its aesthetic evaluations.

Keywords Spatial complexity · Visual complexity · Landscape aesthetics · Betti numbers · Geographic Complexity · Complexity and aesthetics · Complexity and art

16.1 Visual Complexity

> "This art is born; not taught"
>
> (Seneca, 4b.C.-65a.D., discussing the art of spiders in building their webs in his "Moral Letters to Lucilius"
>
> Letter 121.CCXXI.22-23, "On Instinct of Animals")

Complexity is one of the determinants of aesthetics (Winsor 2004) and a crucial factor in consumer preference (Creusen et al. 2010), or preference in general (Stamps 2004, Galbraith 2001). Necklaces, necktie knots (the Windsor knot, Pratt knot etc.)

F. Papadimitriou, *Spatial Complexity*, https://doi.org/10.1007/978-3-030-59671-2_16

can be analyzed in terms of knot theory (Fink and Mao 1999; Fink and Mao 1999a) and indeed, much of the aesthetics of our garments or jewelry are based on artistic ways of handling forms of spatial complexity. As in attires, so in all fields of art and aesthetics of things that can be visually perceived, the degree to which spatial complexity determines aesthetic evaluations is as interesting as it is difficult to fathom. Take for instance, the famous ornamental *celtic knots*. They are topologically interesting due to their exquisite braiding and knotting (Meehan 1991; Trilling 1995) and their design and outfit may be desirable for many, but they may also seem too "complex" to appeal to the aesthetic taste of others.

"Visual complexity" is a commonly used term in aesthetics and psychology (Vitz 1966; Aitken 1974; Saklofstke 1975) and denotes the degree of complexity of an image or shape, as perceived visually. Assessing it has been (and still is) a quite difficult problem, since there is no widely accepted method for measuring it, even by employing as indicators any of its parametres, magnitudes and properties, such as disorder, number of colors etc. (Eisenman 1967; Stamps 2002; Donderi 2006). Visual complexity has been found to affect users' preference for webpages (Deng and Scott Poole 2012; Sun et al. 2018), as well as marketing and buyers' preferences for industrial products (Wu et al. 2016), such as cars (Chassy et al. 2015) and smartphones (Choin and Lee 2012). A particularly characteristic set of researches in visual complexity would certainly comprise those carried out by Berlyne (Berlyne 1963; Berlyne 1970; Berlyne 1971; Berlyne 1974) and colleagues (Berlyne and Peckham 1966; Berlyne and Ogilvie 1974; Berlyne et al. 1968), from which a central finding is that the "visual complexity" of spatial forms is determined by irregularities of their spatial elements.

Experimental psychological research in visual complexity however, has brought evidence of the importance of a symmetry-complexity dipole in viewing preferences (Jacobsen et al. 2006). In fact, both symmetry and complexity of image forms and patterns have been recognized as strong determinants of viewing preference and this has been established in psychological research in aesthetics (Eysenck 1941; Eisenman and Gellens 1968; Jacobsen 2004; Tinio and Leder 2009). Hoping to bridge shape complexity with aesthetics, Birkhoff (1932) proposed an aesthetic measure (M) of a shape, that was linked to "complexity" (C) and order (O) by the (rather overly simple) ratio: M = O/C and Eysenck (1941) derived an empirical linear formula for measuring visual preference for different geometrical shapes, according to their geometric characteristics:

$$20x_1 + 24x_2 + 8x_3 + 7x_4 + 5x_5 + 3x_6 + 3x_7 + 2x_8 + x_9 - 2x_{10} - 8x_{11} - 15x_{12} \tag{16.1}$$

where the parameters from x_1 to x_{12} represent mainly geometric properties, such as symmetries and angles. After empirical research, Eysenck claimed that this formula might explain as much as 80% of the aesthetic appeal that different polygonal shapes had on the subjects of the research.

Adopting a different approach that is reminiscent of topological CW-complexes, Biederman's theory of "*geons*" (abbreviation for geometric icons) highlights the importance of basic geometric structural units of 3d objects (cubes, pyramids, cylinders, handles etc. termed geons) as responsible for their perception of surfaces and objects (Biederman 1987). A basic property of geons is that even if they are viewed from different angles, they are still discernible and perceivable. Being able to decide what an object is on the basis of its geons only, testifies the psychological "principle of componential recovery", which, in turn, explains why we are able to decide what an object is, even as we view it only partially. Evidently, the higher the number of "geons" necessary to identify an object, the higher its spatial complexity.

Various image compression techniques and standards (JPEG, PNG, TIFF, GIF etc.) have resulted from successful efforts to compress image sizes and have revolutionized our information and telecommunication technologies. These are compression techniques that yield some (lossy or lossless) compression of the initial image size. The compression is verified by the reduced memory required to store the image electronically. Although compression techniques and compressed memory are not measures of complexity themselves, reduction in memory size can be helpful to explain why a square map is more or less complex than another. This can better be understood by creating a plot showing how the memory required to store a file (measured in kilobytes) of a small binary image (with only two colours: black and white in this case) changes with increasing participation of black cells in it with respect to spatial complexity (C_{P2} in this case) and with visual complexity. The file size can be measured by a loss-less compression algorithm such as png. The more differentiated an image appears visually, the more spatially complex it should be and the more computer memory size it should require. To test this, consider for instance a 256 × 256 undifferentiated (all white) image in which various blocks of 64 × 64 cells in it are being progressively blackened. By creating an arbitrary number of such black-and-white images (six here, plus the all-white initial image), the different images can be arranged so that each one of them appears increasingly more complex in comparison to its preceding one (Fig. 16.1).

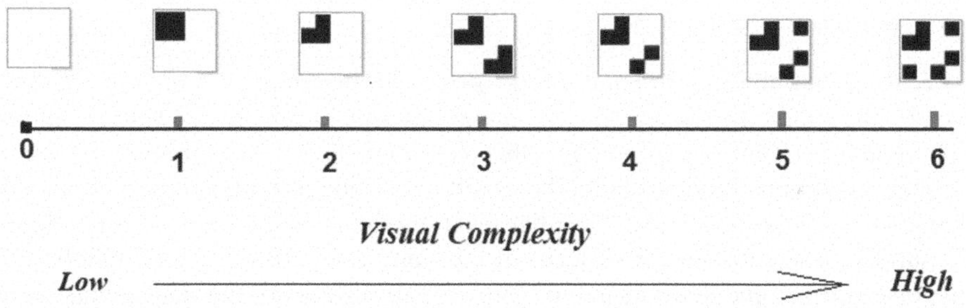

Fig. 16.1 A sample of seven images (numbered), all of the same size, with increasing visual complexity (perceived from the increased spatial differentiation), from 0 = undifferentiated to 6 = most differentiated, with different black-and-white allocations on them. Visually, spatial complexity increases from the 0th image (no complexity at all) to the 6th image (most complex of them all)

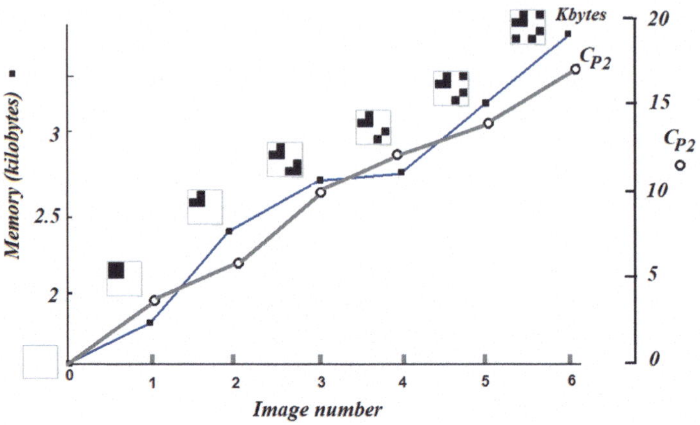

Fig. 16.2 Increasing image size (measured in kilobytes—left vertical axis) with increasing spatial complexity as perceived visually, in six 256 × 256 binary (black and white) images, from absolutely simple (all white—image 0) to various disparate black and white allocations (images 1–6) plotted against spatial complexity C_{P2} (right vertical axis)

The storage capacity requires progressively larger computer memory with spatial complexity, increasing from the 0th image (completely white) to the 6th (with 7 black 64 × 64 cells in it) and thus, image sizes follow an increasing trend with increasing spatial differentiation (Fig. 16.2), matching the visual assessment of the complexity of each one of them.

16.2 Visual Complexity Versus Spatial Complexity of 3d Objects

With what fantasy he conferred multiple curvature on space!
(Gaston Bachelard 1994, p. 157)

Recalling that the Euler characteristic $\chi(s)$ is an invariant of a surface s and is given by the simple formula $\chi(s) = n - v + p$ (where n = nodes, v = vertices and p = polygons), it might be a good guess that a combination of areas (polygons), lines (vertices) and points (nodes) should be useful to provide us with an index of spatial complexity of 3d surfaces. As the Euler invariant depends on the overall topology of a surface, it is constant for surfaces of the same genus. But the total number of polygons, vertices and nodes is not the same for all surfaces of the same genus, so it would be plausible to examine whether it might be initially useful to decide whether some 3d convex object has higher spatial complexity than another, on the basis of the three components of the Euler characteristic of a convex surface. So it might be useful to experiment in terms of the Betti numbers b_j of a convex surface s:

$$\chi(s) = \sum_{j=0}^{\infty} (-1)^j b_j = b_0 - b_1 + b_2 - \ldots, \tag{16.2}$$

whether these numbers might be appropriate to evaluate spatial complexity. The rationale is that the more spatially complex a solid convex body is, the more points, edges and faces it is expected to have, and hence an (arbitrarily defined) metric of spatial complexity to experiment with, which may be referred to as $C(s)$ here, might be based on the sum of Betti numbers:

$$C(s) = \sum_{j=0}^{\infty} b_j = b_0 + b_1 + b_2 + \ldots \tag{16.3}$$

Some example calculations are given for the sake of comparison, for characteristic simple solid bodies: for the simplest 3d solid (the tetrahedron), the immediately more complex (the cube) and some other more *visually* complex geometric solids (Table 16.1 and Fig. 7.14). All these 3d shapes have the same Euler characteristic $\chi(s)$, but, as it turns out, different spatial complexities $C(s)$. Further, values of $C(s)$ grow higher with increasing spatial complexity of each geometric solid, as it is visually perceived.

The calculation for the simplest 3d solid, the tetrahedron, yields $C(s) = 14$, so this can be the minimum spatial complexity $C(s)$ of all simply connected 3d objects. Besides this experimental metric's values increasing with the shape's visual complexity, it is noticeable that topologically equivalent shapes such as the cube and the bipyramid have both the same $C(s)$ value. $C(s)$ performs relatively satisfactorily if applied to objects with low values of n, v, p (some further example calculations are

Table 16.1 Calculation of an experimental simple metric of spatial complexity $C(s)$ for some convex simply-connected three-dimensional shapes (see Fig. 16.3)

3D shape	n	v	p	$C(s)$
Tetrahedron	4	6	4	14
Cube	8	12	6	26
Bipyramid	6	12	8	26
Cube-and-Pyramid	9	16	9	34
Two Cubes	12	20	10	42
4 cubes Tau	20	36	18	74
5-cubes Cross	24	44	22	90
7-cubes Cross	32	60	30	122
3x3x3 Cube	56	108	54	218

All these solid shapes have the same Euler characteristic $\chi(s) = 2$, but different spatial complexities $C(s)$. Comparing the results of $C(s)$ shown here with the shapes shows how $C(s)$ increases for shapes that appear to be *visually* more complex

Fig. 16.3 Some three-dimensional shapes for the calculation of the spatial complexity metric $C(s)$. Upper row: tetrahedron, cube, bipyramid, cube and pyramid, two cubes, four cubes tau. Lower row: Five cubes cross, seven cubes cross, $3\times3\times3$ cube

Table 16.2 Some further calculations of $C(s)$ for some archimedean solids of increasing spatial complexity

Solid	n	v	p	$C(s)$
Sphere divided in 8 parts	6	12	8	26
Triangular icosahedron	12	30	20	62
Truncated cube	24	36	14	74
Rhomb-cub-octahedron	24	48	26	98
Great rhomb-cub-octahedron	48	72	26	146

given in Table 16.2), although it is useful for 3d solids with zero genus only. The extent to which $C(s)$ might be a useful as an estimator of the spatial complexity of 3d objects in general is open for future research.

16.3 Landscape Aesthetics and Spatial Complexity

Often on the mountain, under the shade of the old oak,
I sorrowfully sit during the sunset and I walk my gaze at random over the valley
"Souvent sur la montagne, à l' ombre du vieux chêne,
Au coucher du soleil, tristement je m'assieds;
Je promène au hasard mes regards sur la plaine"
(Alphonse de Lamartine, 1790–1869, "L'Isolement")

Spatial complexity can be involved in a range of non-overlapping approaches to landscape interpretation and explanation (Kühne et al. 2018). Perhaps, some of the most interesting complexification processes can be seen and felt in landscapes of postmodernity (Kühne 2012a; Kühne 2018; Kühne 2019), in which zones of fuzziness often appear in cross-border hybrid spatial zones (Kühne 2012: Kühne and Schönwald 2018).

The visual and aesthetic impact of landscapes has repeatedly been examined by several researchers (Tyrvainen et al. 2003; Cooper 2009; Saito 2010; Papadimitriou 2010; Papadimitriou 2012). The identification of quantitative determinants of visual aesthetic forms seems to be particularly difficult, despite the known connection of landscape structure to aesthetics (Nohl 2001; Franco et al. 2003).

Diversity is a prominent measure of landscape structure (the landscape's arrangement of its spatial elements), reflecting the relative participation of each one of the landscape's elements in the whole. If a landscape has many land types, ecosystems, land uses, land covers, then it has a high diversity. Some researchers have relied on the use of landscape diversity (as measured by Shannon's formula) as an assessor of landscape complexity (Barrett et al. 2009; Ode et al. 2010), but others (Otahel 1999) have considered landscape complexity as a separate criterion of landscape aesthetics than landscape diversity (Coeterier 1996; Sevenant and Antrop 2009; Sevenant and Antrop 2010). Empirical research (Lindemann-Matthies et al. 2010) has shown that alpine landscapes characterized by high landscape diversity (Fig. 16.4) are rated by viewers as more aesthetically appealing; research in landscapes of Norway (Strumse 1994) and Switzerland (Lindemann-Matthies and Bose 2007; Junge et al. 2009)

Fig. 16.4 A snapshot of the alpine landscape of Innsbruck, Austria. High landscape diversity usually implies higher aesthetic appeal, as empirical researches have discovered in alpine landscapes

revealed that species-rich landscapes are more preferred than species-poor (species-rich means higher number of landscape types, higher diversity, and hence higher spatial complexity).

Expectedly, the assessment of complexity of visual forms depends on the viewer and the object viewed. Experimenting with objects of varying complexity, Phillips et al. (2010) found that subjects preferred either very simple or very complex objects, so the complexity of objects can be a predictor of preference. In this context, simple forms coexisting with complex forms within the same spatial setting may also be aesthetically appealing (Fig. 16.5).

With an interesting connection to the properties and behaviors of complex systems, several well known architects have strived to represent self-organizing spatial systems, flocking, swarming and other forms of complex spatial behavior. Just to mention a few names only, Cache (2003) used the term "associative architecture" for complex interlacings of architectural elements in space, Hensel et al. (2006) proposed a complex "morphogenetic design" method, Jenks (1997) introduced "nonlinear architecture". Also, a variety of architectural creations by i.e. Stan Allen, Zaha Hadid, Greg Lynn (among others) appear to have adopted concepts of complexity and/or complex systems theory in their architectural designs.

Fig. 16.5 Athens: View of the city center with the temple of the Olympian Zeus at the forefront and Lycabettus hill at the backdrop. The spatial complexity of the urban landscape in the middle area of the picture coexists with hefty areas of spatially less complex areas (urban green), along with the ancient temple's simplicity. The coexistence of *both* spatially complex *and* simple shapes within the *same* setting may result in higher aesthetic appeal than if there were *either* complex or simple shapes

16.4 Spatial Complexity and Aesthetic Evaluations

> Neither admiration, nor victories but simply to be accepted as part of an undeniable reality like stones and trees
>
> "No admiraciones ni victorias
>
> sino sencillamente ser admitidos
>
> como parte de una realidad innegable,
>
> como las piedras y los árboles"
>
> (Jorge Luis Borges, 1899–1986, "Llaneza"/Simplicity)

Some simple designs have been considered aesthetically pleasant throughout the centuries by many different people. The typical ancient greek ionic style is more complex in its design than the austere dorian style; in both these styles however, simple symmetries account for most of the aesthetic appeal. While symmetry may be more preferrable in the western cultural hemisphere (Gombrich 1979), it is nevertheless a human characteristic to yearn and to search for regularities (Popper 1972). Chen et al. (2011) found that symmetric patterns are more aesthetically appealing than asymmetric ones. But there may be a different reason: some experimental evidence (Krupinksi and Locher 1988; Locher and Nodine 1989; Palmer 1991) has suggested that lower visual complexity can be associated with more appealing forms. Expectedly however, simplicity alone does not guarantee aesthetic appeal and this is why most people would prefer a picture rich in chromatic diversity instead of another with similar but less colors.

Delplanque et al. (2019, p. 146) noticed that "an object that is too simple would lead to boredom, while too much complexity would cause distraction". The source of the problematic relationship between spatial complexity and aesthetic appeal should be sought in the absence of some measure of spatial complexity that might be used to evaluate the complexity of spatial objects and surfaces. The so-called "*aesthetic middle*" investigated by Deplanque et al. (2019) recalls earier experiments (Day 1967), in which it was shown that *neither too complex nor too simple patterns are more preferable.*

The complexity of the viewer's experience of an artwork is often a function of the artwork's spatial complexity. The former has been estimated (Krejtz et al. 2014) by monitoring individuals' gaze patterns using a Markov model and Shannon's formula (Atkins et al. 2010). In other studies however (Kuper 2015), visual complexity and preference did not correlate completely. Indeed, the Shannon formula would probably yield a high figure for an image as disordered as that of Fig. 16.6 and more than few viewers would consider this assortment rather pleasant.

Another fact to consider is the level of spatial detail (that is the level of spatial complexity at small resolutions), which may or may not be indicative of the possible aesthetic appreciation of the image as a whole. Consider for instance, an image (Fig. 16.7), which may be aesthetically appealing to some.

If any part of this image is viewed in detail, it may not be as appealing. Indeed, there are cases where spatially complex forms may be highly precise and with a high

Fig. 16.6 Disorder and high spatial complexity may not necessarily be unpleasant

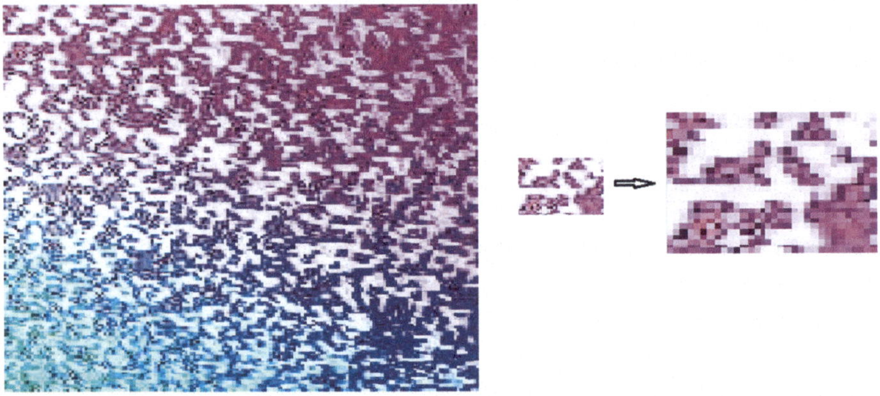

Fig. 16.7 Aesthetic appeal can be scale-dependent on spatial complexity. If the upper left corner area of the grand image (left) is enlarged (right), the resulting image may not be as aesthetically pleasant as the original large image. The same applies to all parts of the grand image, which, nevertheless, many would find aesthetically appealing, despite its highly disorderly allocation of colors

level of "complicatedness", but they may not be aesthetically pleasant, *unless* they constitute themselves parts of some larger spatial arrangement that is aesthetically appealing only if viewed from afar, or at large (Fig. 16.8). Images containing many different geometric shapes (i.e. squares, rectangles) are more spatially complex than images with less such shapes and it was experimentally found that the former were

Fig. 16.8 Branches of a plant interwoven in a dense complex fabric at the ceiling of an arcade, may not be aesthetically pleasant at a close examination, but they may be so if viewed as part of a whole (at the scale of the arcade in its entirety): spatially complex settings of lower aesthetic appeal are combined together within larger settings which give a higher aesthetic appeal

preferred to the latter (Timio and Leder 2009). Besides, as Morris (1995) noticed, the simplicity of the aesthetic experience does not necessarily stem from the simplicity of the viewed shapes.

Perhaps, this explains (Liu and Luo 1996) why some fractals are widely considered as aesthetically pleasant forms as they constitute typical examples of spatially complex objects (Fig. 16.9). In the case of fractals, there is a direct scale-dependence of aesthetic appeal on spatial complexity. As Crawther (1991) argued, mathematical approaches to appreciations of beauty make us recall the Kantian category of the "sublime". The aesthetics of fractal objects may relate to the fractal dimension characterising them, although it may also relate to the branching processes they may present. However, the relationship between the fractal dimension and aesthetic appeal of a fractal object remains an open problem. In fact, the repercussions of mathematical approaches to estimations of beauty are not only philosophical, but also practical in the case of fractals. For instance Li et al. (2007) suggested that fractality and complexity can be basic parts of an aesthetically pleasant design.

Whatever the spatial complexity of the observed object may be, aesthetic evaluations are highly dependent on cognitive psychological determinants. The "*oblique effect*" is one such example: it is the fact that humans perceive horizontally and vertically more and better than along other orientations, as has been experimentally discovered (Campbell et al. 1966; Appelle 1972) and this highlights the preference for square grids instead of other spatial partitions. However, the oblique effect does not necessarily imply the ability to recognize the spatial complexity of a surface or

Fig. 16.9 A fractal object. Should its aesthetic appeal be attributed to the fact that it encapsulates some interesting mathematical results relating to complexity?

object if seen from different viewpoints following the "*viewpoint invariance*" as is called in psychology (Williams et al. 2009), since spatial complexity is not necessarily *perceived* to be the same if the surface or object is viewed from a different perspective (Fig. 16.10).

Fig. 16.10 The spatial complexity of a surface is not necessarily *perceived* as the same if the surface or object is viewed from a different perspective

Fig. 16.11 Repeating similar shapes with the same orientation is usually aesthetically acceptable. But if one shape had a strikingly different angle compared to the other ones, it would probably spoil the aesthetics of the whole

While orientation plays a prominent role in aesthetics, in the context of spatial complexity there are surprising twists in the turn (literally). For instance, same shapes following the same orientations are aesthetically acceptable in general (Fig. 16.11), but different shapes and with different orientations may not be as appealing. Perhaps differences in orientation increase the sense of disorder, may be even more than the geometric differences among the spatial elements themselves, although same shapes with different orientations may receive positive aesthetic evaluations (Fig. 16.12). In fact, the japanese art of *Ikebana* (the art of flower arrangement) is based precisely on the aesthetic appeal of arranging flowers and branches with different orientations and asymmetric slanting.

Besides orientation, patterns of branching, dispersal and expansion into space are also interesting in terms of spatial complexity (i.e. Figure 16.13), as are patterns of aggregates (Fig. 16.14). In these cases, the characteristics of spatial complexity of patterns of dispersal or accumulation play the most important role in aesthetic evaluations.

But geometric relationships between objects are not the only determinants of the relevance of spatial complexity to aesthetics. *Minimalist* features in *surrealist* paintings are very interesting in that they present a significant class of visual material with low spatial complexity which nevertheless is shaped so as to provoke a major intellectual or emotional impact on the viewer. The sculptures of Man Ray

Fig. 16.12 Similar shapes, albeit with different orientations may nevertheless be aesthetically pleasant

Fig. 16.13 Expansion in space, in which similar shapes grouped together but with different orientations in space can be equally appealing

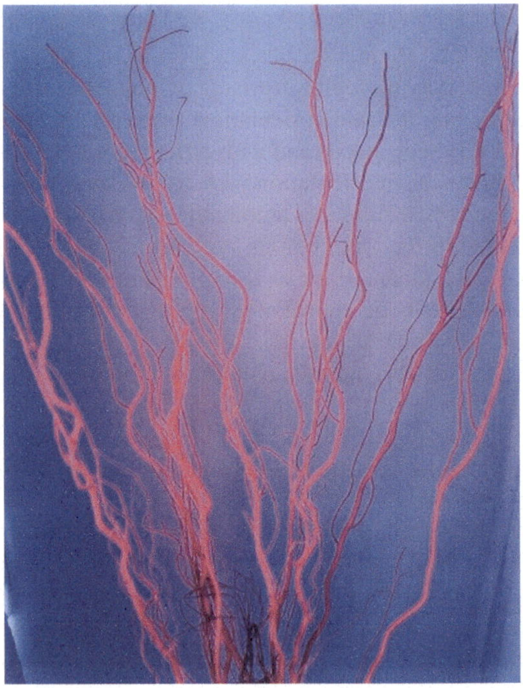

Fig. 16.14 Spatial aggregates (clouds in this case): complex forms are often derived from aggregation of same objects (as in this case) or dissimilar ones

(1890–1976) convey intense messages to the viewer despite their minimal spatial complexity. Similarly, the paintings of René Magritte (1898–1967) usually present forms of minimal spatial complexity but perplexing the viewer with impossible situations. Might (notwithstanding their historical and aesthetic value), such paintings be considered "aesthetically pleasant" today?

Whether the shape is a square or another simple shape, minimalist art eventually tests aesthetics at the lower levels of spatial complexity. But claiming that the elimination of *all* spatial complexity increases aesthetic appeal can hardly be accepted unanimously.

Shifting emphasis from low to high spatial complexity, visual artworks (i.e. paintings) can be made more spatially complex by increasing chromatic diversity and detail. In the context of *neo-impressionism* for instance, details were meticulously added by a kind of "pixelization" i.e. as in the works of George Seurat (1859–1891). On the antipodes of simplicity, some well-known examples are the chaotic compositions in the works of Jackson Pollock (1912–1956) and the assortments of hundreds of different spatial elements in many of the works of Wassily Kandinsky (1866–1944). With curvature being a geometric element contributing to the increase of spatial complexity, paintings of Franz Marc (1880–1916) highlight the contrast between curves and straight lines. It is probably not accidental that artistic and architectural styles such as *baroque* and *art nouveau* reflected precisely the need to increase spatial complexity by enhancing one of its main determinants, curvature.

References

Aitken, P. P. (1974). Judgments of pleasingness and interestingness as functions of visual complexity. *Journal of Experimental Psychology, 103,* 240–244.

Appelle, S. (1972). Perception and discrimination as a function of stimulus orientation: the "oblique effect" in man and animals. *Psychological Bulletin, 78,* 266–278.

Atkins, D. L., Klapaukh, R., Browne, W. N., Zhang, M. (2010). Evolution of aesthetically pleasing images without human-in-the-loop. IEEE World Congress on Computational Intelligence, WCCI 2010—2010 IEEE Congress on Evolutionary Computation, CEC 2010, 5586283.

Bachelard, G. (1994). *The poetics of space.* Boston MA: Beacon Press.

Barrett, T. L., Farina, A., & Barrett, G. W. (2009). Aesthetic landscapes: An emergent component in sustaining societies. *Landscape Ecology, 24*(8), 1029–1035.

Berlyne, D. E. (1963). Complexity and incongruity variables as determinants of exploratory choice and evaluative ratings. *Canadian Journal of Psychology, 17,* 274–290.

Berlyne, D. E. (1970). Novelty, complexity, and hedonic value. *Perception and Psychophysics, 8,* 279–286.

Berlyne, D. E. (1971). *Aesthetics and psychobiology.* New York: Appleton-Century-Crofts.

Berlyne, D. E. (1974). Novelty, complexity, and interestingness. In D. E. Berlyne (Ed.), *Studies in the new experimental aesthetics: Steps toward an objective psychology of aesthetic appreciation* (pp. 175–180). Washington, DC: Hemisphere Publishing Corporation.

Berlyne, D. E., & Ogilvie, J.C. (1974). Dimensions of perception of paintings. In D. E. Berlyne (Ed.), *Studies in the new experimental aesthetics: Steps toward an objective psychology of aesthetic appreciation* (pp. 181–226). Washington, DC: Hemisphere Publishing Corporation.

Berlyne, D. E., Ogilvie, J. C., & Parham, L. C. C. (1968). The dimensionality of visual complexity, interestingness, and pleasingness. *Canadian Journal of Psychology, 22,* 376–387.

Berlyne, D. E., & Peckham, S. (1966). The semantic differential and other measures of reaction to visual complexity. *Canadian Journal of Psychology, 20,* 125–135.

Biederman, I. (1987). Recognition-by-components: A theory of human image understanding. *Psychological Review, 94*(2), 115–147.

Birkhoff, G. D. (1932). *Aesthetic measure.* Cambridge, MA: Harvard University Press.

Cache, B. (2003). Philibert de l' Orme Pavillon: Towards an associative architecture. *Architectural design,* March–April.

Campbell, F. W., Kulikowski, J. J., & Levinson, J. (1966). The effect of orientation on the visual resolution of gratings. *Journal of Physiology, 187,* 427–436.

Chassy, P., Lindell, T. A. E., Jones, J. A., Paramei, G. V. (2015). A relationship between visual complexity and aesthetic appraisal of car front images: An eye-tracker study. *Perception,* 0(0), 1–13.

Chen, C.-C., Wu, J.-H., & Wu, C.-C. (2011). Reduction of image complexity explains aesthetic preference for symmetry. *Symmetry, 3,* 443–456.

Choin, J. H., & Lee, H.-J. (2012). Facets of simplicity for the smartphone interface: A structural model. *International Journal of Human-Computer Studies, 70,* 129–142.

Coeterier, J. F. (1996). Dominant attributes in the perception and evaluation of the Dutch landscape. *Landscape and Urban Planning, 34,* 27–44.

Cooper, D. (2009). Humans in the land: The ethics and aesthetics of the cultural landscape. *British Journal of Aesthetics, 49*(2), 188–191.

Creusen, M. H., Veryzer, R., & Schoormans, J. P. L. (2010). Product value importance and consumer preference for visual complexity and symmetry. *European Journal of Marketing, 44*(9), 1437–1452.

Crowther, P. (1991). Kant's analytic of the sublime: Fro the preliminary sections to the mathematical mode. *The kantian sublime,* March, 78–108.

Day, H. (1967). Evaluations of subjective complexity, pleasingness and interestingness for a series of random polygons varying in complexity. *Perception and Psychophysics, 2,* 281–286.

Delplanque, J., De Loof, E., Janssens, C., & Verguts, T. (2019). The sound of beauty: How complexity determines aesthetic preference. *Acta Psychologica, 192,* 146–152.
Deng, L., & Scott Poole, M. (2012). Aesthetic design of e-commerce web pages—Webpage Complexity, Order and preference. *Electronic Commerce Research and Applications, 11,* 420–440.
Donderi, D. C. (2006). Visual complexity: A review. *Psychological Bulletin, 132,* 73–97.
Eisenman, R. (1967). Complexity-simplicity: I. Preference for symmetry and rejection of complexity. *Psychonomic Science, 8*(4), 169–170.
Eisenman, R., & Gellens, H. K. (1968). Preference for complexity–simplicity and symmetry–asymmetry. *Perceptual and Motor Skills, 26,* 888–890.
Eysenck, H. J. (1941). The empirical determination of an aesthetic formula. *Psychological Review, 48,* 83–92.
Fink, T. M., & Mao, Y. (1999a). Designing tie knots by random walks. *Nature, 398,* 31–32.
Fink, T. M., & Mao, Y. (1999b). Tie knots, random walks and topology. *Physica A, 276,* 109–121.
Franco, D., Franco, D., Mannino, I., & Zanetto, G. (2003). The impact of agroforestry networks on scenic beauty estimation—The role of a landscape ecological network on a socio-cultural process. *Landscape and Urban planning, 62*(3), 119–138.
Galbraith, S. (2001). Beauty from complexity. *Contemporary Physics, 42*(5), 323–325.
Gombrich, E. (1979). *The sense of order: A study in the psychology of decorative art.* Oxford: Phaidon press.
Hensel, M., Menges, A., & Weinstock, N. (2006). Techniques and technologies in morphogenetic design. *Architectural Design, 76,* 78–87.
Jacobsen, T. (2004). Individual and group modelling of aesthetic judgment strategies. *British Journal of Psychology, 95,* 41–56.
Jacobsen, T., Schubotz, R., Hofel, L., & Cramon, D. Y. (2006). Brain correlates of aesthetic judgment of beauty. *NeuroImage, 29,* 276–285.
Jenks, C. (1997). Nonlinear architecture. New science—New architecture. *Architectural Design, 67,* 6–9.
Junge, X., Jacot, K. A., Bosshard, A., & Lindemann-Matthies, P. (2009). Swiss people's attitudes towards field margins for biodiversity conservation. *Journal of Nature Conservation, 17,* 150–159.
Kandinsky, W. (1931). Reflexions sur l' art abstrait. *Cahiers d' Art, 7*(8), 351–353.
Krejtz, K., Szmidt, T., Duchowski, A.T., Krejtz, I. (2014). Entropy-based statistical analysis of eye movement transitions. Eye Tracking Research and Applications Symposium (ETRA), 159–166.
Krupinski, E., & Locher, P. (1988). Skin conductance and aesthetic evaluative responses to nonrepresentational works of art varying in symmetry. *Bulletin of the Psychonomic Society, 26,* 355–358.
Kuper, R. (2015). Preference, complexity, and color information entropy values for visual depictions of plant and vegetative growth. *Horticulture Technology, 25*(5), 625–634.
Kühne, O. (2012a). Urban Nature between modern and Postmodern Aesthetics: Reflections based on the social constructivist approach. *Questiones Geographicae, 31*(2), 61–70.
Kühne, O. (2012a). *Stadt—Landschaft—Hybridität. Ästhetische Bezüge im postmodernen Los Angeles mit seinen modernen Persistenzen.* [City - Landscape—Hybridity. Aesthetic references in postmodern Los Angeles with its modern persistence]. Wiesbaden: Springer VS.
Kühne, O. (2018). *Landscape and power in geographical space as a social-aesthetic construct.* Dordrecht: Springer.
Kühne, O. (2019). *Landscape theories. A Brief Introduction.* Wiesbaden: Springer VS.
Kühne, O. & Schönwald, A. (2018). Hybridisierung und Grenze: Das Beispiel San Diego/Tijuana. [Hybridization and Borders: the example of San Diego/Tijuana]. In M.Heintel, R.Musil, & N.Weixlbaumer (Eds), *Grenzen. Theoretische, konzeptionelle und praxisbezogene Fragestellungen zu Grenzen und deren Überschreitungen* (pp. 401–417). Wiesbaden: Springer VS.
Kühne, O., Weber, F. & Jenal,C. (2018). Der Begriff ‚Landschaft' sowie essentialistisch und positivistisch orientierte Zugänge. [The term 'landscape', essentialist and positivist orientated

approaches] In O. Kühne, F. Weber, & C. Jenal (Eds.), *Neue Landschaftsgeographie. Ein Überblick* (pp. 5–10). Wiesbaden: Springer VS.

Li, C. L., Yeung, Y. C., Chiu, W. K., & Yu, K. M. (2007). Modelling of Complex Fractal Objects for Aesthetic Product Development. *International Journal of Product Development, 4*(3–4), 207–224.

Liu, H., & Luo, J. (1996). A method for generating super large fractal images useful for decoration art. *Communications in Nonlinear Science and Numerical Simulation, 1*(3), 24–27.

Lindemann-Matthies, P., & Bose, E. (2007). Species richness, structural diversity and species composition in meadows created by visitors of a botanical garden in Switzerland. *Landscape and Urban Planning, 79,* 298–307.

Lindemann-Matthies, P., Briegela, R., Schopbach, B., & Junge, X. (2010). Aesthetic preference for a Swiss alpine landscape: The impact of different agricultural land-use with different biodiversity. *Landscape and Urban Planning, 98,* 99–109.

Locher, P., & Nodine, C. (1989). The perceptual value of symmetry. *Computers & Mathematics with Applications, 17,* 475–484.

Meehan, A. (1991). *Celtic design: Knotwork.* New York: Thames and Hudson.

Morris, R. (1995). Notes on Sculpture. In G. Battock (Ed.), *Minimal art: A critical anthology.* Berkeley, CA: University of California Press.

Nohl, W. (2001). Sustainable landscape use and aesthetic perception- preliminary reflections on future landscape aesthetics. *Landscape and Urban Planning, 54,* 223–237.

Ode, A., Hagerhall, C. M., & Sang, N. (2010). Analysing visual landscape complexity: Theory and application. *Landscape Research, 35*(1), 111–131.

Otahel, J. (1999). Visual Landscape Perception: Landscape pattern and Aesthetic Assessment. *Ekologia Bratislava, 18*(1), 63–74.

Palmer, S.E. (1991). On goodness, gestalt, groups, and garner: local symmetry subgroups as a theory of figural goodness. In G. Lockhead, & J. Pomerantz (Eds.), *The perception of structure: essays in honor of Wendell R. Garner*. Washington, DC: American Psychological Association.

Papadimitriou, F. (2010). Conceptual Modelling of Landscape Complexity. *Landscape Research, 35*(5), 563–570.

Papadimitriou, F. (2012). Modelling landscape complexity for land use management in rio de janeiro. *Brazil. Land Use Policy, 29*(4), 855–861.

Phillips, F., Norman, J., & Amanda, M. (2010). Fechner's Aesthetics Revisited. *Seeing and Perceiving, 23*(3), 263–271.

Popper, K. (1972). *Objective knowledge: An evolutionary approach.* Oxford: Clarendon press.

Saito, Y. (2010). Future directions for environmental aesthetics. *Environmental Values, 19*(3), 373–391.

Saklofske, D. H. (1975). Visual aesthetic complexity, attractiveness and diverse exploration. *Perceptual and Motor Skills, 41,* 813–814.

Sevenant, M., & Antrop, M. (2009). Cognitive attributes and aesthetic preferences in assessment and differentiation of landscapes. *Journal of Environmental Management, 90,* 2889–2899.

Sevenant, M., & Antrop, M. (2010). The use of latent classes to identify individual differences in the importance of landscape dimensions for aesthetic preference. *Land Use Policy, 27,* 827–842.

Stamps, A. E. (2002). Entropy, visual diversity, and preference. *The Journal of General Psychology, 129,* 300–320.

Stamps, A. E. (2004). Mystery, complexity, legibility and coherence: a meta-analysis. *Journal of Environmental Psychology, 24,* 1–16.

Strumse, E. (1994). Environmental attributes and the prediction of visual preferences for agrarian landscapes in western Norway. *Journal of Environmental Psychology, 14,* 293–303.

Sun, L., Yamasaki, T., & Aizawa, K. (2018). Photo aesthetic quality estimation using visual complexity features. *Multimedia Tools and Applications, 77*(5), 5189–5213.

Tinio, P. P. L., & Leder, H. (2009). Just how stable are stable aesthetic features? Symmetry, complexity, and the jaws of massive familiarization. *Acta Psychologica, 130,* 241–250.

Trilling, J. (1995). Medieval interlace ornament: The making of a cross-cultural idiom. *Arte Medievale, 9,* 59–86.

Tyrvainen, L., Silvennoinen, H., & Kolehmainen, O. (2003). Ecological and aesthetic values in urban forest management. *Urban Forestry and Urban Greening, 1*(3), 135–149.

Vitz, P. C. (1966). Preference for different amounts of visual complexity. *Behavioral Science, 11,* 105–114.

Williams, N.R., Willenbockel, V., & Gauthier, I. (2009). Sensitivity to spatial frequency and orientation content is not specific to face perception. *Vision Research, 49,* 2353–2362.

Winsor, P. (2004). Complexity in the experimental audio/visual arts. *Chaos, Solitons & Fractals, 20,* 45–53.

Wu, K., Vassilev, J., Zhao, Y., Noorian, Z., Waldner, W., & Adaji, I. (2016). Complexity or simplicity? Designing product pictures for advertising in online marketplaces. *Journal of Retailing and Consumer Services, 28,* 17–27.

Chapter 17
Geophilosophy and Epistemology of Spatial Complexity

We rather prefer to look after many things and to link ourselves to many things... and so linked with many we feel ourselves heavy and dragged down by them.
"Μᾶλλον θέλομεν πολλῶν ἐπιμελεῖσθαι
καὶ πολλοῖς προσδεδέσθαι...
...ἅτε οὖν πολλοῖς προσδεδεμένοι
βαρούμεθα ὑπ' αὐτῶν καὶ καθελκόμεθα»
(Epictetus, 50–135 a.D. "Discourses", 14–15)

Abstract Adding difference to an undifferentiated surface or object is the first step towards increasing its spatial complexity (this difference can be thematic, geometric, topological etc). The philosophy of Gilles Deleuze appears particularly suited to explain spatial complexity and, to a lesser extent, some philosophical ideas of Leibniz and Nicolai Hartmann. The process of creation of spatial complexity from simple square maps begins with number partitions and entropy, followed by topological differentiations (boundaries, spatial arrangements etc.) and, eventually, with the symmetric multiplication of generic forms. Recognizing that curvature and spatial perspective are crucial to our perception (and hence, interpretation) of anything spatial, we are led to a re-appraisal of the role of subjectivity in spatial complexity. As for the research, analysis and exploration of spatial complexity, experimentation is indispensable and this points to the possibility to consider as more suitable an "experimental philosophy". An epistemological problem is that we can only partially manage the immense realm of possible spatial configurations. This leads us to consider that combinatorial explosions induce limits to spatial analysis while studying spatial complexity at very large (i.e. planetary) scale.

Keywords Spatial complexity · Gilles Deleuze · Geocomputation · Spatial Computing · Geography and philosophy · Geophilosophy · Big geospatial data

F. Papadimitriou, *Spatial Complexity*, https://doi.org/10.1007/978-3-030-59671-2_17

17.1 From Difference to Complexity

So numerous and so perplexed
"tam numerosae atque perplexae"
(Quintilian, 35–100 a.D., "Institutio Oratoria", 8.2)

The pair complexity versus simplicity has puzzled philosophers and mathematicians alike for centuries. Geometric simplicity ("Einfachkeit") may have been just a formality for Kant, but for Hilbert it certainly was not: his 24th problem was to state criteria of simplicity, or, essentially, to prove the simplicity of proofs (to decide the simplest proof among many).

In the philosophy of Nicolai Hartmann, space constitutes a special "category" and there are paired categories consisting of opposite pairs, i.e. discreteness-continuity, form-matter, element-complex. While space is an "inorganic" existence ("anorganisches sein"), the "complex" ("gefüge") has an inside structure while its outside relates it to other "complexes" or is itself part of the internal structure of another "complex" (Hartman 1975). The elements of each "complex" and their functions define its internal structure and its resilience (Hartmann 1940). The internal stability of a "complex" is assured by the law of connectivity ("Gesetz der Verbundenheit").

The universe evolved from its primordial unstructured and lifeless state by increasing its complexity, progressively creating living beings, from which emerged intelligence, ethics, aesthetics, justice and affection for other beings. It seems more than likely, that local modifications of entropy have been a key process in this increase in complexity. It took ample time for this chain of emergence to assume visible form, incessant efforts and infinite persistence by nature and, in the later stages, by humankind. And yet, not even half of the story has unfolded, since the universe is still not "old" enough. In these processes, relatively small sets of objects and *simple rules* about them have proved adequate to create spectacularly complex spatial forms.

The undifferentiated space may be perceived like Leibniz's "monad", in the sense that it "has to have simple substances, as the composed is nothing else but an amassment, an aggregate (aggregatum) of simple things" (Leibnitz 1714, p. 2). Such aggregates, composed from simple spatial constituents, can be best examined in the context of the Deleuzian philosophical framework, in which the "mechanosphere" that comprises the biosphere and the noosphere is replete with "machinic assemblages", hybrid forms, and "rizhomatous open systems" (Fig. 17.1), which may include the "promise for a negentropic future" for humankind (Ansell-Pearson 1997, p. 167). Why the "*rhizome*" (the greek word «ρίζωμα» that is a root system) is a philosophical metaphor for spatial complexity is exemplified by two "principles" of rhizomes: the "principle of cartography" and the "principle of connection and heterogeneity". The latter suggests that "any point on a rhizome can be connected with any other" (Deleuze and Guattari 1983, p. 11). Its spatial (map-like) extent makes it different than a simple tree-like hierarchical structure composed of linear elements. In fact, the rhizome is a map, whence also the suggestion (Deleuze and Guattari 1983, p. 25): "make maps, not tracings".

In the philosophy of Deleuze, the differentiation of space ad infinitum inevitably entails "a transportation of difference, a diaphora (difference) of diaphora, until a final difference" (Deleuze 2010, p. 40) and, in this way, natural selection is a process of endless differentiation with aim to enhance the capabilities of survival: "the leitmotiv of the origin of species is: we do not know what individual difference is capable of!" (Deleuze 2010, p. 310). It is in this Deleuzian context, that De Landa (2002, p. 25) wrote that the differentiation process (eventually leading to complex forms) begins from an undifferentiated space which "progressively differentiates, eventually giving rise to extensive structures (discontinuous structures with definite metric properties)". The simplest difference on a 2d squared surface can be created by introducing one "black" cell in a simple undifferentiated (i.e. all-white) square map and convert it into a binary map. In the case of Arnold Cat Maps, these differences can spread all over the image, interchange positions in it and eventually fall back at the same place where they originally were located, so as to reproduce the original image unchanged back again. However, entropy alone does not suffice to explain spatial complexity. Higher entropy only indicates higher spatial complexity in small binary maps, but not always so; topology is probably the other most significant determinant of spatial complexity. The central role of topological properties in the understanding of space reiterates in the context of various philosophical approaches, both ancient and modern, from Aristotle to Wittgenstein, Whitehead, Minkowski, Prior, Heidegger, as has been identified by several researchers (da Costa et al. 1997; Simons 2006; Peters 2008; Sean 2009).

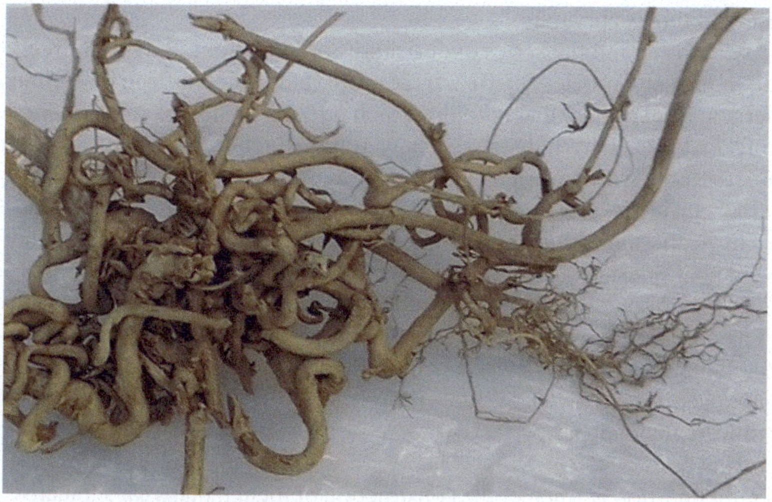

Fig. 17.1 A real "rhizome" (a complex root system) with its knots, links, writhes, variable curvature, intersections, etc.: the epitome of natural spatial complexity?

17.2 Curvature and Subjectivity

> It is among the philosopher's fields, we think, to consider also geographical surveys, as we intend to do here
>
> "Τῆς τοῦ φιλοσόφου πραγματείας εἶναι νομίζομεν,
>
> εἴπερ ἄλλην τινά, καὶ τὴν γεωγραφικήν,
>
> ἣν νῦν προῃρήμεθα ἐπισκοπεῖν"
>
> (Strabo, 62 b.C.–23 a.D., Geography, Chap. 1)

There is not a single philosophical approach to the geographical space. In fact there is a vast array of possible philosophical perspectives, i.e. positivist, essentialist (Kühne et al. 2018), postmodern (Kühne 2012a, b; Kühne 2018) among several other ones, of which a detailed analysis can be found in Kühne (2019). But philosophers were not the only ones to be concerned with space. The geographer Hagerstrand (1982) suggested the mapping of personal 4d place-times, the geographer Virilio (1994, p. 7) a "teletopology", of spaces that are better understood when viewed from a distance, and the science fiction author Stanislav Lem proposed a "toposophy" (Broderick 2001, p. 324), while the new latest generation of geospatial technologies enables the user to create subjective maps and share them with other users (Papadimitriou 2010a, b). These characteristic and many other geophilosophical approaches reverberate the problem of subjectivity in the perception of geographical space. How is subjectivity related to the geographical space has long been an issue of debate among geographers, but there is a hint from the point of view of spatial complexity analysis. And this is curvature.

"The world is an infinite series of curvatures or inflections and the entire world is enclosed in the soul" (Deleuze 2012, p. 26). Curvatures and inflections define a key characteristic of existence that is folds, which are one feature of spatial complexity. Following Deleuze, labyrinths have many folds which can also lead to complex internal structures with caverns and porosity and thus, with a metaphysical extension, "God plays tricks, but also furnishes the rules of the game" (Deleuze 1993, p. 71). Given the current state of human knowledge, a philosophical rather than a mathematical question is: might bi-dimensionality (spatial extent) be a "trick of nature" to convert one-dimensional randomness into non-randomness?

Back to curvature now, its combination *with* perspective is central in the perception of spatial complexity. The viewer's perspective of a spatial extent is critical in the way that space is perceived by the viewer: a seemingly unordered allocation of spatial objects may appear perfectly ordered if viewed from a different perspective (Fig. 17.2).

A central question is whether the construction of mental maps (or subjective maps in general) results as one of the inevitable consequences of the human inadequacy, *not only to perceive, but also to represent* the geographical space (manually or otherwise) without error and this inevitably leads us to introduce "subjective" perceptions, interpretations, theorizations and explanations of the geographical space. It is in this

Fig. 17.2 Light spots on the ceiling of an auditorium. If viewed from the left or the right side of the auditorium, they appear unordered, but if viewed from the center of the auditorium, a perfect alignment surprisingly emerges

epistemological context, that subjective geographies were first proposed by the "situationists" movement, in a way that "the situationists' division of the city echoes the psychologists' topology of the brain" (Bukatman 2005, p. 169). How this problem can be tackled quantitatively can be enlightening as it reveals the underlying spatial complexity.

The distance of point (x_1, y_1) from point (x_2, y_2) on a 2d Euclidean space is:

$$D = \sqrt{(x_1 - x_2)^2 + (y_1 - y_2)^2}, \tag{17.1}$$

but the distance between two points (φ_1, θ_1) and (φ_2, θ_2) on the surface of a sphere (in polar coordinates φ = longitude and θ = latitude) is:

$$D = R[\arccos\{\sin\varphi_1 \sin\varphi + \cos\varphi_1 \cos\varphi_2 \sin(\theta_1 - \theta_2)\}] \tag{17.2}$$

As we shift our point of focus from maps of small geographical regions to maps of large areas (i.e. showing parts of continents or even the entire globe), distances are best calculated by Riemannian metrics. The Riemannian length of a "*path*" h is:

$$M_R(h) = \int_a^b \|(P(t))\|_R \mathrm{d}t \tag{17.3}$$

and the Riemannian distance between two points u and v is:

$$D_R(u, v) = \inf_{p \in W_h} M_R(h) \tag{17.4}$$

where W_h contains all the paths that join u and v. On the basis of these considerations, the total error of a mental map with respect to the reality of the geographic space has been calculated (Tobler 1965; Meshcheryakov 1965) as:

$$\frac{\int_{-\pi}^{\pi}\int_{0}^{2\pi}\int_{0}^{2\pi}\left(\frac{dS}{ds}-1\right)^2\cos\varphi d\theta d\varphi}{2\pi\int_{-\pi}^{\pi^2}\int_{0}^{\pi}\cos\varphi d\theta d\varphi} \tag{17.5}$$

where

$$\frac{dS}{ds} = \sqrt{(g_x^2\cos^2 a + 2g_{xy}\sin a\cos a + g_y^2\sin^2 a} \tag{17.6}$$

is a measure of the distance on the surface of the earth of $S(u, v)$ from $s(x,y)$ expressed in Riemannian metrics.

It is self-evident even from the high number of calculations necessary to compute such distances, that humans have to tolerate imprecisions in their estimations of area, angle, distance, location. Might this suggest that subjective geographies are inevitable?

17.3 A Philosophical Ladder to Spatial Complexity

> The (phenomena of nature) are marvelously complex and interwoven to one another, as will be evident to him who investigates them
>
> "Τούτοις τοῖς παθήμασιν πρὸς ἄλληλα συμπλεχθεῖσιν τεθαυματουργημένα τῷ κατὰ τρόπον ζητοῦντι φανήσεται"
>
> (Plato, 428–348 b.C., "Timaeus", 80c)

The (neo-Platonic) philosopher Plotinus suggested (Enneads, IV, 4xxxii) that "sympathy" may have been the driver for the aggregation of simple things together, the underlying force which pushes them close to one another and this "pervades the entire universe". Thus, in a neo-Platonic sense, clumpiness (B) is a result inevitably emerging from the work of "sympathy" in the world. And, as the (also neo-Platonic) philosopher Porphyry warned (in "The Cave of the Nymphs, 6), the universe may appear charming from "outside", but obscure and dark from "inside": a remarkable observation that may apply to spatial complexity as well: charming as one observes it, but obscure and difficult to decipher. Plotinus questioned also ("Enneads", V, 2i) how everything might have been derived from some "Single One", without that "One" being endowed with either duality or diversity. Translating this thought to twenty-first century science, it means that increasing entropy is precisely the next step after the differentiation of the uniform "One" and, in this sense, 3 × 3 square binary maps are a very characteristic example documenting how changes in spatial complexity rapidly increase from entropy class 0 to class 1. The spatial analysis of 3 × 3 maps may turn out to be very special: neurophysiological experiments on the perception of numbers seem to confirm that the numbers 1, 2, 3 are perceived differently by humans than other numbers, with high-accuracy measurements suggesting that they are perceived as many as 200 milliseconds faster than other numbers (Revkin et al. 2008; Hyde and Spelke 2009).

Having examined how spatial complexity can be created by simple square maps, we saw how and why *entropy is a necessary but not sufficient condition for the generation of spatial complexity* and the same applies to topology (due to the topologically inequivalent positions of the map's square cells).

As the experimentation with small maps revealed, the ladder to creation of spatial complexity from small maps begins with partitions which define "entropy class". Entropy, as a measure of disorder, may also be vested with the meanings of discord, disagreement, chaos, redundancy etc. Disorder is perceived to appear in tandem with bad things. In James 3:16 the Bible reads "Where you have envy and selfish ambition, there you find disorder and every evil practice" ("Ὅπου γὰρ ζῆλος καὶ ἐριθεία, ἐκεῖ ἀκαταστασία καὶ πᾶν φαῦλον πρᾶγμα"). At this point, it is worth remembering Deleuze (2010, p. 299) in his writing that we live in "a world the very ground of which is difference, in which everything rests upon disparities, upon differences". Just one more step up on the ladder linking entropy with complexity will reveal that the combination of *both entropy and topology* seems to be a necessary *and* sufficient condition for the creation of generic binary maps. And there enters symmetry, whereas symmetric transformations of generic maps create the total number of possible map configurations and hence, a general "formula" describing the generation of complex spatial forms from simple square maps is:

$$\begin{aligned}&\text{Partitions(or Entropy)} + \text{Topology} + \text{Symmetries}\\&=> \text{Total number of Possible Map Configurations}\end{aligned}$$

Quite possibly, this formula may be applicable to square maps only and to no other types of spatial arrangements. Yet, it may serve as a model for creating spatial complexity. The number of possible configurations can therefore be considered at three distinct levels, with increasing number of possible configurations at each level: (a) partitions, (b) generic maps and (c) total number of configurations. Each level is characterized by a prevailing property and is endowed with the properties of the level preceding it: adding topological differences to entropy-only differences of configurations, the level of generic maps is attained and adding at this level the range of possible symmetries (algebraic differentiation) eventually yields the total number of possible configurations (Fig. 17.3).

Spatial arrangements corresponding to number partitions is the primary cause driving the whole process of spatial differentiation in this ladder linking spatial simplicity with spatial complexity, so it would make sense to ask whether the possible partitions follow any pattern themselves. An insight can be sought in the "*Ramanujan congruences*" which reveal that the partition function $P(n)$ may follow some rules according to the form of the integer n, for instance:

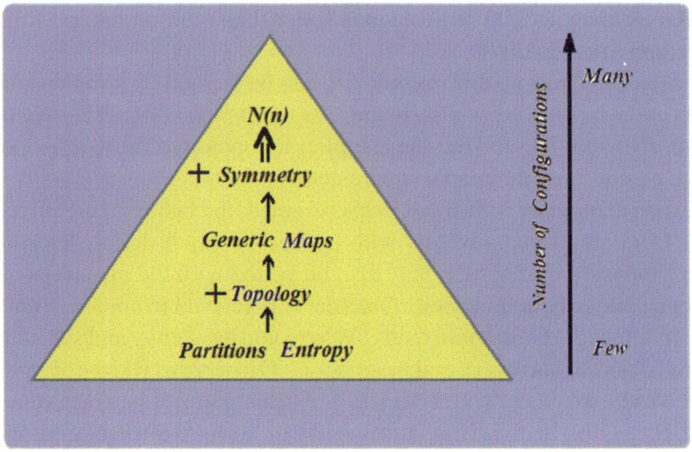

Fig. 17.3 Adding topological differences to configurations derived from partitions results in generic maps. Adding symmetry to generic maps results in the total number of square map configurations $N(n)$

$$\begin{aligned} P(5n+4) &= 0 \bmod 5 \\ P(7n+5) &= 0 \bmod 7 \\ P(11n+6) &= 0 \bmod 11 \end{aligned} \tag{17.7}$$

Following the *Dyson conjecture*, such congruences can be explained by means of the length and the rank of a partition, as proved by Atkin and Swinnerton-Dyer in 1954.

At the upper steps of the ladder, repeating patterns on the basis of symmetries are reminiscent of the concept of the "*ritournelle*" (the refrain), which is an important theme of Deleuzian philosophy. As described in Deleuze's (and Guattari's) "Mille Plateaux", the "refrain" is a block of content, and constitutes the nucleous of a spatial extent which, as Deleuze describes, is perceived as "territory". Metaphorically, the "refrain" is, more or less, equivalent to the λ block used here to calculate the C_{PI} complexity of small strings. The binarization and coding of the spatial lattice takes place in the Deleuzian "striated" space, the Euclidean, a space of the "actual". In Deleuzian terms, this contrasts the "smooth space" ("lisse") that is a "space of intensities", where the "virtual" world of qualities dwells and thus "binary machines" are created, which "grow more complex as they intersect or collide with one another, confront each other, and cut us up in every direction" (Deleuze and Guattari 1983, p. 77).

Some of the syllogisms about spatial complexity however, would not have been possible to make without the use of *square* maps. It seems that, even more than symmetry, orthogonality may be the prominent geometric property enabling us to analyze and understand spatial complexity. Salvador Dali, in his "Disintegration of the persistence of Memory" (1952), influenced by quantum mechanics, illustrates

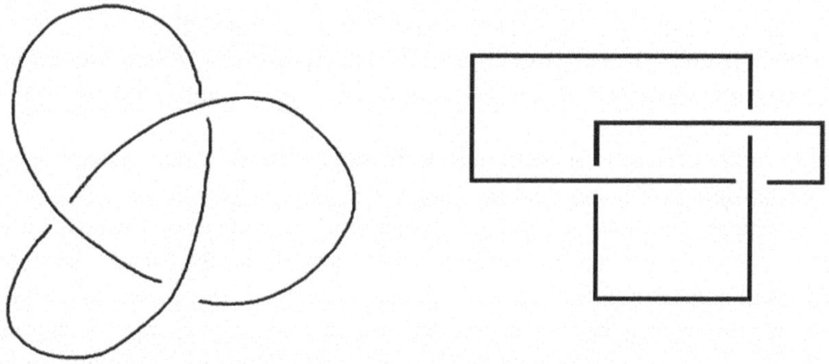

Fig. 17.4 The Morse diagram of a simple knot (left) and its Dynnikov "arc-presentation" (right)

how the fabric of space-time is discrete at quantum scales. In this work, the fundamental discreteness of space at quantum scales is represented by arrays of disjoint orthogonal bricks. Even curves in R^3 can in many ways be better understood if they are represented by vertical and horizontal lines. For instance, *Dynnikov diagrams* with vertical and horizontal lines can be used to represent and solve knots; these are called "arc-presentations" and their complexity is equivalent to the number of the vertical lines of the diagram and, following a theorem by Dynnikov (2006), every knot has an arc-presentation (Fig. 17.4).

17.4 Experimenting in the Spatium Numerorum

"I think therefore I err"
(Gigerenzer 2005, p. 1)

Several centuries ago, the ancient greek philosopher Parmenides first posed the problem of deciding what it means for a spatial extent to have a property differentiating it from another. Plato discussed in "Philebus" the aesthetics of geometric shapes (Philebus, 51d), Aristotle (in his "Metaphysics") attributed beauty to order and symmetry, while Pythagoras was concerned with "figurative numbers" (i.e. triangular, square, pentagonal numbers). Following Kant, all humans have a common basic form of spatial understanding inherited "apriori" (Godlove 1996), but, according to Hume, we can better understand discrete quantities in space (De Pierris 2012). Might this discreteness imply computability problems beyond human reach? And, as Levy aptly observed (Levy 2004, p. 133), "the relevance of space in human life is increasing. Individuals have become actors of their own spatiality as well as of the spatiality of others". Such subjective "spatialities" might explain the rise in non-postitivist and "idiographic" approaches in geography (Hallisey 2005) over the last decades. Sometimes, spatialities can not be understood without elaborate experimentation and spatial complexity may be very characteristic.

Epistemologically, adopting experimentation as a central method to explore spatial complexity appeals to the current philosophical perspective of "*Experimental Philosophy*" an emergent trend aiming to (Alexander 2012, p. 2): "apply the methods of the social and cognitive sciences to the study of philosophical cognition", by shifting the focus of emphasis from what particular philosophers think about something (as has been the traditional philosophical practice for ages) to what ordinary people think. Thus, it often poses problems, carries out quantitative analyses on the basis of which it then derives philosophical interpretations, which often allow it to come up with conclusions that were previously unanticipated. Clearly, all physical science has deep roots in experimentation, but that does not mean the experimental approaches always follow the practices of experimental philosophy or that they are eventually subjected under philosophical scrutiny. But computer-assisted explorations may prove indispensable in analyzing spatial complexity. Yet, whether we are able to understand spatial complexity by using *only* mathematical methods or not is a different issue.

The perspective of "*mathematical formalism*" suggests that mathematics is no more than a meaningless world of symbols and hence, processing strings of symbols of cover types on surfaces is equally "meaningless". In such a case, our estimates of spatial complexity may be well-structured or even accurate, but not necessarily producing meanings for us. It remains to be seen how and if this might lead us further, i.e. to *nominalism* or *intuitionism*. Although reaching at a satisfactory "explanation" of a phenomenon implies its "understanding" according to some philosophers (De Regt 2009a, b), it does not imply the same to others. Following Lipton (2009) for instance, "understanding" is possible even without explanation. Whatever the case, experimentation (as in the case of 3×3 binary maps) is a potentially useful approach to tackle spatial complexity, particularly with the adoption of mathematical methods, models and simulations towards this end. In mathematics, despite the fact that we are still short of an analytic proof of the Riemann hypothesis thus far, the experimentation hypothesis has been verified computationally for up to ten trillion zeroes. Other examples of experimentalist approaches abound in mathematics: another case is a theorem proven by "*collaborative mathematics*", an effort initiated by T. Gowers, and the *Hales-Jewett theorem* (Hales and Jewett 1963) which is a multidimensional generalization of the tic-tac-toe arrangements and generalization of van der Waerden's theorem from Ramsey theory, essentially proving that high-dimensional objects must be expected to exhibit some combinatorial structure and that it is impossible to encounter complete randomness in them.

With a strong experimentation-and-simulation method, the study of spatial complexity of square maps might also be interpreted as a "*philosophical animation*"; a term coined by Mullarkey (2000, p. 179) to describe the use of diagrams and shapes for philosophical thinking. The use of square maps for deriving results about spatial complexity may thus serve as an explanation of the ways by which spatial complexity is generated and how it changes over surfaces, as well as a general model for "understanding" spatial complexity.

17.5 Large-Scale Spatial Complexity: Limits to Spatial Analysis?

"Human mind can not bear very much reality"

(T.S. Elliot, 1888–1965, "Burnt Norton")

Spatial combinatorial "explosions" generate large numbers of possible map configurations very quickly. Computer processing of the resulting spatial datasets is one problem; stretching the human mental capacities enough to figure them out is another. Some human communities are accustomed to using parts of the human body for counting and only count small numbers. The Oksapmim in New Guinea use 28 parts of the body (face hands, shoulders) to count (Saxe 2014). However, despite the fact that different types of magnitudes are encoded by different neurons, but the brain areas responsible for spatial information processing and for numerical processing overlap in the fronto-parietal quantity network (Nieder 2019).Tricks to shorten our way to calculations of "explosively" large numbers may appear in tandem with technological advances in computing. For instance, multiplication as taught at school requires n^2 operations, but converting numbers to be multiplied to polynomials with complex roots by using fast Fourier transforms (Rittaud 2013) requires a significantly lower number of operations, specifically of the order of $n\log n[\log(\log(n))]$.

Even such compressions however, can not always help us perceive the very big numbers emerging from spatial combinations. In fact, the problem of perceiving and coping with such numbers has apparently been known since the antiquity. Archimedes (287–212 b.C.), in his short work "The Sand Reckoner" («Ψαμμίτης», or «Ἀμμου Καταμέτρης») estimated that as many as $10^{8\times10^{16}}$ "grains of sand" might fill the entire space of the universe: a myriad of myriads and this raised to the power of the myriad of myriads and all this to the power of a myriad of myriads ("μία μυριὰδα μυριὰδων εἰς τήν μυριοστή μυριὰδα καὶ ὅλο εἰς τήν μυριοστή μυριὰδα"); a number that turns out to be much bigger than the number of baryons in the universe. Unsurprisingly therefore, some philosophers have repeatedly highlighted the inadequacy of human mind to imagine very large numbers (i.e. McTaggart 1927; Norcross 1997; Broome 2004). In view of this challenge, the philosophical currents of "epistemic modality" and "fictionalism" are particularly relevant: by examining modal expressions in languages (like "might", "could" etc.), "epistemic modality" focuses on possible worlds (Egan and Weatherson 2011). In this way, $N(n)$ can be interpreted philosophically, as a "modal" expression of possible combinations of n objects.

Yet, spatial complexity may also be perceived through the philosophical lenses of "fictionalism", which accepts the plurality of worlds under certain conditions (Liggins 2008; Woodward 2008). In response to these philosophies, "modal realism", in a synthesis of epistemic modality and fictionalism, accepts both possible and impossible worlds as "real". Modal realism begun with a work on pragmatic and semantic problems arising from pluralities of worlds by Lewis (1986) and progressed with Yagisawa's (2010) philosophy of "possible" worlds. The overwhelming computational complexity that easily emerges from problems involving large scale spatial combinations and possible spatial outcomes crops up in spatial decision-making and agrees with empirical evidence suggesting that when humans are asked to evaluate uncertain spatial outcomes, they accept uncertainty, but "remain profoundly divided over what strategies to adopt" (Retaillé and Walther 2011, p. 85).

Indeed, the epistemological impact of calculations of $N(n)$ can be more far-reaching if considered at the level of practice, for they unavoidably suggest there may be possible limits to geographical knowledge and practice. This raises a serious epistemological problem for spatial scientists, with geographers and cartographers among them most prominent: Might the "exploding" numbers of possible map configurations force researchers to accept subjective interpretations as indispensable to spatial analysis?

It is not difficult to explore the orders of magnitudes resulting from such "explosions". Take, for instance, a square binary map of size 100×100. How many binary map configurations are possible from the 10,000 cells of this map up to maximum entropy? The answer is that this map can host precisely 100,891,344,545,564,193,334,812,497,256 different binary square maps. This figure is approximately tantamount to 100 billion billion billion, or 100 octillion of possible binary map configurations.

Equivalently, a map of size 1 km $\times$ 1 km may host as many as

$$2.702 \times 10^{299}$$

possible different binary maps of size 1 square meter each (2, 7 followed by 300 digits) up to maximum entropy class, or, more precisely,

270 288 240 945 436 569 515 614 693 625 975 275 496 152 008
446 548 287 007 392 875 106 625 458 705 522 193 898 612 483
924 502 370 165 362 606 085 021 546 104 802 209 750 050 679
917 549 894 219 699 518 475 423 665 484 263 751 733 356 162
464 079 737 887 344 364 574 161 119 497 604 571 044 985 756
287 880 514 600 994 219 426 752 366 915 856 603 136 862 602
484 428 109 296 905 863 799 821 216 320

What might the geographical implications of these magnitudes be? Consider that the total surface area of the earth is approximately 5.1×10^8 km^2. If, instead of a sphere, the area of the terrestrial surface stretched over onto a planar square map of equal area, this map would have a side equal to the square root of 5.1×10^8, that is 22,583 km. Consequently, if up to $r = 225{,}000{,}000$ black cells were defined on this planar map, the number of all possible 1 km $\times$ 1 km binary map configurations would be:

$$N(n) = 10^{10^{8.817116294672106}}$$

How big is this number? To compare with, there are only about 3×10^{80} cubic metres volume of space in the observable universe, while a commonly used estimate of the number of baryons in the observable universe is of the order of 10^{80} and the number of atoms in the universe ranges between 4×10^{78} and 6×10^{79}. Also, for the sake of comparison, the (notoriously high) number of all valid positions in a chess game ranges between 10^{43} and 10^{50}. It is easy to see that $N(n)$ exceeds all these numbers:

$$10^{10^{8.3}} > 10^{80} > 10^{79} > 10^{50}$$

Hence, if the entire earth's (planar) surface were subdivided into square cells of 1 km^2 size each and all possible allocations of black cells in this map were calculated (even *only* up to maximum entropy class), the total number of possible map configurations would be higher than the number of atoms and higher than the number of baryons in the entire (known) universe.

Would it be an exaggeration to consider this simple calculation as an indication of the limits to our geographical knowledge and understanding? Evidently, this would also limit the potential of spatial planning alternatives.

And if the earth's surface were stretched out as a square planar map and if it were divided into cells of 1 square meter size each, then the number of possible binary square maps up to maximum entropy would be:

$$10^{10^{10^{1,105837751009754}}}$$

(10 to the power of 10 to the power of 10 to the power of 1.106). This number can be expressed as approx. 1 "googol" to the power of 1.106 (1 googol equals 10^{100}); a number clearly beyond comprehension for the human mind (1 followed by 101 zeroes):

10,000,000,000,000,000,000,000,000,000,
000,000,000,000,000,000,000,000,000,000,
000,000,000,000,000,000,000,000,000,000
000,000,000,000

The naming "google" was introduced arbitrarily, by the 9-years old Milton Sirotta in 1920. Although rather useless in calculations so far, googol is sometimes used to benchmark the border between the very large and the physically uncountable (Kasner 2001).

Such magnitudes are reminiscent of the "oceanic feeling of limitlessness and boundlessness" (Lambert et al. 2006, p. 479) which has long inspired, disappointed (and motivated) geographers and navigators throughout the ages. We are thus in position to calculate only a very small number of possible spatial configurations of the earth's surface and, therefore, *we can only partially manage (computationally) the vast realm of possible spatial configurations.* It won't be too far from the truth to claim that there are limits not only to our understanding, but may be even to our very imagination of how an adequately large piece of space might possibly look like. If more types of spatial allocations were considered in the geographical space (i.e. not only binary, but multi-colored allocations also), then the number of possible configurations would increase intractably for relatively "small" *n*. The temptation to suggest that *potential* geographies imply limits to spatial computability seems hard to resist. When Garry Kasparov lost to IBM"s "Deep Blue" in 1997, many doubted whether human mental power might be up to the computational challenge posed by supercomputers. But, as was pointed out (Rasskin-Gutman 2012), a player looking eight moves ahead in chess is already presented with as many possible game evolutions in chess as there are stars in the galaxy.

So how many "moves ahead" can a spatial planner or a geographer make? How many possible spatial allocations could we possibly reflect upon? How long will it take before being able to decide that a spatial scenario or a spatial plan has really examined *all* the possible spatial outcomes? And, eventually, will we ever be able to cope with spatial analysis at such large spatial scales?

We so far have no definite answers to these questions and we may never have. In as much as there are many possible interesting chess games which we do not know of (because they have never been played), there are (almost) innumerable ways for humans to consider possible alternative spatial arrangements on the surface of the earth that have never been considered and there may not even exist adequate time available to examine them.

References

Alexander, J. (2012). *Experimental philosophy. An introduction.* Cambridge: Polity Press.

Ansell-Pearson, K. (1997). *Viroid life: Perspectives on Nietzche and the Transhuman condition.* London: Routledge.

Broderick, D. (2001). *The spike.* New York: Forge.

Broome, J. (2004). *Weighing lives.* Oxford: Oxford University Press.

Bukatman, S. (2005). *Terminal identity: The virtual subject in postmodern science fiction.* Durham: Duke University Press.

Da Costa, N. C. A., Bueno, O., & French, S. (1997). Suppes predicates for space-time. *Synthese, 112*(2), 271–279.

De Landa, M. (2002). *Intensive science and virtual philosophy.* London: Continuum.

Deleuze, G. (1993). *The fold.* London: Bloomsbury.

Deleuze, G. (2010). *Difference and repetition.* London: Continuum.

Deleuze, G. (2012). *The logic of sense.* London: Continuum.

Deleuze, G., & Guattari, F. (1983). *On the line.* New York: Semiotext(e).

De Pierris, G. (2012). Hume on space, geometry, and diagrammatic reasoning. *Synthese, 186*(1), 169–189.

De Regt, H. W. (2009a). The epistemic value of understanding. *Philosophy of Science, 76,* 585–597.

De Regt, H. W. (2009b). Understanding and scientific explanation. In H. W. De Regt, S. Leonelli, & K. Eigner (Eds), *Scientific understanding* (pp. 21–42). Pittsburgh: University of Pittsburgh Press.

Dynnikov, I. A. (2006). Arc presentations of links-Monotonic simplification. *Fundamenta Mathematicae, 190,* 27–96.

Egan, A., & Weatherson, B. (Eds.). (2011). *Epistemic modality.* Oxford: Oxford University Press.

Gigerenzer, G. (2005). I think therefore I err. *Social Research, 72*(1), 1–24.

Godlove, T. F., Jr. (1996). Is "space" a concept" Kant, Durkheim and French neo-Kantianism. *Journal of the History of the Behavioral Sciences, 32*(4), 441–455.

Hagerstrand, T. (1982). *Diorama, path.* Tijdschrift voor Economische en Sociale: Project.

Hales, A. W., & Jewett, R. (1963). Regularity and positional games. *Transactions of the American Mathematical Society, 106,* 222–229.

Hallisey, E. J. (2005). Cartographic visualization: An assessment and epistemological review. *Professional Geographer, 57*(3), 350–364.

Hartmann, N. (1940). *Der Aufbau der realen Welt. Grundriss der allgemeinen Kategorien lehre.* Berlin: De Gruyter.

Hartmann, N. (1975). *New ways of ontology.* Connecticut: Greenwood Press.

Hyde, D. C., & Spelke, E. S. (2009). All numbers are not equal: An electrophysiological investigation of large and small number representations. *Journal of Cognitive Neuroscience, 21,* 1039–1053.

Kasner, E., & Newman, J. R. (2001). *Mathematics and the imagination.* New York: Dover.

Kühne, O. (2012a). Urban nature between modern and postmodern aesthetics: Reflections based on the social constructivist approach. *Questiones Geographicae, 31*(2), 61–70.

Kühne, O. (2012b). *Stadt – Landschaft – Hybridität. Ästhetische Bezüge im postmodernen Los Angeles mit seinen modernen Persistenzen.* [City—Landscape—Hybridity. Aesthetic references in postmodern Los Angeles with its modern persistence]. Wiesbaden: Springer VS.

Kühne, O. (2018). *Landscape and power in geographical space as a social-aesthetic construct.* Dordrecht: Springer.

Kühne, O. (2019). *Landscape theories. A brief introduction.* Wiesbaden: Springer VS.

Kühne, O., Weber, F., & Jenal, C. (2018). Der Begriff ‚Landschaft' sowie essentialistisch und positivistisch orientierte Zugänge. [The term 'landscape', essentialist and positivist orientated approaches]. In O. Kühne, F. Weber, & C. Jenal (Eds.), *Neue Landschaftsgeographie. Ein Überblick* (pp. 5–10). Wiesbaden: Springer VS.

Lambert, D., Martins, L., & Ogborn, M. (2006). Currents, visions and voyages: Historical geographies of the sea. *Journal of Historical Geography, 32,* 479–493.

Leibnitz, G. W. (1714). *Monadologie.*

Lévy, J. (2004). A geographical turn I [Eine geographische Wende]. *Geographische Zeitschrift, 92*(3), 133–146.

Lewis, D. (1986). *On the plurality of worlds.* Oxford: Blackwell.

Liggins, D. (2008). Modal fictionalism and possible worlds discource philosophical studies. *International Journal for Philosophy in the Analytic Tradition, 138,* 151–160.

Lipton, P. (2009). Understanding without explanation. In H. W. De Regt, S. Leonelli, K. Eigner (Eds) *Scientific understanding* (pp. 43–63). Pittsburgh: University of Pittsburgh Press.

McTaggart, J. M. E. (1927). *The nature of existence.* Cambridge: Cambridge University Press.

Meshcheryakov, G. (1965). The problem of choosing the most advantageous projections. *Geodesy and Aerophotography, 4,* 263–268.

Mullarkey, J. (2000). *Bergson and philosophy.* Notre Dame: Notre Dame University Press.

Nieder, A. (2019). *A brain for numbers: The biology of the number instinct.* Cambridge, MA: MIT University Press.

Norcross, A. (1997). Comparing harms: Headaches and human lives. *Philosophy and Public Affairs, 26,* 135–167.

Papadimitriou, F. (2010a). A “neogeographical education”? The geospatial web, GIS and digital art in adult education. *International Research in Geographical and Environmental Education, 19*(1), 71–74.

Papadimitriou, F. (2010b). Introduction to the complex geospatial web in geographical education. *International Research in Geographical and Environmental Education, 19*(1), 53–56.

Peters, M. A. (2008). Wittgenstein as Exile: A philosophical topography. *Educational Philosophy and Theory, 40*(5), 591–605.

Rasskin-Gutman, D. (2012). *Chess metaphors. Artificial intelligence and the human mind.* Cambridge, MA: The MIT Press.

Retaillé, D., & Walther, O. (2011). Spaces of uncertainty: A model of mobile space in the Sahel. *Singapore Journal of Tropical Geography, 32*(1), 85–101.

Revkin, S. K., Piazza, M., Izard, V., Cohen, L., & Dehaene, S. (2008). Does subitizing reflect numerical estimation? *Psychological Science, 19,* 607–614.

Rittaud, B. (2013). Nombres complexes pour calculs d’ ecolier. *Les Dossiers de la Recherche, 5,* 30–32.

Saxe, G. B. (2014). *Cultural development of mathematical ideas: Papua New Guinea Studies.* New York, NY: Cambridge University Press.

Sean, R. (2009). Heidegger’s topology: Being, place, world. *Australasian Journal of Philosophy, 87*(1), 169–171.

Simons, P. (2006). The logic of location. *Synthese, 150*(3), 443–458.

Tobler, W. R. (1965). Computation of the correspondence of geographical patterns. *1964 Papers Regional Science Association, 15,* 131–139.

Virilio, P. (1994). *The vision machine.* London: British Film Institute.

Woodward, R. (2008). Why modal fictionalism is not self-defeating. *Philosophical Studies, 139,* 273–288.

Yagisawa, T. (2010). *Worlds and individuals—Possible and otherwise.* Oxford: Oxford University Press.

Chapter 18
Spatial Complexity and the Future

Some of the legends are simple and some are complex
"Εἰσὶ δὲ τῶν μύθων οἱ μὲν ἁπλοῖ οἱ δὲ πεπλεγμένοι"
(Aristotle, 384–322 b.C., "Poetics", 1452a)

Abstract Three sets of determinants of spatial complexity have been identified: (i) probabilistic (entropy, randomness), (ii) geometrical (symmetries, intersections, orthogonality, curvature), (iii) topological (boundaries, genus, dimension, knottedness, braiding, knotting, linking, braiding, writhing). Spatial complexity is an inter-disciplinary subject and a spatial complexity observatory might need to be established, to monitor developments and research results related to spatial complexity from across all disciplines. The future of humankind highly depends on the degree to which it is able to understand, exploit and manage spatial complexity at large (planetary) scales. Even in planning a utopia, would we prefer maximum spatial complexity? Or just maximum spatial diversity? Or a co-presence of different "compossible" (as Leibniz named them) spatial configurations? Might quantum computation and/or spatial computing help us explore spatial alternatives? No doubt, crossing borders into the amazing realm of spatial complexity poses challenging problems at both theoretical and practical levels.

Keywords Spatial complexity · Spatial computing · Geography and complexity · Quantum computation · Complexity and inter-disciplinary · Mens Spatii · Kardashev

18.1 Key Determinants of Spatial Complexity

Our knowledge is only partial and our prophecies are also partial
"ἐκ μέρους δὲ γινώσκομεν καὶ ἐκ μέρους προφητεύομεν"
(St.Paul, A' Corinthians, 13)

It is now appropriate to take stock of what we know to try to identify general fields, axes, criteria, or, better, "determinants" of spatial complexity. As research

F. Papadimitriou, *Spatial Complexity*, https://doi.org/10.1007/978-3-030-59671-2_18

from disparate research fields has revealed, spatial complexity is a genuinely interdisciplinary field of research. This is because spatial contrasts, disparities and inequalities in variables, parameters, intensities, populations, quantities and qualities are ubiquitous, cutting across most (if not almost all) domains of scientific enquiry. We live in a world of incessant and unlimited spatial differentiations, so we have developed science and technologies to explore them. Whether we use satellite imagery or MRI, data from field observations or 3d representations from physical and chemical experiments, or even higher-dimensional spatial analyses, we are prompted to deal with innumerable types and forms of more or less spatially complex objects and surfaces. As a matter of fact, it often passes unnoticed that questions and problems relating to particular determinants of spatial complexity essentially conceal the overall problem of defining and assessing spatial complexity itself. For instance, when a physicist examines the "spatial entropy" of a chemical process, or when a landscape ecologist calculates the "landscape diversity" of a landscape, they are both referring to some determinants of spatial complexity. And when a textile designer draws celtic braids and knots on the canvas and a fluid mechanics engineer examines braided flows of liquids, they essentially deal with the same determinant of spatial complexity (braiding).

Spatial complexity is a property of every spatial object or allocation in space; it may be trivially low or intractably high. Whether we deal with two, three, or higher-dimensional entities, it nevertheless depends on some key factors such as quantity of spatial elements, quantity of non-spatial elements allocated in space, randomness, symmetries, geometry type, topology, genus, dimension, knottedness, and, possibly, qualities. Defining the degree of complexity or simplicity of a spatial object or surface is tantamount to calculating the difficulty involved in this definition, whether it is i.e. a nano-surface or the large-scale structure of the universe. We deal with spatial complexities, we suffer from them, we need to rid of them, but we also need to create them with our technologies and we enjoy taking aesthetic pleasure from observing them in various forms of visual artworks.

As concerns the psychological context of spatial complexity, there are strong indications that the mechanisms defining the perception of spatial complexity may not be as quantitatively tractable as we would like them to be. However, transferring such difficulties to categories of mathematics for which we have well-crafted tools may be encouraging.

Our current mathematical armature does not guarantee complete success in conquering the summits of spatial complexity analysis. Quite the contrary; it sometimes only serves to prove that some spatial problems easily lead to combinatorial explosions or intractable situations. But knowing the limits of human knowledge and the limits of human understanding in relation to spatial problems is not a minor issue, whatever the particular field of one's scientific expertise may be. And yet, there is another, brighter side: the fact that in most cases, we generalize spatial data, and through these generalizations we are efficient in counting, evaluating and handling the levels of spatial complexity we need. And in this respect, possessing easy-to-use metrics such as C_{P1} and C_{P2}, helps in assessing the spatial complexity of "small" maps on (admittedly vaguely defined) "not-too-complex" surfaces. Extending these indices (and any other ones that may appear in the future) to three and more dimensions and adjusting them to also apply to non-euclidean geometries and to higher dimensional objects is clearly a big challenge, but certainly not the only one. The study of spatial complexity is replete with open problems and addressing them effectively will significantly impact the way we perceive the world around us.

Spatial complexity apparently depends on three key sets of determinants: entropy/randomness, geometry and topology. All these determinants should eventually yield some algorithmic and/or computational expressions defining spatial complexity. The manifestations of each one of these sets of determinants are manifold. Thus, with respect to entropy, the higher the number of cover types and the more random the allocation, the more complex the object is. Conversely, if there are symmetries and simple shapes, the less complex it is. And, if the space is non-euclidean, complexity increases. But neither entropy nor geometric properties are adequate indicators of spatial complexity; topology is indispensable, by offering us further descriptors of complexity (whether the space examined has high or low genus, the number of spatial dimensions, the number of boundaries in it etc). All in all, the spatial complexity of a surface or object therefore seems to depend on the following factors (Fig. 18.1):

(A) Entropy and randomness

- Richness in non-spatial categories/classes/covers: the higher the number of distinctively different categories, the higher the spatial complexity.
- Entropy: the more equiprobable the number of non-spatial categories/classes/covers, the more complex the object.
- Randomness of allocation of these spatial or non-spatial categories: generally, but not always, the higher the randomness of the allocation of the categories, the higher the spatial complexity, or, otherwise put, less patterning means higher spatial complexity.

(B) Geometry

- Symmetries: a less symmetric object or form is expectedly a more complex one.
- Intersections: more intersections among spatial elements imply a more complex spatial setting.
- Orthogonality: intersections at right angles result in less complex objects.
- Geometry type: it is more computationally demanding to describe an object that obeys a non-euclidean geometry.

(C) Topology

- Boundaries: more boundaries among the different spatial elements means the object is more complex.
- Genus: the higher the genus of an object, the more complex it is.
- Dimension: The higher the spatial dimension of an object, the more complex.
- Knottedness, braiding, linking, writhing: the presence of knots, braids, links and writhes make an object more complex.
- Immersions and singularities.

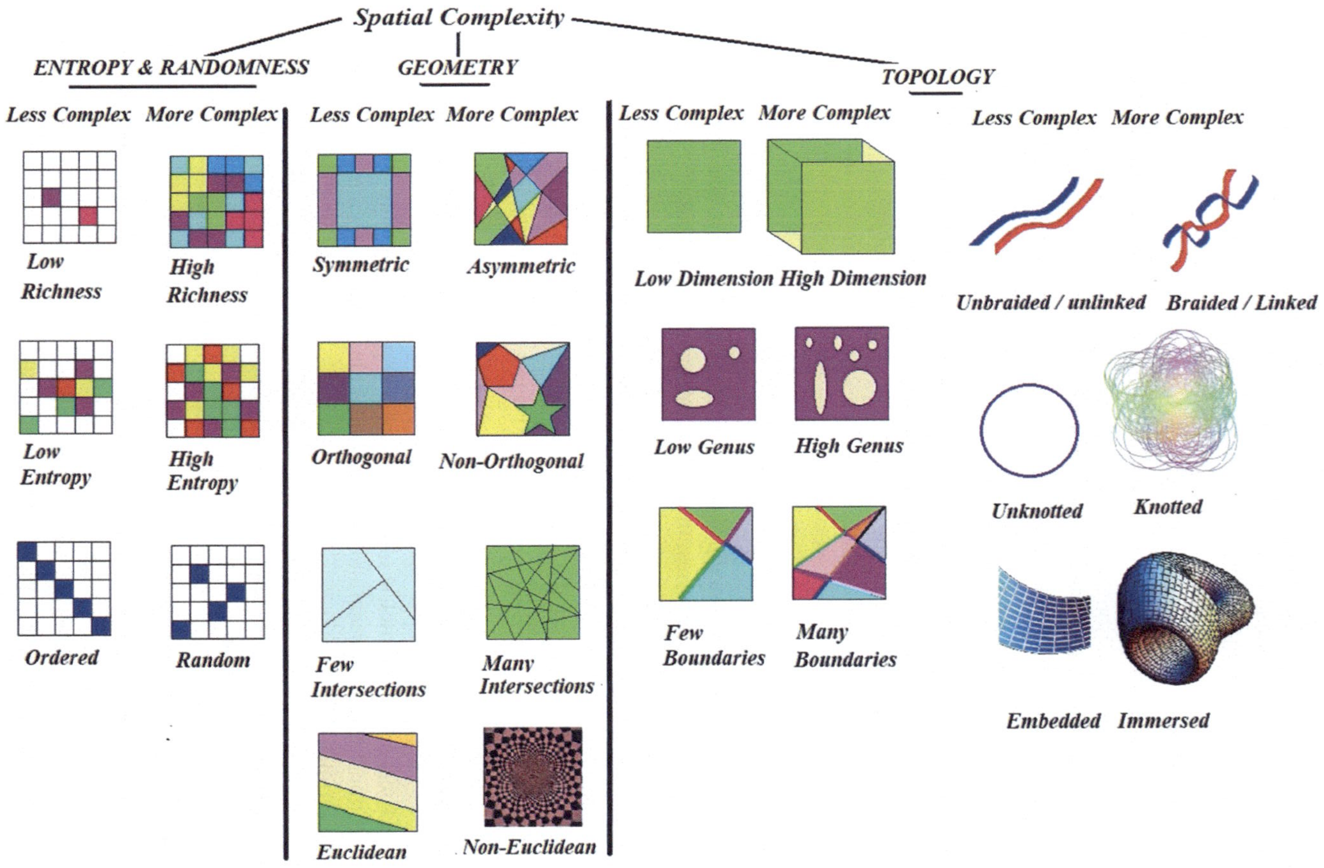

Fig. 18.1 Three key sets of determinants of spatial complexity: entropy and randomness, geometry, topology (see text for explanations)

The identification of key determinants of spatial complexity however, does by no means imply that the riddle of spatial complexity has been solved. There are limits to the quantitative assessments of several of these determinants. And besides these, there lies the unfathomable domain of qualitative and subjective interpretations of spatial complexity, which entails an enormous complexity of its own. It would be far too ambitious to claim that such a literally "knotty" subject might easily be "unknotted" within a single book. Yet, one can at least hope that the importance of spatial complexity will be highlighted as a central factor in the vast range of scientific explorations, as it already is and as it deserves to be. As we are not alone on the face of the earth, we should not loose sight from the fact that the ways by which other living beings perceive and assess spatial complexity is largely unknown. For humans however, it seems that spatial complexity is intricately related to the very essence of human identity: we, as a species, are able both to *generate* immensely complex spatial forms, as well as to *de-complexi-fy* those spatial forms that are necessary to maintain and improve our livelihoods. Furthermore, assessing spatial complexity is practically useful, from identifying melanomas to producing more efficient electronic devices, designing intellectually challenging spatial games, exploring aesthetics, etc. Truly, the bulk of this book is essentially a mathematical approach to spatial complexity, but this might make one wonder: is it inevitable to resort to mathematics in order to explain spatial complexity? The answer is: "Yes. Mostly", because it seems nearly impossible to derive any reliable assessment of spatial complexity without resorting to quantified criteria which have to do with measurable quantities (number of intersections, bounds to combinatorial complexity, genus, number of knot crossings etc.), as well as with mathematical but not necessarily measurable quantities (i.e. the geometry type of the setting that is considered). May be, it is precisely this combination of *measurable* (and "assessable" more generally) properties with *not easily measurable* properties that defines the overall spatial complexity of a surface or object. Possibly, with some degree of generalization, it might make sense to assume that any qualitative or subjective estimates of spatial complexity may also be partially attributed to such combinations. Plausibly, one can not easily prove or disprove this conjecture without compelling experimental evidence at hand. But we are still very far from that.

18.2 Spatial Complexity, "Com-Possible" Worlds and Quantum Computation

And they said to all the Israelites:
The land which we spied on is very very beautiful
"καὶ εἶπαν πρὸς πᾶσαν συναγωγὴν υἱῶν Ἰσραὴλ λέγοντες·
ἡ γῆ, ἣν κατεσκεψάμεθα αὐτήν, ἀγαθή ἐστι σφόδρα σφόδρα"
(The Bible, Numbers, 14.7)

In a letter dated 23 November 1697, Leibniz expressed his thought that "out of the infinite combinations of possibles and possible series, there exists one with the greatest amount of essence or possibility and this is brought into existence". Indeed, for Leibniz (in the 1923 edition of Samtliche schiften und briefe". Vol.A VI iv 2232), "possible things have no existence at all and therefore have not the power to exist". Further, not all possible combinations are "*com-possibles*". That is, some possible things are not compatible with each other, and thus can not be realized simultaneously and syn-chorically: "there are many possible universes, each collection of compossibles making up one of them" (Leibniz ed. 1969, L662), but, as he wrote in a letter to Louis Bourguet (5 August 1715) "God chooses the best possible" (Leibniz ed. 1890, G III 583). Amazingly, Leibniz appreciated that the whole process of com-possibles ressembles spatial games where cells on the board are either filled or left empty according to the game's rules and the player's strategy (Leibniz 1890 ed., G IV 405-6, from a letter sent to Fracois Pinsson in June 1701). Hence, to the question "Which one of the worlds is the best possible?", Leibniz answers that it is the one containing "as many things as possible" (Strickland 2006,p. 30), or the set in "which most things co-exist" (op,cit,195). With this thought, Leibniz essentially concluded to a principal criterion for the determination of which world would be the best possible: the one ensuring the highest variety. This condition, interpreted in modern terms, means a world endowed with the highest diversity possible (or maximum entropy).

In vol.VI of his "Interpretation of Nature", Denis Diderot (1754) wondered about possible explanations of the imaginary situation that "Eternity" wrote the "universal mechanism" in the pages of an enormous book, and posed the question: Would the book be more understandable than the universe itself? ("Si l'Éternel, pour manifester sa toute-puissance plus évidemment encore que par les merveilles de la nature, eût daigné développer le mécanisme universel sur des feuilles tracées de sa propre main, croit-on que ce grand livre fût plus compréhensible pour nous que l'univers même?"). This seemingly paradoxical question, by modern standards, ehoes the very concept of algorithmic complexity and can be restated as follows: Would the complexity of the book be equal to the complexity of the physical universe?

Indeed, the worlds of "possibles" have proven a very fertile ground to non-philosophers and non-scientists and the existence of alternative worlds has been a source of inspiration for several authors. Simply consider J. R. R. Tolkien's "Lord of the Rings" and "Silmarillion", H. Rh. Lovecraft's "Kthulu", G. Orwell's "1984", T.More's "Utopia" and so on, among innumerable others. Whether completely fantastic, or created on the basis of modern scientific thinking as in Francis Fukuyama's "posthuman future" (Fukuyama 2002), these worlds so far remain nothing more than merely imagined worlds of human fiction, appearing only in printed paper and films; never in reality.

Nonetheless, the existence of "Parallel worlds" and alternative possible worlds is not precluded, as they constitute significant parts of certain interpretations of quantum mechanics. Although their examination is beyond the scope of this work, it is worth considering that, following the Ghirardi-Rimini-Weber theory of quantum mechanics (Ghirardi et al. 1986), the entire world can be considered as eventually one physical object only: the wave function of the universe. Following this line of thought, all realizations in this world result from vibrations of Schrödinger's wave function. This may seem too simplistic or too theoretical an approach, but the oncoming quantum computers may overcome many of the obstacles standing in the way towards massive spatial computation, eventually paving the way to the computation of "alternative worlds" for very large map sizes. This is because quantum computers are exponential, non-deterministic Turing machines, with extremely high computational capacity, able to process simultaneously overlapping possible eigenstates of quantum systems ("qubits"). A qubit comprises several possible states of combinations of 0 and 1, simultaneously. It takes two eigenstates: $|1>$ and $|0>$ and these are derived from the superposition of two quantum states $|\psi> = a|1> + b|0>$ (where a and b are complex numbers, with $a + b = 1$). For instance, an 8-bit set of mutually exclusive states is 256 such states if processed on a classic computer, but a quantum computer is expected to be able to process in polynomial time all 2^8 combinations simultaneously. In particular, according to Heisenberg matrix mechanics, the quantum state vector $|\psi>$ in the Hilbert space is:

$$|\psi\rangle = a_1 \begin{pmatrix} 1 \\ 0 \\ 0 \\ \cdots \\ 0 \end{pmatrix} . + a_2 \begin{pmatrix} 0 \\ 1 \\ 0 \\ \cdots \\ 0 \end{pmatrix} + a_3 \begin{pmatrix} 0 \\ 0 \\ 1 \\ \cdots \\ 0 \end{pmatrix} + \ldots + a_n \begin{pmatrix} 0 \\ 0 \\ 0 \\ \cdots \\ 1 \end{pmatrix}$$

so the 2-qubit states appear as column vectors:

$$|00\rangle = \begin{pmatrix} 1 \\ 0 \\ 0 \\ 0 \end{pmatrix}, \ |01\rangle = \begin{pmatrix} 0 \\ 1 \\ 0 \\ 0 \end{pmatrix}, \ |10\rangle = \begin{pmatrix} 0 \\ 0 \\ 1 \\ 0 \end{pmatrix}, \ |11\rangle = \begin{pmatrix} 0 \\ 0 \\ 0 \\ 1 \end{pmatrix}$$

Quantum computation effectuated on the basis of qubits m is based on the power 2^m, so a calculation with m=6 qubits gives 64 possible routes. Consequently, with quantum computers, a number of routes as high as 1,9x10^{81} can be calculated, with only m=270 qubits; this number of routes is so big that approximates the total amount of particles in the universe.

So, quantum algorithms, such as "*Shor's algorithm*" may be promising in shortening calculation times. While factorizing a number of n digits long by means of classical arithmetic requires a number of operations rising exponentially with n, Shor's quantum algorithm requires a growth of only $O(n^3)$. The same algorithm can be used to rapidly factorize numbers to primes. To avoid using complex numbers, instead of q-bits, the u-bits have been introduced, with even more surprising properties: a u-bit is essentially capable of interacting with all other bits of information in the universe.

Consequently, given the very rapid growth in computational power, it may be too early to define the ultimate limits to our computational capability, and hence the limits to how spatial complexity can be perceived, understood and explained, either by human intelligence alone or aided by machines.

18.3 Large-Scale Spatial Complexity, Spatial Computing and Planetary Futures

"You're on Earth. There's no cure for that"

(Samuel Beckett, 1906–1989, "Endgame", 1957)

Were a super-powerful planner given the chance to re-plan the earth, what should he need to maximize? Homogeneity? Diversity? Symmetry? Spatial complexity? As the reader may correctly have suspected, this question reverberates the centuries-old problem of utopia and utopism, from Plato (his "Politeia") to Thomas Moore. The naturalist Richard Jefferies wrote in his novel "After London" (1885) about a future in which, after a disaster, the population of London is eliminated and the formerly busy city is colonised by plants and has reverted into wilderness: an appalling dystopia or a green utopia? Similarly, William Morris in his "News from Nowhere" (1890) described a post-industrial world without transport networks or industry, where woodlands have replaced buildings in a clearly rural "utopia", that might be summarised by the known motto "Et in Arcadia ego". But the urban development across the world has proven that what humans eventually create on the face of the earth is quite the opposite: a rapid complexification of planetary proportions. For Jameson (1990), the greatest danger in utopian thinking is the regression to uniformity. In short, *maximum order is not utopia* and, as a matter of fact, the major challenge for the future is to achieve a stable complex society (Whitehead 1926, p. 90) instead of a society with maximum order.

The world has become immensely more complex and the earth's spatial complexity has attained heights that none could ever expect at the time of the first industrial revolution in which the utopists lived. This *visible* spatial complexity is convoluted with an equally immense *invisible* one, created by all sorts of cyber-physical systems, AR, VR, networks of networks and "hypernetworks" surrounding the globe, connecting the Internet with sensors and cableless networks (Papadimitriou 2010a, b), forming vast arrays of "spatial computing".

Where does this exploding spatial complexity lead us to? Some decades ago, Freeman Dyson and Nikolai Kardashev classified civilizations in the universe (provided there exist other ones aside of ours') into three types, according to their energy management techniques (Dyson 1979; Kardashev 1964): "Type I civilizations" are those who have become capable of using all forms of their planet' s energy. By virtue of a globally established synergy, they have avoided major military conflicts, have constructed planetary telecommunication networks, and have developed methods to take advantage of their planet's energy (i.e. wind energy, geothermic

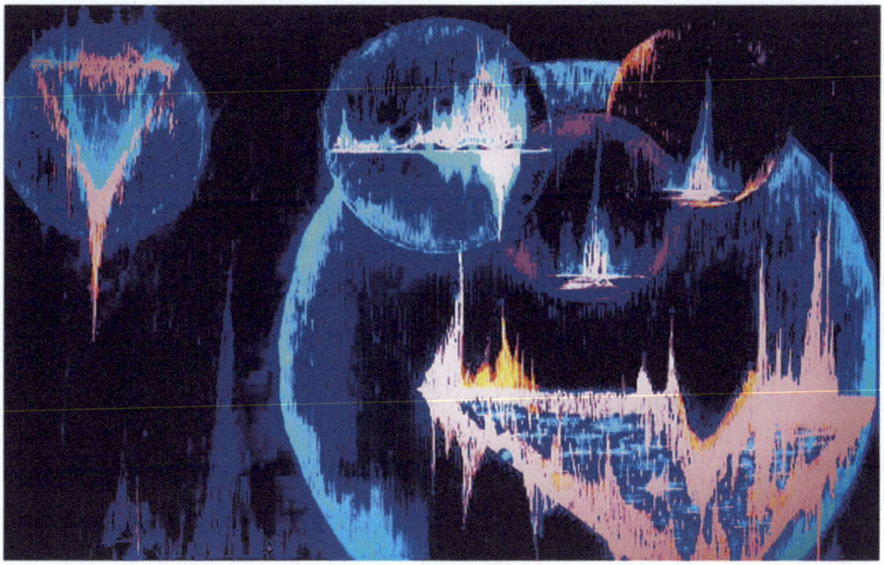

Fig. 18.2 Imagining a possible galactic Type-II civilization and "Dyson spheres" (created by the author).

fields). "Type II civilizations" have mastered stellar energy sources and have thus become able to use energy from stars, which they use to colonize neighboring stellar systems (Fig. 18.2)."Type III civilizations" are able to use energy from stars of many stellar systems and thus colonize their galaxy. This classification also relates to the "Dyson spheres" which sometimes astronomers have mistakenly thought to have discovered. Following Kaku (1997), the "United Federation of Planets" that is the central entity in the series "Star Trek" is an example of the "Type II" status, whilst the "Borg" extraterrestrial societies qualify for the "Type III" status. The imaginary cubic Borg colonies are advanced human-cybernetic hybrids that have assimilated individualities and transformed them into a unified system of life and intelligence, embracing every living being in their own system of spatial intelligence. But we are not there, since humankind still does not even qualify for the "Type I" status: a "planetary intelligence" is still in the making.

As suggested by Doyle et al. (2011), the entropy and complexity of a message of extraterrestrial origin may determine whether the signal is complex enough to have been emitted by some intelligent agent and (op.cit., p. 417) "a direct indication of the measure of the intellectual capabilities and possibly the complexity of the social structure of the senders". Such a message may be transmitted as a string of symbols (in a linear format), but its decoding and/or interpretation may as well be spatial and the complexity of the signal emitted would most likely be an indication of the civilization's complexity level. A spatially-encoded intelligent message might as well entail a universally understandable mathematical pattern. Indeed, the famous message emitted from the earth to the outer space, the "Arecibo message", was characteristically spatial. Emitted on the 16th of November 1974, the Arecibo radio telescope in Puerto Rico sent the "Arecibo Message" to the star cluster M13, with a frequency of 238 MHz and a power of 1000 kW. The message consisted in 1679 binary digits. This number is the product of two prime numbers ($23 \times 73 = 1679$) and hence the message was a binary map of 23 columns by 73 rows. If read column-wise (23×73), it does not make any sense, but if read row by row (that is 73×23), it contains information about the atomic numbers of the chemical elements H, C, N, O and P, which make up the basis of life on earth (the DNA), the basic forms of DNA nucleotides. The same spatial arrangement makes sense in other ways also, as it conveys information about the form of a human body, the DNA helix, the graph of the solar system and a graph of the Arecibo radio telescope (the antenna dish). The message was sent to star cluster M13, some 25000 light years away, hoping that any intelligent beings that would receive it would also be able to decode it and understand the information it conveys. Essentially, this is tantamount to hoping that their aptitudes in decoding and deriving meanings from spatial complexity will be matching our own.

Most likely, two of the hallmarks of our unique position in the universe are our ability to decode the complexity of the natural world and our capacity to create our own complexity. Both these exist regardless of the chemical, physical or biological substances involved or the processes generating them. Most of our fellow humans live in a world of high complexity (both natural and human-made), although some still live in a world of more natural than human-made complexity. Yet, others have decided to abandon the complexity of the "civilized" life to indulge in the lures of a less complex life. In the Dhammapada for instance, it is written (25:22) that "The monk, joyful and fulfilled by Buddha's doctrine, achieves the blisssing of the state of equanimity, and has let aside the world's complexities". The majority of us however, prefer to live in a complex world, with complex lives full of complex networks, complex technologies, complex economies, etc. In either case, humans make choices of simplicity over complexity and complexity over simplicity on a daily basis. How many of these choices, livelihood types and interactions are spatial? No doubt, a large share of them, from choosing how to present food on a dish, to how to combine shapes and colors of garments and how to arrange clothes in the cupboard. This complexity-related decision making is deeply rooted in our nature and differentiates us from other species, whose choices are considerably more limited.

Meanwhile, in a world that gets increasingly more complex, we experience a frenzying "spatialization" of digital data, from GPS and locational data to selfies. As we walk down the streets of a big city, we see innumerable signs and labels. All these are spatial objects. We seldom bother to see them all in their true 3d context; we most often read them as 2d images and signs. Understanding, decoding and managing spatial complexity is one of the major future challenges for science and technology.

That's because we need to simplify the unnecessary spatial complexity and we only choose a very small portion of this huge inflow of spatial information to process for future reference (which includes advertisements, posters, vast numbers of colors of cars, attires of other people, building facades etc). And all these are spatial data, spatial forms, spatial randomness and spatial patterns. We live in a realm of natural and human-made spatial complexity, in a world of "*global complexity*" (Urry 2003). And our future very much depends on what we do with the natural spatial complexity that the planet has bequeathed to us and what we do with the spatial complexity we have created ourselves. At much larger spatial scales however, increasing complexification (through information and communication technologies, complex transport networks etc) might consist a potential vulnerability of human civilization. Managing wisely the large-scale spatial complexity of the planet is an obviously extremely "complex" undertaking, with unfathomable implications for everyone and everything on the face of the earth.

18.4 *Mens Spatii*: Towards an Observatory of Spatial Complexity

"The battle against Chaos and Old Night,
which is our one truly human activity"
(Bertrand Russell 1926, p. 267)

As Fernando Pessoa wrote in his poem "Ulisses", "myth is the nothing that is everything" (O mito é o nada que é tudo). Thousands of years ago, in ancient greek mythology and according to Hesiod, the goddess Mitis (*Μῆτις*), daughter of the Ocean and Tythys (the Land), born together with Eros (love) and Aether (spirit) and was the first wife of Zeus (the leader of the 12 gods of the ancient greek mythology). She was the primary goddess of wisdom and mother of the more widely known goddess of wisdom, Athina to whom she bestowed her wisdom. Athina gave her name to the city of Athens, but her mother's capacities may have been even greater, since she also was the goddess of cleverness, able to cleverly *weave* things and processes ("*μῆτιν ὑφαίνειν*" as written in the Iliad H. 324). Complexity is the result of things interwoven together. The greek verb pleko (πλέκω) means to weave. Had the ancient greeks known about the modern concept of complexity, it is likely they might have attributed to it one of their foremost and primary divinities.

In geographical analyses nowadays we need maps to analyze the geographical space and to perceive the world from images, but we also need our spatial data to be easily updatable in real time. As imagined by Cubitt (1998, pp. 50–51) we might eventually need an "Imperial Encyclopedia" created from satellite maps, eventually fulfilling "the dream of Cartography: the endlessly updatable map".

Although this may be sounding like a very remote possibility, we hopefully however are in position to approximate an initial coarse classification of determinants of spatial complexity. The set proposed here may not be exhaustive in details; future research findings might suggest regroupping these determinants otherwise or even adding new ones. In any case, the "philosophers' stone" of spatial complexity should be the unification of all the various determinants of spatial complexity into a unified system of spatial complexity assessment, which would render a single quantitative value of complexity for a surface or object, knowing the values for each one of the determinants given here. It is still uncertain whether concepts of homology should be used, or of group-theory, or algebraic geometry or category theory or indeed, of just any other branch of mathematics, in order to bring together *all* the determinants of spatial complexity together at some higher level of generalization. What can we say, for instance, about the overall spatial complexity of a setting of interwoven branches of plants with intersections *and* braiding *and* curvature? And would the complexity be higher if the branches were all straight lines or if the they were with less intersections? So, combining assessments of spatial complexity on the basis of different criteria (geometric and topological in this case) simultaneously presents one of the foremost important challenges for future research in spatial complexity.

It would probably be erroneous to assign any one single determinant of spatial complexity some higher status in comparison to others. Although we need huge efforts to get close to that stage of knowledge, there seems to be enough evidence, theoretically and practically derived, enabling us to suggest that if any such overall assessment is indeed possible at all, it should be expected to have the form of some kind of algorithmic assessment, since in examining the spatial complexity of a surface or an object, any geometric, topological or probabilistic formalism will eventually have to be translated into algorithmic length and/or computational cost. With this thought, we uncomplicatedly revert back to traditional algorithmic and computational complexity. There is no reason really to assume that it might have been feasible to proceed otherwise and this is encouraging, considering that some algorithmic tools for measuring simple cases of spatial complexity are already available. Indeed, the length of description of a spatial object or surface is firmly set at the very heart of spatial complexity analysis. Succeeding to convert the values of the determinants of spatial complexity to a final algorithmic or computational description seems to be a rather remote possibility given our current state of knowledge. But, at least, we might establish some landmarks on our way towards it. Thus, future research does not have to divert into unknown pathways; it simply needs to follow the relevant domains of research that are already established (mathematics, geography, earth sciences, ecology, social sciences, engineering, topography, cosmology, computer science, biomedical sciences, spatial computing etc.) and just be sensitive to all upcoming research results that might be relevant to spatial complexity.

Consequently, due to the inter-disciplinary nature of spatial complexity analysis, it would probably be best to monitor, accumulate and evaluate results relevant to spatial complexity as they come out from the different scientific domains. This will ensure the robustness of the advancing knowledge about spatial complexity and will also help to avoid the danger of biasing this essentially inter-disciplinary subject to a single scientific discipline. For this reason, it may be better to envisage our theoretical knowledge of spatial complexity as advancing in pace with our knowledge from experiments and practice in various disciplines. And "practice" in this case may also entail artifacts of human creativity, constructions, artistic production, engineering applications etc. In short, we need a "*Spatial Complexity Observatory*", which might be named "*Mens Spatii*" (the mind of space). In this way, besides the theoretical achievements, the exploration of spatial complexity may gradually develop into a truly fascinating experience along the way.

We all dwell in a vast realm of spatial complexity, but we probably haven't realized its importance for us. But, as we progressively advance its exploration, we will also sharpen our sensitivity to appreciate it to the extent it deserves, because it is precisely spatial complexity that reflects the plurality and diversity of colors, forms, shapes and features, that the fabric of the world is made of. The moment we think we grasped it, the next it has already changed; and yet, we carry on. Because, as the Dhammapada (8.14) reads, "it is more worthy to live even one single day by grasping the impermanence of all things, than one thousand years in ignorance".

The realm of spatial complexity is vast, overwhelming, dangerous, tricky, playful, deceiving, puzzling, dazzling, elusive, and may be, above all, charming.

Quite possibly, attempting to cross its borders is an endeavor resembling that of the great navigators of the past who embarked to discover new continents.

Embarking to discover new realms has always been challenging and…complex. But handsomely rewarding also.

References

Cubitt, S. (1998). *Digital Aesthetics*. London: Sage.

Diderot, D. (1754). *Pensées sur l'interprétation de la nature.*

Doyle, L. R., McCowan, B., Johnston, S., & Hanser, S. F. (2011). Information theory, animal communication and the search for extraterrestrial intelligence. *Acta Astronautica, 68,* 406–417.

Dyson, F. (1979). *Disturbing the Universe.* New York: Harper and Row.

Fukuyama, F. (2002). *Our Posthuman Future: Consequences of the Biotechnology Revolution.* New York: Farrar Strauss and Giroux.

Ghiradri, G., Rimini, A., & Weber, T. (1986). Unified dynamics for microscopic and macroscopic systems. *Physical Review D, 34,* 470–491.

Jameson, F. (1990). *Late Marxism: Adorno, or the persistence of the Dialectic.* London: Verso.

Leibniz,G. (1890). *Die Philosophischen Schriften von Gottfried Wilhelm Leibniz.* (ed. C.I. Gerhardt), Berlin: Weidmann.

Leibniz,G. (1923). *Samtliche schiften und briefe. Vol.VI.* Berlin: Akademie der Wissenschaften.

Leibniz, G. (1969). *Philosophical papers and letters.* (ed.and transl. Leroy E.Loemker). Dordrecht: D.Reidel.

Kaku, M. (1997). *Visions: How Science will Revolutionize the 21st Century.* New York: Anchor books.

Kardashev, N. (1964). Transmission of information by extraterrestrial civilizations. *Soviet Astronomy, AJ, 8,* 217–221.

Papadimitriou, F. (2010a). A "neogeographical education"? the geospatial web, gis and digital art in adult education. *International Research in Geographical and Environmental Education, 19*(1), 71–74.

Papadimitriou, F. (2010b). Introduction to the complex geospatial web in geographical education. *International Research in Geographical and Environmental Education, 19*(1), 53–56.

Russell, B. (1926). *Education and the Good Life.* London: Allen and Unwin.

Strickland, L. (2006). *The shorter Leibniz texts: A collection of new translations.* London: Continuum.

Urry, J. (2003). *Global Complexity*. Cambridge: Polity.
Whitehead, A.N. (1926/1974). *Process and Reality. An essay in Cosmology*. New York: The Free Press.

Index

A
Adjacency, 163, 164, 169, 171
Aesthetic evaluations, 243, 244, 251, 253, 255
Aesthetic middle, 251
Albers, Joseph, 133
Alexander polynomial, 69, 70
Alexander's wild knots, 195
Algorithmic complexity, 56, 58, 59, 86, 87
Alzheimer's disease, 29
Anosov diffeomorphism, 139
Antoine's necklace, 195, 202
Apollonian gasket, 202
A priori complexity, 222
Archimedes, 273
Architecture, 250
Arecibo message, 287, 288
Aristotle, 265, 271
Arithmetic topology, 218
Arnold Cat Map, 127, 137–140, 265
Art, 127, 131–133
Arte programmata, 133
Artificial Life, 136
Artin word, 72
Art nouveau, 257
Asymmetry, 7, 15

B
Banach-Tarski paradox, 195, 197, 198
Baroque, 257
Beknottedness, 67
Bellard's formula, 188
Bennett's "logical depth", 57
Bessel function, 148
Betti groups, 102
Betti numbers, 246, 247
Binary maps 2 x 2, 158, 159
Binary maps 3 × 3, 163, 165–175
Binary maps 4 x 4, 215, 224
Binary maps 7 X 7, 47, 174, 184, 188, 189
Binary maps of primes, 179, 186
Binary maps of transcendentals, 179
Bio-medical sciences, 19, 28
Bipyramidal astroid, 105
Birkhoff aesthetic measure, 244
Blocks of symbols, 87
Bolyai, 197
Border cells, 173, 174
Borel-Cantelli lemma, 213
Borg, 287
Borsuk's conjecture, 197
Boundaries, 63–66, 75, 76, 172, 173
Boy's surface, 200
Braid complexity, 71, 72
Braiding, 63, 71, 72
Braids word problem, 72
Brownian stochastic motion, 57
Bruinier & Ono formula, 143, 148
Buffon's needle, 209–211
Burnside lemma, 143, 155
Burnside–Polya theorem, 155

C
Cantor gasket, 130
Cantor, Georg, 187
Cantor's theorem, 201
Cats, 137
Cellular automata, 24–26
Celtic knots, 244
Chaitin-Levin complexity, 57
Chartres cathedral, 134
Chebyshev map, 139

F. Papadimitriou, *Spatial Complexity*, https://doi.org/10.1007/978-3-030-59671-2

Checkers, 127, 134, 135
Chess, 127, 134–136
Chiavatti, Gianfranco, 133
China, 134, 176
Clifford Tori, 115, 118
Clumpiness, 163, 164, 169–171
Cobordism theory, 120
Cognitive limitations, 238
Cohomology, 118
Colin de Verdiere number, 47
Combinatorial complexity, 46, 47
Combinatorial explosion, 160, 263
Complex functions (binary maps from), 196
Complex systems, 5, 19, 25
λ-compressed string, 87
Computational complexity, 13, 14
Conchoids, 104
Connectedness, 201
Connected sums, 118, 119
Contagion, 54, 55
Continuous fraction, 189
Conway polynomial, 69
Corpus Hypercubus, 116
Cosmic web, 19
Cosmology, 19
Covering problem, 136
C_{P1}, 81, 86–88, 90, 91, 93–98, 101, 103–107, 109, 164–170, 174, 175, 179, 186, 189, 195, 197–200, 202–205, 221–224, 238, 270
C_{P2}, 81, 90, 91, 93, 101, 103–107, 164–166, 168–170, 174, 175, 179, 184, 189, 197–200, 202–204, 206, 221, 223, 245–246
C_R, 110–112
Creating spatial complexity, 143
Crossing number of braids, 71
$C(s)$, 247, 248
Cubical complex, 115, 118
Curvature, 44, 46, 257, 263, 265, 266
Curve shortening flow, 64

D
Dali, Salvador, 116
Dedekind eta function, 149–151
Dehn, Max, 197
Dehn's classification theorem, 76
Dekeract, 116
Deleuze, Gilles, 263–266, 269, 270
Dendrites, 31, 32
De Stijl movement, 132
Dhammapada, 291
Diabolical square, 132
Diderot, Denis, 284
Diffeomorphic, 121
Differential structure, 119
Dilogarithm function, 46
Dimension, 46, 48, 63, 64, 74–76
Disorder, 4, 8, 9, 15
Diversity, 53, 163, 174
Divisor function, 150, 151
DNA topology, 29
Dodekeract, 116
Dominoes, 47, 48
Doxel, 13, 115–117
Durer, Albrecht, 131
Dynnikov diagrams, 271
Dyson conjecture, 270
Dyson spheres, 287

E
Ecological complexity, 19
Edge of chaos, 27
Edit distance, 85
Effective homology, 118
Eilenberg — MacLane spaces, 201
Einstein-Bose condensates, 182
Eisenstein series, 149–151
Electronic devices, 283
Elliptic integral of the first kind, 151
Elliptic modulus, 150–152
Enigmas (of spatial complexity), 195
Enneper's surface, 200
Enneract, 116
Enriques-Kodaira theorem, 120
Entropy class, 144, 153, 154, 158, 159, 163, 165–169, 171, 172, 174, 175
Entropy encoding, 91, 92, 108
Entropy of a curve, 55
Epistemic modality, 273
Epistemology, 176
Erdös- Szekeres theorem, 215
Eternity-II puzzle, 134
Euler characteristic, 110
Euler-Mascheroni constant, 152, 182
Euler-Poincaré characteristic, 118
Even-numbered maps, 179, 184, 186
Exotic spheres, 115, 118, 120, 121
Expected complexity, 222
Experimental philosophy, 263, 272
EXPSPACE class, 13
EXPTIME class, 13, 70
EXPTIME-complete, 135

F

Fary-Milnor theorem, 66
Fenchel's theorem, 66
Fibonacci sequence, 128, 144
Fictionalism, 273
Fivos Papadimitriou C_{P1} measure, *see* C_{P1}
Fivos Papadimitriou C_{P2} measure, *see* C_{P2}
Fivos Papadimitriou C_R measure, *see* C_R
Fivos Papadimitriou *C(s)* measure, *see* *C(s)*
Flatland, 132
Floer homology, 74
Focus-dependent, 236
Ford circles, 195, 202, 203
Four colors problem, 47
Fourier Transform, 83
Fractal, 19, 23, 24, 26, 29, 223
Future, 279, 281, 284, 286, 288–290

G

Gage-Hamilton-Grayson theorem, 63, 64
Game of Life, 136
Gamma function, 148, 151, 152
Garden of Eden, 136
Gardens Shopping Mall, 134
Gaussian curvature, 45
Gaussian-type functions, 58
Gauss linking number, 73
Gelfand, 220
Generalization, 229, 231, 233–235
Generic maps, 143, 154, 155, 163, 165, 166, 168, 171, 172
Genus, 63, 64, 70, 74
Geographical complexity, 23
Geographical Information Systems (GIS), 21, 22, 82
Geometric complexity theory, 218
Geometrization, 195, 196
Geons, 245
Geophilosophy, 263
Geospatial, 21–23, 27
Gerstner, Karl, 133
Gestalt psychology, 229, 230, 233
Global complexity, 289
Glome, 121
Go (game), 127, 134, 135
Golden section, 128, 138, 189
Googol, 275
Grope, 118
Grosberg-Nechaev complexity, 63, 70
Grothendieck constant, 220
Grothendieck's inequality, 209, 218, 220

H

Hales-Jewett theorem, 215, 272
Hamiltonian cycle, 115, 117
Hamiltonian path, 101, 108, 117
Hardy-Ramanujan formula, 143, 146
Hartmann, Nicolai, 263, 264
Hausdorff, Felix, 197
Heegard complexity, 103
Hendekeract, 116
Hepteract, 116
Hermann grids, 230
Hermitian metric, 120
Heron's formula, 40, 75, 76
Hesiod, 175
Hexeract, 116
Hilbert, David, 132, 195, 197, 201, 264
Hilbert space, 220
Hilbert space-filling curve, 202
Homeomorphic, 64–66, 76, 121
Homogeneous map, 195, 204
Hyperbolic geometry, 46, 196
Hypercubes, 115–118
Hypersphere, 120, 121

I

I-Ching, 176
Ikebana, 255
Immersion, 63, 76, 77, 195, 197, 200
Impairment of meaning, 235
Infinity, 195, 201, 202, 204
Information-richness, 23
Intersections, 39, 41–44
Intuitionism, 272
Isoperimetric inequality, 44, 45

J

Jacobi theta function, 151, 152
Jones polynomial, 70
Jordan Curves, 63
Jordan Curve theorem, 64
Jung's theorem, 209

K

Kaaba cube, 131
Kähler metric, 120
Kandinsky, Wassily, 243, 257
Kanenobu knots, 70
Kant, Immanuel, 253, 264, 271
Kardashev, Nikolai, 286
Kasparov, Gary, 276
Kauffman polynomial, 70

K-cobordism, 120
Khinchin's bound, 209
Khovanov homology, 70
King's neighborhood, 25, 175
Klein surface, 77, 135, 196
Kloosterman sum, 147, 148
Knight's tour problem, 135
Knot complexity, 29, 30, 67, 70
Knottingness, 63, 67
Knotting of proteins, 19
Koch curve, 24
Kodaira embedding theorem, 120
Kolmogorov complexity, 57, 86
Kronecker delta, 148
Kühne, Olaf, 249, 266

L
Labyrinths, 127, 134
Lambert function, 218
Landau-Ginzburg equation, 30
Landscape aesthetics, 249
Landscape complexity, 19, 21, 53
Landscape diversity, 22
Landscape fragmentation, 22
Landscape patchiness, 22
Language (of space), 82
Large-scale spatial complexity, 286, 289
Latin squares, 127, 134
Law of familiarity, 231
Law of good continuation, 231, 232
Law of nearness, 231, 232
Law of simplicity, 231, 232
Leibniz, Gottfried, 264, 279, 284
Lem, Stanislav, 266
Levenshtein distances, 84, 90, 91
Le Witt, Sol, 177
Lhuilier's formula, 110
Lie group, 196
Linear array scanning, 28, 92
Linking, 63, 72–74
Link topology, 29
Lissajous curves, 105
Lorenz attractor, 26

M
Mach bands, 230
Magic, 132, 137
Magic square, 131, 132
Magritte, René, 257
Malevich, Kazimir, 133, 177
Marc, Franz, 257
Mathematical formalism, 272
Matisse, Henri, 132
Matousek, 180
Matveev complexity, 66, 101, 102, 118
Maze-generating algorithms, 134
Mazes, 127, 134
Maze-solving algorithms, 134
Meaning, 229, 233, 235, 236, 239–241
Melanoma, 19, 28
Mens Spatii, 289, 290
Milnor spheres, 120–121
Minimalism, 127, 131, 132
Minos' labyrinth, 134
Misperception (of spatial complexity), 3–11, 13, 14, 16
Mitis (Μῆτις), 289
Möbius band, 76, 101, 103–105
Modal realism, 273
Mondrian, Piet, 133
Monomino, 48, 49
Moore neighbourhood, 25
More, Thomas, 284, 286
Morphogenesis, 30–32
MRI imagery, 29, 107, 280
Multiplicity of a generic map, 166
Multiply-connected, 101, 103
Multiverse, 20

N
Neo-impressionism, 257
Neo-Platonic, 175, 176, 268
Network analysis, 209
Networks, 22, 23
Neurophysiology, 230
Nil geometry, 196
N(n), 146, 147, 158–160
Noether's theorem, 216
Nome, 149–152
Nominalism, 272
Non-Euclidean geometries, 44–46
NP class, 13, 68, 103
NP-complete, 13, 117, 135
NP-hard, 14, 134, 135, 220
NPSPACE class, 13
Nuclear spaces, 220
Number of crossings, 67, 68, 71, 72
Numbers defining spaces, 180

O
Octachoron, 116
Octeract, 116
Odd-numbered maps, 184–186

Oncology, 29
Optical art, 233
Optoelectronic systems, 28
Orientability, 76, 77
Orientation changes, 87
Orthogonality, 39, 40
Otherness, 163, 174
Over-simplification, 235

P
Pac-Man, 134
Papentin complexity, 86–87
Paracelsus, 28
Partitions, 143–148, 153, 155, 158
Patchiness, 163, 164, 169–171
P class, 13
Penteract, 116
Perception of spatial complexity, 6, 14, 16
Perception of spatial randomness, 236
Perelman, Grisha, 196
Perspective, 263, 266, 272
Pessoa, Fernando, 289
Phase space, 26, 27
Philosophical animation, 272
Physics, 19, 23, 24, 27
Pick's theorem, 63, 65, 66, 75
Planetary intelligence, 287
Plato, 271, 286
Plotinus, 268
PNG compression, 245
Poincaré conjecture, 116, 121
Poincaré recurrence theorem, 139
Pollock, Jackson, 257
Polyominoes, 39, 46–48
Porphyry, 268
Precision, 188
Prime manifolds, 119
Principle of uniform connectedness, 233
Problem of perceptual segregation, 233
Prostate cancer, 29
Pseudosphere, 104
PSPACE class, 13
PSPACE-complete, 134, 135
Pyramidal number, 129, 130
Pythagoras, 175
Pythagoreans, 175–176

Q
QR encodings, 27
QR technologies, 19
Qualitative complexity, 14, 229, 239, 240
Quantum computation, 279, 284, 285
Quantum state, 220
Qubits, 285

R
Rademacher formula, 143, 146
Ramanujan congruences, 269
Ramanujan's formula, 144, 151, 219
Ramsey theory, 215, 216, 272
Random graphs, 220
Random image, 59
Randomness, 51, 56–59
Random string, 51, 56, 57, 86
Rasterization/digitization, 82, 83
Raster scanning, 92
Ray, Man, 257
Redfield polynomials, 217
Reeb graphs, 101, 110
Reeve's formula, 75
Reidemeister moves, 67, 68, 74
Reinhardt, Ad, 177
Relative complexity, 223
Relative primes, 148
Rhizome, 264, 265
Riddled system, 27
Riemann, Bernhard, 115, 119, 182–183, 196, 267–268, 272
Riemann surface, 105
Riemann zeta function, 149–151, 182, 183
Ring of Gaussian integers, 129
Rodchenko, Alexander, 132
Rosette scanning, 28, 92
Ruggedness of a surface, 59
Rule of thirds, 177
Russian constructivism, 132

S
Salience maps, 236
Samsø Labyrinten, 134
Scale-dependence, 233
Scanning, 28, 81, 88, 91, 92
Schönbrunn Palace, 134
Schwenk's theorem, 135
Scrabble (game), 135
Semantic complexity, 239
Semantic generalization, 69
Semidefinite programming, 220
Seurat, George, 257
Shannon's formula, 51, 249, 251
Shape complexity, 5
Shiva, 131
Shor's algorithm, 285

Sierpinski square, 130
Simplicial complexes, 101, 102
Singularities, 195, 199, 201
Snakes and Ladders (game), 135
Sol geometry, 196, 197
Space complexity, 3, 5
Spatial aggregates, 257
Spatial combinatorics, 46
Spatial entropy, 51
Spatial games, 127, 134–136
Spatial generalization, 84
Spatial inhomogeneity, 20
Spatial partitions, 127, 128
Spatial randomness, 229, 236
Spatial stochasticity, 57
Spatium Numerorum, 143, 271
Species richness, 51, 54
Spencer-Welch theorem, 136
Spherical geometry, 196
Spherical tetrahedron, 46
Spherical triangle, 40, 41
Sphero-cylindrical curve, 106
Spiral scanning, 92
Steenrod squares, 118
Stefan-Boltzmann law, 182
Steinhaus entropy, 55
Subjective complexity, 14, 236
Subjectivity, 263, 266
Sudoku, 134, 135
Sudoku method, 221, 223
Supercoils, 29
Suprematism, 133
Surrealist, 255
Swallowtail catastrophe, 199
Symmetric maps, 143
Symmetry, 143, 154–158
Symplectic geometry, 119
Szemerédi regularity theorem, 216

T
Tesseract, 115, 116
Tetraktys (τετρακτύς), 175, 176
Tetris, 135
Tetrominoes, 47, 48
Texture gradient, 230, 231
Thematic simplification, 235
Theorema Egregium, 45
Thurston's geometrization theorem, 195, 196
Tibetan mandalas, 131
Tic-tac-toe, 127, 135
Topological complexity, 3, 5
Topological differentiation, 11, 195, 204, 205
Toposophy, 266
Torus, 106
Transfinite numbers, 202
Transition rules, 24
Trefoil, 67, 70
Trominoes, 48, 49
Turing equations, 30

U
Ulam spiral, 186
Undifferentiated map, 205
Universe, 19–21
Unknotting, 238
Utopia, 279, 284, 286

V
Van der Waerden-Erdős–Szekeres theorem, 215
Van Doesburg, Theo, 132
Varignon's theorem, 211
Vasarely, Victor, 233
Video games, 127
Vishnu, 131
Visual complexity, 16, 243–247, 251
Von Mises, 56–57
Von Neumann neighbourhood, 25
Voronoi polygons, 84, 85
Voxel, 11, 13, 107, 108, 110
Voxelization, 107, 108

W
Wada lakes, 47
Weierstrass' elliptic function, 105
Weierstrass function, 199
Whitney embedding theorem, 120
Writhing, 63, 72

X
XP-hard, 70

Y
Yugong, 176

Z
Zermelo-Fraenkel theory, 202

The manufacturer's authorised representative in the EU is Springer Nature Customer Service Centre GmbH, Europaplatz 3, 69115 Heidelberg, Germany. If you have any concerns regarding our products, please contact ProductSafety@springernature.com

Printed and bound by CPI Group (UK) Ltd, Croydon, CR0 4YY
30/09/2026
02225697-0008